CHANGING FAUNAL ECOLOGY IN THE THAR DESERT

Editors

Dr. B.K. Tyagi
MSc, DPhil, FRES, FISCD, FISP, FAZRA
Deputy Director (Senior Grade),
Head, Division of Insecticide and Vector Control
Centre for Research in Medical Entomology (ICMR)
Madurai 625002, TN, India
&
President, Jodhpur Natural History Society
(Formerly the Officer in-Charge, Desert Medicine Research Centre, Jodhpur;
Member in-Charge in South Asia, Odonata Specialist Group of the Species
Survival Commission, IUCN)

Dr. Qaiser H. Baqri
MSc, PhD, PG Nematol (Belgium), DSc, FZSI, FNAAS, FNASc
Additional Director & Officer in-Charge, Desert Regional Station,
Zoological Survey of India (Rtd.), Jodhpur 342005, Rajasthan, India

SCIENTIFIC
PUBLISHERS

Published by:

SCIENTIFIC PUBLISHERS (INDIA)
5A, New Pali Road, P.O. Box 91
Jodhpur 342 001, India
E-mail: info@scientificpub.com
Website: www.scientificpubonline.com

Print: 2023

ISBN: 978-93-86652-06-5
eISBN: 978-93-87307-50-6

Printed in India

Dedicated to

Professor Dr. Ishwar Prakash

D.Sc., FNA, FNASc, FNAAS

The doyen of rodent sciences and The Father of Indian Rodentology

CONTENTS

FOREWORD

It is not always that one comes across books devoted to regional ecological impacts and the volume under review on the *'Changing Faunal Ecology of the Thar Desert'* is exceptional, and is of special significance for reasons more than one. Firstly, the Thar desert is a unique ecosystem which in recent years has been subjected to anthropogenic idiosyncracies resulting in considerable changes in the biota. Ecologists and environmentalists engaged in studies on biological diversity have necessarily to compare the past with the present scenarios. This volume is, therefore, of special significance and goes a long way in demonstrating how a change in environment can bring about diverse changes in animal and plant adaptations, not to mention of changes in ecological succession. The second aspect of significance relates to the dedication of this volume to the late Dr. Ishwar Prakash, an untiring crusader in the cause of studies on Indian Desert Fauna, notably on rodents. An acknowledged expert in this field, Dr. Ishwar Prakash was a very humane, humble, soft-spoken gentleman, noted for serious involvement in matters of research. I had the privilege of knowing him for well over three decades with intimate involvements as task force members of the DST and as Co-Fellows of the Indian National Science Academy.

The fifteen articles on diverse aspects of Thar desert faunal biodiversity are very significant contributions, from soil bio-resources to the impact of the Indira Gandhi Canal on climate change, followed by the changing faunal diversities of amphibia, reptiles, birds and mammals, besides that of several insect groups, notably coleoptera, lepidoptera and Diptera. The last article on the emergence of vector-borne diseases in the context of the changed climate appears very relevant.

It has been my pleasure to have known the authors (*i.e.*, Dr. Tyagi and Dr. Baqri) all through their professional career and am happy that they have collaborated to produce this exceptional volume which has great relevance to the impact of changing ecology on the faunal dynamics of the Thar desert.

10.2. 2005

Prof. T.N. Ananthakrishnan
PhD, DSc, FNA, FASc, FNAAS
Emeritus Scientist
(*Former* Director, Zoological Survey of India, &
Founder Director, Entomology Research Institute, Chennai)

PREFACE

The Thar Desert region, whose geographic limits are generally understood to be between the irrigated lands of the Indus and Sutlej, and the western edge of the Aravalli Hills (*i.e.*, approximately between 24°-30° N and 68°-78° E), had been the abode for the once existing historic Harappa civilization in Kalibanga along the extinct holi river Saraswati, traversing across the Great Indian Desert, together with Yamuna and Ghaggar rivers. At present only the relic Ghaggar marks its presence during monsoon, while Yamuna has in course of time changed its direction, now flowing far away from Rajasthan's north-eastern border into the neighbouring State of Uttar Pradesh. The Thar Desert spreads over as many as four states along the north-western flank of the country, though its very major portion lies in Rajasthan The entire Thar Desert region in Rajasthan is organized in 12 districts, spread over nearly 196,150 km^2 (62% of the total land of the State), and harbouring a population of 12.8 million people (39% of the population of the State). The seasonal Luni river divides the desert into two parts; in the north a vast stretch of sand dunes or levees and in the south an expansive basin rimmed by flat-topped hills of sandstone and limestone. For centuries, Thar Desert continued to exist as a vast mass of monotonous physiography dominated by sandy levees, though constantly subjected to imposing environmental and anthropogenic vicissitudes. Yet, the greatest change in the Thar's ecology became clearly discernible with the advent of extensive canalization brought about by the triad of Gang canal (1928), Sirhind-feeder canal (1955) and, probably one of the world's largest canal systems of its type in a desert ecosystem, the Indira Gandhi canal (1958). The more or less monotonous character of various different faunal groups, both invertebrates and vertebrates, was affected a great deal under their impact and soon species began to change in their composition, demography, population dynamics, product-outputs, and interaction with surrounding abiotic and biotic environments, including human health.

During past nearly 7-8 decades the Thar has consequently experienced profound changes in its complexion arising primarily from greatly intensified use of natural assets and an exploitative management, both dictated by greatly increased human pressure. The traditional, conservative management has lost ground to the expediency of making a living. Though people are by large having a better living condition that have been made possible by improved means of transport and communication, drinking water and public health schemes, nevertheless, overexploitation of available resources and/or inadvertently ushered mismanagement of Thar's recent and most precious aquatic resource – the canal waters, caused about some devastating impacts on the life of its fauna.

This book is a result of keen observations generated from a mélange of animal groups, both as tiny as protozoans and as giant as mammals, in the xeric environmental settings of the Thar Desert, in principle. The underlying philosophy has been that the resultant cadence in animal life under the impact of constantly changing Thar Desert ecology should be highlighted for a better understanding. There are totally 15 contributions, each one by a specialist or a group of specialists in his/their respective field. In addition, an updated bibliography together with an authentic biographic note of Dr. Ishwar Prakash is also presented. Invariably all of these contributions deal with the life forms of a particular type, be it protozoans, termites, butterflies, rodents, frogs, reptiles or mammals, under the changing climatic stress.

It is hoped that the book will offer plenty of novelties on the subject of changing faunal ecology in as fast altering xeric ecosystem as the Thar Desert in north-western India that could well form a model for similar explorations elsewhere in the world. We feel obliged in dedicating this book to the fond memory of Prof. Dr. Ishwar Prakash - *the doyen of rodent sciences* and *the father of Indian rodentology.*

Finally, preparation of this commemorative volume in the everlasting memory of Dr. I. Prakash has brought us several pleasant experiences. This book as such is the first attempt of its type in offering dedication to a legend of zoological sciences in the Thar Desert region. In accomplishing this rare task we always continued to have God's blessings, for which we humbly offer our heartfelt gratitude to Him.

Dr. B.K. Tyagi
Dr. Q. H. Baqri

ACKNOWLEDGEMENT

In compiling this commemorative volume we have received enormous help and support from a large number of professional colleagues, particularly those who have associated with Prof. I. Prakash during his long scientific career, in various scientific organizations both in India and abroad and our heartfelt thanks are due to all of them, even though citation of individual benefactor's name might not be possible here.

Dr. Mohd. Idris, CAZRI, Jodhpur was particularly generous in helping with the compilation of Prof. I. Prakash's authentic and updated bio-bibliographical account, for which we extend our cordial gratitude.

We also wish to thank Mr. Muralidharan, V. Stenographer Gr.II, DRS/ZSI, Jodhpur for his assistance in assembling the manuscripts submitted for the commemorative volume.

Such a major and rare task would not have been possible without constant caring, support and encouragement of a few chosen near and dear ones, and to this effect we foremost remain grateful to our families.

Last, but not the least, we are thankful to Shri Pawan Kumar, Publisher, Scientific Publishers (India), for his unflinching confidence and support, and bringing out the book in time.

LIST OF CONTRIBUTORS

(in alphabetical order)

1. **Dr. V.C. Agrawal**
Zoological Survey of India,
M-Block, 535-New Alipore,
Kolkata - 700053

2. **Dr. Anil Ahuja**
Associate Professor,
Dept. of Clinical Veterinary Medicine
College of Veterinary and Animal Science,
Bikaner (Raj.)- 334 001
E-mail: anil_ahuja@mail.com

3. **Dr. Q.H. Baqri**
Addl. Director & Officer in-Charge,
Desert Regional Station, Z.S.I.,
Jhalamand Chauraha, New Pali Road,
Jodhpur- 342 005 (Raj.) India
E-mail: qaiserbaqri@rediffmail.com

4. **Dr. S. Chakraborthy**
Zoological Survey of India,
M-Block, 535-New Alipore,
Kolkata - 700053 (W.B.) India

5. **Dr. R.P. Dhir**
Principal Scientist and Director,
CAZRI (Rtd), 498 Defence Colony,
Jodhpur 342009 (Raj.) India
E-mail: dhirrp@sancharnet.in

6. **Dr. Shalini Gaur**
Desert Regional Station,
Zoological Survey of India,
Jhalamand Chauraha, New Pali Road,
Jodhpur- 342 005 (Raj.) India
E-mail: drshalini@yahoo.co.in

7. **Dr. G.P. Gupta**
Entomology Division,
Defence R & D Establishment,
Jhansi Road,
Gwalior – 474002, India

8. **Dr. Mohd. Idris**
Central Arid Zone Research Institute,
Jodhpur-342003 (Raj.) India
E-mail: Idris_m@cazri.raj.nic.in

9. **Dr. S.I. Kazmi**
Desert Regional Station,
Zoological Survey of India,
Jhalamand Chauraha, New Pali Road,
Jodhpur- 342 005 (Raj.) India
E-mail: drs_zsi@datainfosys.net

10. **Dr. Madhumita Mandal**
Desert Regional Station,
Zoological Survey of India,
Jhalamand Chauraha, New Pali Road,
Jodhpur- 342 005 (Raj.) India
E-mail: drs_zsi@datainfosys.net

11. **Dr. M.J. Mendki**
Entomology Division,
Defence R & D Establishment,
Jhansi Road, Gwalior – 474002, India

12. **Dr. Shri Prakash**
Entomology Division,
Defence R & D Establishment,
Jhansi Road,
Gwalior – 474002, India

13. **Dr. N.S. Rathore**
Desert Regional Station,
Zoological Survey of India,
Jhalamand Chauraha, New Pali Road,
Jodhpur- 342 005 (Raj.) India
E-mail: drs_zsi@datainfosys.net

14. **Dr. A.S. Rao**
Central Arid Zone Research Institute,
Jodhpur-342 003 (Raj.) India
E-mail: asrao@cazri.raj.nic.in

15. **Prof. H.S. Rose**
Lepidoptera Laboratory,
Dept. of Zoology, Punjabi University,
Patiala - 147 002, India
E-mail: profhsrose@yahoo.com

16. **Dr. B.M. Sharma**
Department of Zoology,
J.N.V. University,
Jodhpur-342 001, India
E-mail: bmsharmara@rediffmail.com

17. **Dr. R.C. Sharma**
Desert Regional Station,
Zoological Survey of India,
Jhalamand Chauraha, New Pali Road,
Jodhpur- 342 005 (Raj.) India
E-mail: drs_zsi@datainfosys.net

18. **Dr. B.K. Tyagi**
Centre for Research in Medical Entomology (ICMR),
4-Sarojini Street, Chinna Chokkikulam,
Madurai – 625 002 (TN), India
E-mail: bk_tyagi@sify.com

19. **Dr. Vijay Veer**
Entomology Division,
Defence R & D Establishment,
Jhansi Road,
Gwalior – 474002, India
E-mail: vijayveer50@yahoo.com

20. **Dr. K C. Verma**
Station Health Organization (SHO),
Jodhpur Cantt
Jodhpur, India

Changing Faunal Ecology in Thar Desert, pp. 1-25, 2005.
Eds.: B.K. Tyagi and Q.H. Baqri

1

DR. ISHWAR PRAKASH: POLYMATH AND RODENTOLOGIST – A BIO-BIBLIOGRAPHY (1931-2002)

B.K. Tyagi*, Mohd. Idris and Qaiser H. Baqri*****

*Centre for Research in Medical Entomology (ICMR), 4-Sarojini Street, Chinna Chokkikulam, Madurai - 625 002 (TN); **Central Arid Zone Research Institute, Jodhpur-342003 (Raj.); ***Desert Regional Station, Zoological Survey of India, Jhalamand Chauraha, New Pali Road, Jodhpur- 342 005 (Raj.), India

PART I
BIOGRAPHY: GLIMPSES INTO LIFE AND ACHIEVEMENTS OF DR. ISHWAR PRAKASH

"The woods are lovely, dark and deep.
But I have promises to keep,
And, miles to go before I sleep,
And miles to go before I sleep."

The life of Dr. I. Prakash, an indefatigable genius *par excellence*, truly reflected the spirit radiated from these famous lines of Robert Frost. With an ever growing child-like inquisity in eyes and a rock-like colossal determination in mind, he became one of the greatest rodentologists in the world in modern times.

Professor Dr. I. Prakash, popularly known as **IP** among his friends and colleagues, was born on 17th December, 1931 at Mathura in Uttar Pradesh. He completed his early schooling at Mt. Abu, Pilani and Jaipur in Rajasthan and later obtained his B.Sc. (Biology group) and M.Sc.,with Zoology, from Rajasthan University, Jaipur. He was awarded with the Ph.D. degree in Zoology in 1957; his dissertation being on the ecology of desert mammals carried out under the aegis of an UNESCO Project. In course of time, he specialized on Mammalia, particularly the rodents which earned him the coveted degree of D.Sc. in 1983 for his work on *'Ecology and management of desert rodents'*.

Following a brief stint as a lecturer in Zoology in Rajasthan Education Service and the University of Jaipur, Dr. I. Prakash joined his *alma mater* - the Central Arid Zone Research Institute (CAZRI), Jodhpur - in 1961 as Animal Ecologist. At CAZRI he worked in various capacities till his suprannuation on 31st December, 1991. During his long scientific career he firmly established the rodent research laboratory engaged in eco-physiological researches of international repute. His creative and innovative mind brought many laurels to his institute as he became the 'founder' Project Coordinator of the All India Co-ordinated Research Project (AICRP) on Rodent Control from 1977-1982. In recognition of Dr. I. Prakash's ingenius research endeavours the Indian Council of Agricultural Research soon bestowed on him the prestigious chair of Professor of Eminence in 1980, which he held till his retirement. In fact, Dr. I. Prakash really never retired from pursuing scientific research as the Department of Science and Technology conferred on him the title of Professor Emeritus, whereby he was able to carry out his research work on small mammals of the Aravally mountain ranges at the Desert Regional Station of Zoological Survey of India, Jodhpur. Due to his life long dedication towards rodent research the Indian National Science Academy, New Delhi selected him as INSA's Senior Scientist, an honour he lived with till his last breath.

Dr. I. Prakash played a long inning as a devoted scientist in as diversified disciplines as vertebrate taxonomy, ecology, biology, behavior, physiology and rodent pest management. He is also credited with the initiation of original research pertaining to biochemical communication amongst desert rodents, under the ICAR's Professor of Eminence Scheme, awarded to Dr. Prakash in 1980. Dr. I. Prakash's team studied olfactory communication through the body odours, and those emanated by ventral scent marking gland and urine in three desert rodents, *viz.*, Desert gerbil, *Meriones hurrianae*, Indian gerbil, *Tatera indica,* and soft furred field rat, *Rattus meltada pallidior.* This serious investigation revealed that both the sexes in *Meriones* and *Tatera* species and only males in *Rattus meltada* possess the ventral scent marking gland. This ventral scent marking gland plays multiple roles in the social organization of these rodents; to signal 'home' to the marking animals in their territory, to enhance their confidence in the environment which is saturated with their own odour to advertise ready-to-mate status by estrous females, genetic upgradation, innate mechanism to regulate population size and food reservation. Dr. I. Prakash had also found that if a very small quantity of sebum secreted by the scent-marking gland or the conspecific female urine was mixed in the poison bait, it mitigated the poison shyness behaviour. Working in this Professor of Eminence Scheme of ICAR, two Research Associates, namely, Dr. Mohd. Idris and Dr. Saroj Kumari were awarded Ph.D. Degree from the then University of Jodhpur and the University of Agra, respectively, under the able supervision of Dr. I. Prakash, both of which are now eminent scientists in rodent sciences.

The Thar Desert region bore a very special place in Dr. I. Prakash's mind and his love for it was well known. Based on his innovative observations on population dynamics of the rodents in the Thar Desert he had ventured to reorganize desert districts under four major habitats, *i.e.*, Sandy, Rocky, Ruderal and Gravel. On over all basis, the desert gerbil, *Meriones hurrianae* formed 26.0% of the total rodent population, whereas, the ratio of other predominant rodents, the Indian gerbil, *Tatera indica* and the soft furred field rat, *Rattus meltada pallidior*, was 24.7% and 13.1%, respectively. Two species, *i.e.*, hairy footed gerbil, *Gerbillus gleadowi*, and little Wagner's gerbil, *G. nanus indus*, solely occupied sandy habitat. However, in rocky habitat, Cutchus rock rat, *Rattus cutchicus cutchicus* (66.6%) and browny spiny mouse, *Mus platythrix sadhu* (9.8%), were found abundantly. Besides, Dr. I. Prakash was always concerned about the future of animal species, and was a keen and staunch advocate of conservation of threatened species. A nature lover, he often lamented extinction of animals like lion and cheetah from the desert late in last century, and worried a great deal about the imposing threat to survival of others like tiger, panther, bear, wild boar, black buck, gazelle and Great Indian Bustard. He had always given emphasis to protect the vast areas of grassland for enhancing the wildlife resources and left no stone unturned for the wildlife management in the Rajasthan desert.

A prolific writer, Dr. I. Prakash published a large number of scientific papers (>350), along with several monographs and edited books (>12) published by renowned publishers like CRC, Dr. Junk of The Hague, Arnold Heimann, ICAR, Scientific publishers (India) *etc.* He worked ceaselessly all his life and, in fact, was very enthusiastically working on one of his last treatises pertaining desert fauna. He was a founding member of the Arid Zone Research Association of India (AZRAI). One of the objectives of the Association was to bring out an inter-disciplinary research journal named *Annals of Arid Zone* of which editorial board he was a member in its formative years. He also served AZRAI as secretary for well over two decades, during which period he organized an international symposium on "*Arid Zone Research and Development*" at CAZRI, Jodhpur in February 1978. He was an active member on several national and international Expert Panel/committees such as those of FAO on Vertebrate Pests, the Rodent Control Expert Group of Government of India and the ICAR, ICMR, DST, UGC, Planning commission, wildlife Institute of India, Man and Biosphere, Ministry of Environment, FAO/DANIDA Panel *etc.* Dr. Prakash was an elected fellow of INSA and Foreign Fellow of Indian Theriological Society, besides being a Fellow of the prestigious National Academy of Scienes, Allahabad and National Academy of Agriculture Sciences, New Delhi. He was also recipient of the prestigious Rafi Ahmed Kidwai Award of ICAR for the biennium 1974-75 and was conferred with the Harswarup Memorial Award of INSA in 1990. He had earned the rare honour of being the Indian Science Congress Associations's Sectional President for Zoology and the Indian National Science Academy's Head of the Local Chapter, Jodhpur in the late 1990s. Dr. I. Prakash had traveled to a large number of countries including Australia, New Zealand, USA, Thailand, Philippines, UK, France, China, USSR, Kuwait and Italy.

A truly creative and competent scientist, Dr. I. Prakash was a hard taskmaster. He was an apostle of honesty and an astute administrator. Above all, he was a hearty, loving and considerate person always ready to help anyone who would seek his advice. He had several qualities which enhanced his ever charming personality and he had special inclination towards music and Urdu poetry. He breathed his last on May 14th, 2002 at Jodhpur. He is survivived by his wife Professor Laxmi, two daughters Subhra and Mukta, and a son Gaurav.

In his passing away the whole scientific community across the country had lost one of the finest, and most disciplined and dedicated scientist - widely recognized as 'The Father of Indian Rodentology". Dr. I. Prakash would however continue to live by his enormous scientific inputs of very high quality which would inspire scores of young researchers to carry forward selflessly the "torch of enlightenment" he had been instrumental to light up through sheer dint of hard work and focused dedication – traits inevitably indispensable for high quality research. He truly lived a *karmayogi*'s life, emulating all along his life the tenets of '*Karma*' as preached in '*Shrimadbhagwadgita*':

"कर्मण्येवाधिकारस्ते मा फलेषु कदाचन।
मा कर्मफलहेतुर्भूर्मा ते सङ्गोऽस्त्वकर्मणि।"

PART II

AN UP-TO-DATE SCIENTIFIC BIBLIOGRAPHY OF DR. ISHWAR PRAKASH

Anonymous, 1988. Thar Desert Biosphere Reserve. Project Document No. 7, Ministry of Environment & Forests, GOI, New Delhi, pp. 1-100. (Dr. I. Prakash was one of the major contributors).

Advani, R., Mathur, R.P. and Prakash, I., 1981. Rodent control and its socio-economic impact in rural environment. *Indian Farming* 32(4): 9-11.

Advani, R., Mathur, R.P. and Prakash, I., 1982. Rodent control in Jodhpur. *Indian Farming* 32(4): 9-11.

Advani, R., Prakash, I. and Mathur, R.P., 1982. Assessment of rodent damages and yield losses in standing wheat crops in a desert village complex. *Zeitschr. Angew Zool.* 89: 257-266.

Advani, R., Prakash, I. and Mathur, R.P., 1982. Chooha Niyantran-Gramin Anchal Mein uska Samajik evem Aarthik Prabhav (In Hindi). *Kheti.* April 1982. 6-7 & 9-13.

Advani, R. and Prakash, I., 1980. Behaviour of the Indian desert gerbil, *Meriones hurrianae* (Jerdon) during recovery from food and water deprivation. *Behav. Process* 5: 75-79.

Advani, R. and Prakash, I., 1981. Neophobia in the Indian desert gerbil, *Meriones hurrianae* (Jerdon, 1867). *Saugt. Mitteil.* 29. 75-78.

Advani, R. and Prakash, I., 1984. Evaluation of zinc phosphide, aluminium phosphide and RH_787 in reducing rodent damage to chilli grown in Rajasthan. *Indian J. Agric. Res.* 54(6): 500-505.

Advani, R., Idris, M. and Prakash, I., 1984. Seasonal changes in feeding rhythm of *Meriones hurrianae* in the Indian desert. *J. Interdiscipl. Cycle Res.* 15: 133-138.

Advani, R. and Prakash, I., 1987. Variations in the rodent populations in response to four anticoagulant rodenticides in the residential habitats of the Indian desert. *Pesticides* XXI (5): 18-22.

Advani, R., Prakash, I. and Mathur, R.P., 1988. Reduction in rodent population through intermittent control operations in the cropping ecosystem of the Indian desert. *Proc. XIII Vertebr. Pest. Conf.* (Eds. A.C. Crabb and R.E. Marsh), University of California, Davis: 119-122.

Agarwal, V.C. and Prakash, I., 1998. Ecological distribution of Indian rodents. In: *Rodents in Indian Agriculture.* (Ed. Ishwar Prakash and P.K. Ghosh) Scientific Publishers, Jodhpur: 1: 1-10.

Barnett, S.A., Cown, P.E. and Prakash, I., 1975. Circadian rhythm of movements of the house rat, *Rattus rattus. Indian J. Expt. Biol.* 13: 153-155.

Barnett, S.A., Cowan, P.E., Radford, G.G. and Prakash, I., 1975. Peripheral anosmia and the discrimination of poisoned food by *Rattus rattus* L. *Behavioural Biology* 13: 183-190.

Barnett, S.A., and Prakash, I., 1975. *Rodents of Economic Importance in India.* Arnold-Heinemann, New Delhi, & Heinemann, London, 1-175.

Bharadwaj, D. and Prakash, I., 1979. Intensity of poison aversion and taitshyness in *Rattus rattus* due to intermittent and successive exposure to RH-787. *Z. angew. Zool.* 66: 427-436.

Bharadwaj, D. and Prakash, I., 1981. Movements of *Rattus rattus* in an artificial environment. *Indian J. Expt. Biol.* 19: 794-796.

Bharadwaj, D. and Prakash, I., 1982. Discrimination between harmless and harmful stimuli by the house rat, *Rattus rattus rufescens. Indian J. Expt. Biol.* 20: 302-304.

Bharadwaj, D. and Prakash, I., 1982. Mitigation of poison aversion in *Rattus rattus* through acclimatization. *Indian J. Expt. Biol.* 20: 396-398.

Bhardwaj, D. and Prakash, I., 1984. Poison bait efficacy with and without pre-baiting the black rat, *Rattus rattus. Pestology.* VIII (10): 21-24.

Bhardwaj, D. and Prakash, I., 1984. Efficacy of three anticoagulant rodenticides for the control of poison shy *Rattus rattus. XI Vertebrate Pest Control Conference*: 82-84.

Bohra, H.C., Goyal, S.P., Ghosh, P.K. and Prakash, I., 1992. Studies on ethology and eco-physiology of the antelopes of the Indian desert. *Annals of Arid Zone.* 32: 83-96.

Cowan, P.E. and Prakash, I., 1978. House rats from the Indian arid zone. *Z. angew. Zool.* 65: 187-194.

Fitzwater, W.D. and Prakash, I., 1969. Burrows, behaviour and home range of the Indian desert gerbil, *Meriones hurrianae* Jerdon. *Mammalia* 33: 598-606.

Fitzwater, W.D. and Prakash, I., 1989. *Handbook of vertebrate Pest Control* (revised by I. Prakash, III Edition) Indian Council of Agricultural Research, New Delhi: 1-103.

Ghosh, A.K., Baqri, Q.H. and Prakash, I., (Eds.) 1996. *Faunal Diversity in the Thar Desert: Gaps in Research.* Scientific Publishers (India), Jodhpur, 1-410.

Ghosh, P.K., Goyal, S.P. and Prakash, I., 1979. Metabolism and eco-physiology of Rajasthan desert rodents. *J. Arid Environ.* 2: 77-83.

Ghosh, P.K. Purohit, K.G. and Prakash, I., 1962. Studies on the effects of prolong water deprivation. *Environmental Physiology Psychol. in Arid Conditions, pp. 301-306. Proc. Lucknow Symp.* UNESCO, Paris.

Ghosh, P.K. and Prakash, I., (Eds.) 1988. *Ecophysiology of desert vertebrates.* Scientific Publishers, Jodhpur: pp.475.

Goyal, S.P., Ghosh, P.K. and Prakash, I., 1981. Significance of body fat in relation to basal metabolic rate in some Indian desert rodents. *J. Arid Environ.* 4: 59-62.

Goyal, S.P., Ghosh, P.K. and Prakash, I., 1982. Energetic aspects of adaptation in the Indian desert gerbil, *Meriones hurrianae* Jerdon. *J. Arid Environ.* 5: 69-75.

Goyal, S.P., Ghosh, P.K. and Prakash, I., 1983. Estimation of energy flow during winter through a primary consumer (Indian gerbil, *Tatera indica,* Hardwicke) in a desert ecosystem. *Proc. Indian Natl. Sci. Acad.* B49: 295-302.

Goyal, S.P., Bohra, H.C., Ghosh, P.K. and Prakash, I., 1988. Role of *Prosopis cineraria* pods in the diet of two Indian desert antelopes. *J. Arid Environ.* 14: 285-290.

Gupta, B.K. and Prakash, I., 1975. The landscape and land forms. Chapter I in "*Environmental Analysis of the Thar Desert*" (Eds. R.K. Gupta & I Prakash) pp. 11-21, English Book Depot, Dehradun.

Gupta, R.K. and Prakash, I., (Eds.) 1975. *Environmental Analysis of the Thar Desert.* English Book Depot, Dehradun, 1-484.

Gupta, R.K. and Prakash, I., 1975. The origin and geomorphic evolution. Chapter II in "*Environmental Analysis of the Thar Desert*" (Eds. R.K. Gupta & I. Prakash). 22-37. English Book Depot., Dehradun.

Gupta, R.K. and Prakash, I., 1978. Management of range resources in the Indian arid zone with reference to rodent control. *Proc. Indian natnl. Sci. Acad.* 44 B: 312-319.

Idris, M and Prakash, I., 1982. Behavioural responses of the Indian gerbil, *Tatera indica* to conspecific subum odour of the ventral scent marking gland. *Proc. Indian Acad. Sci.* 9: 259-265.

Idris, M. and Prakash, I., 1983. Response of male Indian gerbil, *Tatera indica* to urine from females. *Indian J. Expt. Biol.* 21: 97-98.

Idris, M. and Prakash, I., 1985. Hormonal control of scent marking behaviour in Indian gerbil. *Acta Theriol.* 30: 267-274.

Idris, M. and Prakash, I., 1985. Social and scent marking behaviour in Indian gerbil. *Biol. Behaviour* 10: 31-39.

Idris, M. and Prakash, I., 1986. Influence of odours of Indian gerbil, *Tatera indica* on the social and scent marking behaviour of sympatric desert gerbil, *Meriones hurrrianae. Proc. Indian Nat. Sci. Acad.* 52: 333-340.

Idris, M. and Prakash, I., 1986. Scent marking in the Indain gerbil, *Tatera indica* in response to conspecific odour. *Proc. Indian Acad. Sci.* 95: 89-96.

Idris, M. and Prakash, I., 1987. Scent marking activity in the Indian gerbil, *T. indica* in relation to population density. *Animal Behaviour* 35: 920-922.

Idris, M. and Prakash, I., 1987. Scent marking behaviour and social organization in the Indian soft furred field-rat. *Acta Theriol.* 32: 11-20.

Idris, M. and Prakash, I., 1988. Perception of dominance status through conspecific scent in the desert gerbil, *Meriones hurrianae. Indian J. Expt. Biol.* 26: 479-480.

Idris, M. and Prakash, I., 1991. Scent marking behaviour in desert gerbil in relation to social hierarchy. *Proc. Indian Natl. Acad. Sci.* (B). 57 (3&4): 263-266.

Idris, M. and Prakash, I., 1992. Shyness behaviour. In *Rodents in Indian Agriculture* (Ed. Ishwar Prakash and P.K. Ghosh). Vol. 1, pp. 433-442 Scientific Publishers, Jodhpur.

Jain, M.B. and Prakash, I., 1974. Bird damage in relation to varietal difference in bajra crop. *Ann. Arid Zone* 13: 139-144.

Jain, A.P. and Prakash, I., 1978. The organ, body weight relationships in the Indian desert hare, *Lepus nigricollis dayanus Blanford. Z. angew. Zool.* 63: 31-36.

Jain, A.P., Prakash, I. and Rana, B.D., 1974. Baits for the control of the soft-furred field-rat, *Rattus meltada pallidior. Ryley. Z. angew. Zool.* 61(2): 183-190.

Jain, A.P., Prakash, I. and Rana, B.D., 1975. Baits for the control of the cutch rock rat, *Rattus cutchicus cutchicus* (Wroughton). *Proc. All India Rodent Seminar,* Ahmedabad. Pp. 179-186.

Jain, A.P., I. Prakash, and B.D. Rana, 1975. Morphological variation in the field rodents in the Indian desert. III. *Rattus c. cutchicus* (Wroughton) 1912 and *Rattus meltada pallidior* (Rylay), 1914. *Acta Theriologica* 20: 133-140.

Jain, A.P. and Prakash, I., 1981. Population ecology of Indian desert rodents. *Proc. Symp. Animal Ecol. Zool.* Survey, India 1: 5-13.

Jain, A.P. and Prakash, I., 1983. Laboratory evaluation of poison-base for the control of Indian gerbil, *Tatera indica. Indian Acad. Sci.,* 92(3): 253-261.

Krishna, D. and Prakash, I., 1955. Hedgehogs of the desert of Rajasthan. I. Distribution and fossorial habits. *J. Bombay Nat. Hist. Soc.* 53: 38-43.

Krishna, D. and Prakash, I., 1956. Hedgehogs of the desert of Rajasthan. II. Food and feeding habits. *J. Bombay Nat. Hist. Soc.* 53: 362-366.

Krishna, D. and Prakash, I., 1960a. Hedgehogs of the desert of Rajasthan. III. Food in nature. *Proc. Raj. Acad. Sci.* 7: 60-62.

Krishna, D. and Prakash, I., 1960b. A note on mammals of the Rajasthan desert. *Anusandhan Patrika* 3: 203-214 (in Hindi).

Krishnamurthy, K., Prakash, I. and Chaturvedi, G.C., 1977 (Eds.). *Proceedings of All India Rodent Seminar,* Ahmedabad, pp. 1-376.

Kumari, S. and Prakash, I., 1979. Conspecific odour as a phagostimulant for the Indian gerbil, *Tatera indica indica* Hardwicke. *Indian J. Expt. Biol.* 17: 981-982.

Kumari, S. and Prakash, I., 1980. Role of conspecific urine in enhancing food consumption and masking of shyness behaviour in the desert gerbil, *M. hurrianae* (Jerdon). *Indian J. Expt. Biol.* 18: 730-732.

Kumari, S. and Prakash, I., 1980. Discrimination of poisoned food by *Funambulus pennanti* and persistence of the learning behaviour. *Indian J. Rodentology* 2: 11-15.

Kumari, S., Cowan, P.E. and Prakash, I., 1981. The midventral gland of the Indian desert gerbil, *M. hurrianae* Jerdon. *Acta Theriol.* 26: 97-106.

Kumari, S. and Prakash, I., 1981. Behavioural responses of *Meriones hurrianae* (Jerdon) to conspecific sebum odour of ventral sebaceous gland. *Biology of Behaviour,* 6: 255-263.

Kumari, S. and Prakash, I., 1981a. Observations on the social behaviour of the Indian desert gerbil, *Meriones hurrianae. Proc. Indian Acad. Sci.* 90: 463-471.

Kumari, S. and Prakash, I., 1981b. Scent marking behaviour of *Meriones hurrianae* during oestrous. *Animal Behaviour* 29: 1269-1271.

Kumari, S. and Prakash, I., 1983a. Mid ventral marking gland of *Tatera indica* and soft furred field rat, *Rattus meltada pallidier. Acta Theriol.* 28: 425-434.

Kumari, S. and Prakash, I., 1983b. Seasonal variation in the dimension of scent marking gland of three desert rodents and its possible relationship with their reproductive performance. *Proc. Indian Acad. Sci.* 92: 299-304.

Kumari, S. and Prakash, I., 1984a. Association between scent marking density and dominance in the Indian desert gerbil, *Meriones hurrianae. Indian J. Expt. Biol.* 22: 421-423.

Kumari, S. and Prakash, I., 1984b. Perception of oestrus odours by female *Meriones hurrianae. Animal Behaviour* 32: 927-929.

Kumari, S. and Prakash, I., 1984. Relative efficiency of male and female conspecific urine in masking shyness behaviour in Indian gerbil, *Tatera indica. Proc. Indian Acad. Sci.* 93: 431-436.

Mann, H.S. and Prakash, I., 1982. The desert habitat and desertification. In: State of Environment, some aspects. National Committee on Environment. Govt. of India, New Delhi, III: 1-63.

Mann, H.S. and Prakash, I., 1983. Halting the March, Eco-development in the Thar. Deptt. of Environment and W.W.F., New Delhi, 1-36.

Mathur, R.P. and Prakash, I., 1980a. Laboratory evaluation of anticoagulant treated baits for the control of the northern palm squirrel, *Funambulus pennanti* Wroughton. *J. Hyg. Camb.* 85: 421-426.

Mathur, R.P. and Prakash, I., 1980b. New food reaction among desert rodents. *Sauget. Mitteil.* 28: 28-30.

Mathur, R.P. and Prakash, I., 1980c. Acceptability and toxicity of anticoagulants to the palm squirrel, *Funambulus pennanti Wroughton. Z. angew. Zool.*, West Germany 67(4): 477-482.

Mathur, R.P. and Prakash, I., 1981a. Comparative efficacy of three anticoagulant rodenticides against desert rodents. *Prot. Ecol.* Amsterdam, 3: 327-331.

Mathur, R.P., and Prakash, I., 1981b. Evaluation of brodifacoum against *T. indica, M. hurrianae and R. rattus. J. Hyg.* Cambridge, 87: 179-184.

Mathur, R.P. and Prakash, I., 1981c. Laboratory evaluation of Fumarin for the control of *Tatera indica and Meriones hurrianae. Bull. Grain Tech.* 19(1): 26-29.

Mathur, R.P. and Prakash, I., 1982a. Evaluation of warfarin against *T. indica* and *M. hurrianae. Proc. Indian Acad. Sci.* 91B (5): 463-468.

Mathur, R.P. and Prakash, I., 1982b. Laboratory evaluation of chlorophacinone against *Tatera indica* and *Meriones hurrianae. Tropical Pest Management.* London, 28(3): 291-294.

Mathur, R.P., Singh, C.P. and Prakash, I., 1982. Susceptibility of Indian gerbil, *Tatera indica* to three anticoagulant rodenticides. *Indian J. Rodent* 4: 7-14.

Mathur, R.P. and Prakash, I., 1983. Methods used in the field evaluation of anticoagulant rodenticides. *Vertebrate Pest Control and Management ASTM,* 817: 256-261.

Mathur, R.P. and Prakash, I., 1983. Toxicity of coumatetralyl to two gerbils, *Tatera indica* and *Meriones hurrianae. Z. angew. Zool.,* 70: 257-263.

Mathur, R.P. and Prakash, I., 1984a. A field evaluation of brodifacoum, chlorophacinone and coumatetralyl for control of desert rodents. *Tropical Pest Management,* London, 30(1): 59-61.

Mathur, R.P. and Prakash, I., 1984b. Comparative efficacy of some anticoagulant rodenticdes against Indian rodents. *The Organization and Practice of Vertebrate Pest Control,* Hampshire, U.K., 381-393.

Mathur, R.P. and Prakash, I., 1984c. Reduction in the population of Indian desert rodents with anticoagulant rodenticdes. *Indian Acad. Sci. (B)* 93(6): 585-590.

Mathur, R.P. and Prakash, I., 1984d. Toxicity of coumatetralyl against *Rattus rattus rufescens. Pestology,* VIII (4): 15-17.

Prakash, I., 1953a. Cannibalism in hedgehogs. *J. Bombay Nat. Hist. Soc.* 51: 730.

Prakash, I., 1953b. Additions to recorded food items of the bull frog *(Rana tigrina). J. Bom. Nat. Hist. Soc.* 1.

Prakash, I., 1954a. Cannibalism in hedgehogs. *J. Bombay Nat. Hist. Soc.* 52: 922.

Prakash, I., 1954b. Swimming and locomotion of captive hedgehogs. *J. Bombay Nat. Hist. Soc.* 52: 584-585.

Prakash, I., 1955a. Mammalian life in the desert of Rajasthan. *Biologica* 59-64.

Prakash, I., 1955b. Notes on desert hedgehog (*Hemiechinus auritus collaris* Gray). *J. Bombay Nat. Hist. Soc.* 52: 921-922.

Prakash, I., 1955c. A list of mammals of Rajasthan desert. *J. Bengal Natural Hist. Soc.* 28: 1-17.

Prakash, I., 1955d. Additions to the list of mammals of Rajasthan desert. *Ibid.* 28: 169.

Prakash, I. and Sharma, S.C., 1955. Nematode and hedgehog mortality. *J. Bombay Nat. Hist. Soc.* 53: 23.

Prakash, I., 1956a. Studies on the ecology of the desert hedgehogs. *Proc. Raj. Acad. Sci.* 6: 24-30.

Prakash, I. and Sharma, S.C., 1956. Strange behaviour of bats. *J. Bombay Nat. Hist. Soc.* 53: 123.

Prakash, I., 1957. A survey and ecological studies of the mammals of the desert of Rajasthan with special reference to the food and feeding habits of certain insectivores and rodents. Ph.D. Thesis. Univ. of Rajasthan, Jaipur, pp. 1-245.

Prakash, I., 1958a. Extinct and vanishing mammals from the desert of Rajasthan and the problem of their preservation. *Indian Forester* 84: 642-645.

Prakash, I., 1958b. Game in Rajasthan. *Biologica* 1: 33-38.

Prakash, I., 1958c. Study of vertebrates in the Indian desert. *Turtux News* 36: 266.

Prakash, I., 1958d. The breeding season in the monkey *Macaca mulatta* (Zimmerman). *J. Bombay Nat. Hist. Soc.* 55: 154.

Prakash, I., 1958e. *Calotes* lizard occupying birds nest. *J. Bombay Nat. Hist. Soc.*, Bombay.

Prakash, I., 1958f. Behaviour of certain mammals towards snakes. *Indian Psychol. Bull.* 3: 48-51.

Prakash, I., 1959a. Destruction of vegetation by desert animals in Rajasthan. *Indian Fmg.* 25: 251-253.

Prakash. I., 1959b. Foods of Indian desert mammals. *J. Biol. Sci.* 2: 100-109.

Prakash, I., 1959c. Food of the Indian False vampire. *J. Mamm.* 40: 545-547.

Prakash, I., 1959d. Food of certain insectivores and rodents in captivity. *Univ. Raj. Studies* 4: 1-16.

Prakash, I., 1959e. Checklist of mammals of the Rajasthan. *Univ. Rajasthan Stud., Biol. Sci.* 4: 30-56.

Prakash, I. 1959f. Hypertrophy of *Bullae tympanicae* in the desert mammals. *Sci. Cult.* 24: 580-582.

Prakash, I., 1960a. Bats as a storm signal. *J. Bobmay Nat. Hist. Soc.* 57: 216.

Prakash, I., 1960b. Breeding of mammals in Rajasthan desert. *J. Mamm.* 41: 386-389.

Prakash, I., 1961a. Taxonomic and biological observations on the bats of the Rajasthan desert. *Proc. Ind. Mus.* 50: 149-170.

Prakash, I., 1961b. The present status of the caracal. *J. Bombay Nat. Hist. Soc.* 56: 317.

Prakash, I., 1961c. Die saugietiere der Rajasthan wuste in Inden. *Nature and Volk* 91: 443-448.

Prakash, I., 1961d. Mutual assistance between mother langurs, *Presbytis entellus entellus* Dufresne. *J. Bombay Nat. Hist. Soc.* 58: 790-791.

Prakash, I., 1962a. Group organization, sexual behaviour and breeding season of certain Indian monkeys. *Jap. J. Ecol.* 12: 83-86.

Prakash, I., 1962b. Ecology of gerbils of the Rajasthan desert, India. *Mammalia* 26: 311-331.

Prakash, I., 1962c. Taxonomical and ecological account of the mammals of Rajasthan desert. *Ann. Arid Zone* 1: 142-163; 2: 150-161.

Prakash, I., 1962d. Rolling of the desert hedgehog. *Hemiachinus aurtus collaris. Indian J. Anat. Soc.* 10: 19-22.

Prakash, I. and Kumbkarni, C.G., 1962. Ecotoxicology and control of Indian desert gerbil, *M. hurrianae* (Jordon). I. Feeding behaviour, energy requirements and selection of bait. *J. Bombay Nat. Hist. Soc.* 59: 800-806.

Prakash, I., 1963. Destructive rodents of desert grasslands. *Gosamvardhan* 11: 15-18.

Prakash, I. and Ghosh, P.K., 1963. The great Indian bustard in Rajasthan desert. *Newsletter for Birdwatchers* 3: 1-2.

Prakash, I., 1964a. Some vertebrate pests in the Rajasthan desert. *Indian Forester* 90: 107-112.

Prakash, I., 1964b. Ecology of the Indian desert gerbil, *Meriones hurrianae* Jardon, *Ministry of Education & UNESCO Symp.*, Jodhpur, pp. 305-310.

Prâkash, I., 1964c. Eco-toxicology and control of Indian desert gerbil, *Meriones hurrianae* Jerdon. II. Breeding season, litter size and postnatal development. J. *Bombay Nat. Hist. Soc.* 61: 142-149.

Prakash, I., 1964d. Eco-toxicology and control of Indian desert gerbil, *Meriones hurrianae* Jerdon. Pt. IV. Toxicity, relative acceptability and efficiency of certain poisons. *Indian Forester,* 90(8): 518-528.

Prakash, I., 1964e. Cytomorphology of the adrenal gland of the Indian desert gerbil. *J. Anat. Soc.* India 13: 76-79.

Prakash, I. and Ghosh, P.K., 1964. The great Indian bustard breeding in Rajasthan desert. *Newsletter for Birdwatchers* 3: 15-16.

Prakash, I., Kumbkarni, C.G. and Krishnan, A., 1965. Eco-toxicology and control of the desert gerbil, *Meriones hurrianae* Jerdon. III. Burrow temperature. *J. Bombay Nat. Hist. Soc.* 62: 237-244.

Prakash, I., 1966. A review of the studies on the food of Indian gerbils. *Proc. Indian Rodent Symp.*, Calcutta 147-151.

Prakash, I., 1967. Palm squirrel (*Funambulus palmarum* L.) in Australia. *J. Bombay Nat. Hist. Soc.* 64: 108.

Prakasth, I. and Jain, A.P., 1967. Occurrence of *Rattus meltada* and *Gerbillus dasyuurus* in the Rajasthan desert. *Ann. Arid Zone* 6: 235.

Prakash, I, and Purohit, K.G., 1967. Some observations on the hairy footed gerbil, *Gerbillus gleadowi* Murray, in the Rajasthan desert. *J. Bombay Nat. Hist. Soc.* 63: 431-434.

Prakash, I., 1968a. Zoogeography of rodents of Rajasthan desert. *Proc. Symp. Arid Zone Int. Geogr. Congr.* 74-75.

Prakash, I., 1968b. The importance of ecological studies in rodent control. *Proc. Symp. Bionomics and Control of Rodents*, Kanpur, pp. 550-552.

Prakash, I. and Jain, A.P., 1968. Toxicity of norbormide for desert rodents. *Proc. Symp. Bionomics and Control of Rodents,* Kanpur: 146-149.

Prakash, I. and Taneja, G.C., 1968. Geographical distribution of the Indian desert gerbil. Proc. Symp. *Arid Zone Int. Geogr. Congr.* 76-77.

Prakash, I. and Taneja, G.C., 1968. Adrenal weight in relation to duration of pregnancy in the Indian desert hare. *Lepus nigricollis dayanus Blanford. Experientia* 24: 850.

Prakash, I., Kametkar, L.R. and Purohit, K.G., 1968. Home range and territotality of the northern palm squirrel *Funambulus pennanti* Wroughton. *Mammalia* 32: 603-611.

Prakash, I., Purohit, K.G. and Kametkar, L.R., 1968. Intake of seeds of grass, shrub and tree species by three species of gerbils in Rajasthan desert. *Indian Forester* 93: 801-805.

Prakash, I., 1969. Eco-toxicology and control of Indian desert gerbil, *Meriones hurrianae* Jerdon. V. Food preference in the field during monsoon. *J. Bombay Nat. Hist. Soc.* 65: 581-589.

Prakash, I. and Kametkar, L.R., 1969. Body weight, sex and age factors in population of the northern palm squirrel, *Funambulus pennanti* Wroughton. *J. Bombay Nat. Hist. Soc.* 66: 99-115.

Prakash, I. and Taneja, G.C., 1969. Reproduction biology of the Indian desert hare, *Lepus nigricollis dayanus*, Blanford. *Mammalia* 33: 102-117.

Prakash, I, Rana, B.D. and Taneja, G.C., 1969. Adrenal weight in relation to body weight and reproductive activity in the Indian desert hare, *Lepus nigricollis dayanus Blanford. Indian J. Anat. Soc.* 18: 42-47.

Prakash, I., 1970. Rodents in rural areas. *Participant J.* 4: 16-18.

Prakash, I. and Jain, A.P., 1970. Eco-toxicology and control of Indian desert gerbil, *Meriones hurrianae* Jerdon. VI. One-shot baiting technique. *J. Bombay nat. Hist. Soc.* 67(2): 274-278.

Prakash, I. and Rana, B.D., 1970. A study of field population of rodents in the Indian desert. *Z. angew. Zool.* 51: 129-136.

Prakash, I., Fitzwater, W.D. and Jain, A.P., 1970. Toxic chemicals and baits for the control of two gerbils, *Meriones hurrianae* Jerdon and *Tatera indica* Hardwicke. *J. Bombay nat. Hist. Soc.* 66(3): 500-509.

Prakash, I., 1971. Breeding season and litter size of Indian desert rodents. *Zeit. Agnew. Zool.* 58: 441-454.

Prakash, I., 1971. *Testudo elegans* in western Rajasthan. *J. Bombay Nat. Hist. Soc.* 68: 273-274.

Prakash, I. and Jain, A.P., 1971. Some observations on the Wagner's Gerbil, *Gerbillus nanus indus* Thomas, in the Indian desert. *Mammalia* 35: 614-628.

Prakash, I. and Jain, A.P., 1971. Bait shyness of two gerbils, *Tatera indica* and *Meriones hurrianae. Ann. Appl. Biol.* 69: 169-172.

Prakash, I., Jain, A.P. and Rana B.D., 1971. New records from the Rajasthan desert. *J. Bombay Nat. Hist. Soc.* 68: 447-450.

Prakash, I., Jain, A.P. and Purohit, K.G., 1971. Breeding and post-natal development of the Indian gerbil, *Tatera indica indica* Hardwicke (1807) in Rajasthan desert. *Saugetier Mitteil* 27(4): 375-380.

Prakash, I., Gupta, R.K., Jain, A.P., Rana, B.D. and Dutta, B.K., 1971. Ecological evaluation of rodent populations in the desert biome of Rajasthan. *Mammalia* 35: 384-423.

Prakash, I., Taneja, G.C. and Purohit, K.G., 1971. Eco-toxicology and control of Indian desert gerbil, *Meriones hurrianae* Jerdon. VII. Relative numbers in relation to ecological factors. *J. Bombay Nat. Hist. Soc.* 68: 86-93.

Prakash, I., 1972a. Notes on little known lizards from the Rajasthan desert. *J. Bombay Nat Hist. Soc.* 69: 424-428.

Prakash, I., 1972b. Eco-toxicology and control of Indian desert gerbil, *Meriones hurrianae* Jerdon. VIII. Body weight sex ratio and age structure. *J. Bombay Nat. Hist. Soc.* 68: 717-725.

Prakash, I. and Rana, B.D., 1972. A study of field population of rodents in the Indian desert. II. Rocky and piedmont zones. *Zeit. Angew. Zool.* 59: 129-139.

Prakash, I., 1973a. Rodent control in the desert. *Indian Farming.* 23: 41-43 & 47.

Prakash, I., 1973b. Rodent damages in field and their control in prevention of waste and loss of food-grains. *National Productivity Council.* III. 287-295 (also in *Productivity* XIV (4): 413-419, 1971) and *Asian Productivity Organization.* Tokyo, 193-199 (1974).

Prakash, I., 1973c. Rodents in agriculture and development of effective control systems. *Proc. IX Session of Plant Protection Committee for South East Asia and Pacific Region,* FAO: 1-24.

Prakash, I., 1973d. Extension of the range of *Suncus stoliczkanus* in the Rajasthan desert. *J. Bombay Nat. Hist. Soc.* 69: 643-645.

Prakash, I., 1973e. Intensification of vertebrate pest control research in India. FAO Reprot. *Vertebrate Pest Management* (also *Rat Fighter* 8: 58-60).

Prakash, I., 1973f. Eco-toxicology and control of the Indian desert gerbil, *Meriones hurrianae* (Jerdon). IX. Ecological distribution in the Rajasthan desert. *J. Bombay Nat. Hist. Soc.* 69: 510-517.

Prakash, I. and Rana, B.D., 1973. A study of field population of rodents in the Indian desert. III. Sand dunes in 100 mm rainfall zone. *Z. angew. Zool.* 60: 31-41.

Prakash, I., 1974a. The Ecology of Vertebrates of the Indian Desert. Chapter XIII in *Biogeography and Ecology in India (*Ed. M.S. Mani) pp. 369-420. Dr. Junk b.v Verlag, The Hague,

Prakash, I., 1974b. A note on the use of *Salmonella* for the control of rodents in India. *Plant Pot. Bull. Manila,* 3(9): 15-16.

Prakash, I., 1974c. Rodents and their control: Rodent damages in the field and their control. In *Training Manual "Post-harvest prevention of waste and loss of food grains. Asian Productivity Organization.,* Tokyo, 193-199.

Prakash, I., 1974d. Some observations on the birds of great Rann of Kutch. *Newsletter for Birdwatchers* 14: 4-6.

Prakash, I. Jain, A.P. and Rana, B.D., 1974. A white Indian gerbil, *Tatera indica indica* Hardwicke. *J. Bombay Nat. Hist. Soc.* 70: 375-376.

Prakash, I., 1975a. The amazing life in the Indian desert. *Illustrated Weekly (Annual),* 96-121.

Prakash, I., 1975b. Population ecology of rodents in the Rajasthan desert, India. Chapter V in *"Rodents in Desert Environments"*(Eds. I. Prakash & P.K Ghosh), Dr. W. Junk Publishers, The Hague, The Netherlands,.

Prakash, I., 1975c. The ecology and zoogeography of mammals. Chapter XIX in *"Environmental Analysis of the Thar Desert"* (Eds. R.K. Gupta & I. Prakash), English Book Depot, Dehradun, pp. 448-467.

Prakash, I., 1975d. Wildlife ecology and conservation. Chapter XX in *"Environmental Analysis of the Thar Desert"* (Eds. R.K. Gupta & I. Prakash) English Book Depot, Dehradun, pp. 468-480.

Prakash, I., 1975e. Bird watching (or searching) in Mizoram. *Newsletter for Birdwatchers* 14: 4-6

Prakash, I., 1975f. Wildlife ecology and conservation. In: *"Environmental Analysis of the Thar Desert"* (Eds. R.K. Gupta and I. Prakash), pp. 468-480. Dehradun.

Prakash, I., 1975g. Replacement of sympatric rodent species in the Indian desert. *Proc. All India Rodent Seminar,* Ahmedabad, pp. 29-31.

Prakash, I. (Ed.) 1975h. Proceedings of the Summer Institute on Rodentology. Monograph No. 2. 1-315. (mimeo.), Central Arid Zone Research Institute, Jodhpur.

Prakash, I. and Ghosh, P.K. (Eds.) 1975. *Rodents in Desert Environments.* Dr. Junk Verlag b.v. The Hague, 1-628.

Prakash, I., Jain, A.P. and Rana, B.D., 1975a. A study of field population of rodents in the Indian desert. IV. Ruderal habitat. *Z. angew. Zool.* 62: 339-348.

Prakash, I., Jain, A.P. and Rana, B.D., 1975b. Baits for the control of hairy-footed gerbil, *Gerbillus gleadowi* Murray. *Z. angew. Zool.,* 62(2): 349-359.

Prakash, I., Rana, B.D. and Jain, A.P., 1975. Bait shyness in three species of *Rattus. Z. angew Zool.* 62: 89-97.

Prakash, I., 1976a. Rodent Pest Management-Principles and Practices. Monograph No. 4, 1-28, Central Arid Zone Research Institute, Jodhpur.

Prakash, I., 1976b. Declare war on rats. *Illustrated Weekly of India.* October pp. 19, 21 & 23.

Prakash, I., 1976c. Rodent control in rural and urban areas: *Proc. Symp. Recent Trends in Plant Protection,* Allahabad, 46: 156-160.

Prakasth, I., 1976d. Significance of rodent pest management for progressive agriculture. *Pesticides Annual.* 54-57.

Prakash, I. and Gupta, R.K., 1976. An ecological exploration of an extremely arid tract of Rajasthan. *Indian J. Ecol.* 3: 11-21.

Prakash, I., 1977a. Desert rodents and their management. In: *"Desertification and its Control",* pp. 263-268 Indian Council of Agricultural Research, New Delhi.

Prakash, I., 1977b. Biology of the rodents of Rajasthan desert. In: *National Resources of Rajasthan* (Ed. M.L. Roonwal), Vol. 1, pp. 337-352, University of Jodhpur, Jodhpur.

Prakash, I., 1977c. Current status of rodent control in India. *Pestology* 1: 1-3.

Prakash, I., 1977d. Rodent pests of agricultural crops and their control. *Indian Farming.* 27(1): 39-43.

Prakash, I., 1977e. Second Regional Training Seminar on field rat control and research. *Pestology* 1(4): 23-31.

Prakash, I., 1977f. The Amazing Life in the Indian desert. Monograph No. 6, 10-18, Central Arid Zone Research Institute, Jodhpur

Prakash, I., 1977g. Wildlife and its inventory. *Proceedings of the Summer Institute on* Resouce Inventory and Land Use Planning, pp. 269-276. CAZRI, Jodhpur.

Prakash, I., 1977h. Population dynamics of rodents and their control. In: Desert Ecosystem and its Improvement. (Ed. H.S. Mann), Central Arid Zone Research Institute Monogr. 1: 341-349.

Prakash, I., 1978a. Impact of changing land use pattern on the rodent communities in the Indus valley. *Proc. Symp. Land and Water Management in the Indian Basin*, Vol. 2, pp. 481-486.

Prakash, I., 1978b. Rodent ecology: its significance in rodent control. In *Special Bulletin of Plant Protection on Rodent Control, Directorate of Plant Protection,* Govt. of India, New Delhi. Plant Protection Bulletin 26: 13-15.

Prakash, I., 1978c. Wildlife resources and management. CAZRI Souvenir "25 Years of Arid Zone Research", pp. 61-65, CAZRI, Jodhpur.

Prakash, I. and Mathur, R.P., 1978. Efficacy of chlorophacinone for the control of Indian desert rodents. *Pesticides* 13(6): 41-46.

Prakash, I. and Ojha, P., 1978. The phenomenon of poison bait aversion and its possible attenuation in Indian gerbil, *Tatera indica indica* (Hardwicke). *Z. angew. Zool.* 65: 385-395.

Prakash, I., Rana, B.D. and Jain, A.P., 1978. Evaluation of RH-787 as a rodenticide. *Z. angew. Zool.,* West Germany, 65(1): 5-10.

Prakash, I. and Saroj Kumari, 1978. Morphological variation in *Meriones hurrianae* (Jerdon, 1867) inhabiting two bioclimatic zones in the Indian desert. *Saugt. Mittlen.* 40, 26: 281-284.

Prakash, I., Soni, B.K. and Mathur, R.P., 1978. Baits for the control of Indian desert gerbil, *Meriones hurrianae* Jerdon. *Z. angew. Zool.* 65(4): 303-312.

Prakash, I., 1979a. Sparrow control in China. *Newsletter for Birdwatcher* 19: 11.

Prakash, I., 1979b. Rodent research in India. *J. Bombay Nat. Hist. Soc.* 75: 792-799.

Prakash, I., 1979c. Adaptations of Indian rodents for desert survival. *Proc. Symp.* Organisonic Adaptation to Tropical Environments, pp. 1-10, Madurai.

Prakash, I., 1979d Eliminate rodents to prevent food losses. *Loss Prevention News* 1(4): 12.

Prakash, I., 1979e. Cereal crops protection against rodents in India. *Proc. Cong. For Crop Prot.* Barseilla, France. 1-6.

Prakash, I. and Kumari, Saroj, 1979. Occurrence of ventral marking gland in desert rodents. *Saugt. Mitteil.* 27: 315-316.

Prakash, I., 1980a. National Programme for Rodent Pest Management. In: *Indian Pesticides Industry – fact and figures* (Ed. B.V. David). Vishwas Publication, Bombay, pp. 326-327.

Prakash, I., 1980b. Rodenticides and future trends. In : *Indian Pesticide Industry-facts and figures* (Ed. B.V. David), Vishwas Publications, Bombay, pp. 232-233.

Prakash, I., 1980c. Current status of Great Indian bustard in the Thar desert. *Proc. Intl. Symp. Great Indian Bustard*, Jaipur.

Prakash, I., Advani, R., Soni, B.K. and Rana, B.D., 1980. Evaluation of the poison base for the control of common house rat, *Rattus rattus rufescens* (Gray), *Z. angew. Zool.*, West Germany, 67: 213-223.

Prakash, I. and Ghosh, P.K., 1980. Human animal interactions in the Rajasthan desert. *J. Bombay Nat. Hist. Soc.* 75: 1259-1261.

Prakash, I., 1981a. Wildlife conservation in the Thar. *Arid Lands Newsletter* 14: 2-8.

Prakash, I., 1981b. Bordi (*Z. nummularia*) important food centre for Indian desert animals. CAZRI Bulletin No. 13, pp. 48-50.

Prakash, I., 1981c. Current status of rodent control in India. Coúntry report, GTZ Tech. Bullein (Ed. Norbert Weis), West Germany 5: 1-3.

Prakash, I., 1981d. Rodent pest management in arid zone. *Development Policy and Administration Review.* VII(2): 111-117.

Prakash, I., 1981f. Rodent pests and their management. *Proc. Summer Instt. on Agro-forestry.* CAZRI, Jodhpur, 261-268.

Prakash, I., 1981g. Rodent pests in the desert environment and hazards to human health. *Proc. Symp. Deserts, Man and Health,* Jaipur.

Prakash, I., 1981h. Ecology of mammals of Rajasthan Desert with special reference to rodents and their management. D.Sc. thesis. Univ. of Rajasthan, Jaipur, pp. 1-1345.

Prakash, I., 1981i. Ecology of the Indian desert gerbil, *Meriones hurrianae.* Monograph No. 10, 1-87, Central Arid Zone Research Institute, Jodhpur

Prakash, I., Advani, R. and Mathur, R.P., 1981. Population dynamics of rodents in rural residential environment in response to control operations. *Proc. Workshop Farm Level Post Harvest Tech. for Prevention of Food Losses,* New Delhi.

Prakash, I., 1982a. The Indian arid zone-strategies for management. In Alternative Strategies for Desert Development and Management (Eds. M.R. Biswas and A.K. Biswas), Pergaman Press, New York, Vol. 4: 1128-1135.

Prakash I., 1982b. Damage by *Bandicota bengalensis* to growing rice and other cereal crops in India and rodent control programmes to reduce it. *Proc. Third Inst. Theriol. Cong. Acta Zoologica Fennica.*

Prakash, I., 1982c. The living Thar desert. *Sanctuary* 11: 130-137.

Prakash, I. and Idris, M., 1982a. Response to conspecific ordour of the scent marking gland by *Rattus meltada. Z. angew. Zool.* 69: 481-486.

Prakash, I. and Idris, M., 1982b. Scent marking by the female Indian gerbil, *Tatera indica* from two distinct habitats during oestrus. *Indian J. Expt. Biol.* 20: 915-816.

Prakash, I., 1983a. The two threats. In Souvenir: *Desert Wildlife Protection Society, Godavan*, Jodhpur p. 28.

Prakash, I., 1983b. Mammal pests of agricultural crops. In "*Agricultural Entomology*" (Eds. P.D. Srivastava, M.G. Jotwani, R.A. Agarwal, S.R. Wadhi, R.K. Bhanotar and R.K. Bhatnagar), Vol. 2, pp. 370-386. All India Scientific Writers' Society, New Delhi.

Prakash, I., 1983c. Rodents acquire newer territories: Strategies for their management. In *Frontiers of Research in Agriculture* (Ed. S.K. Roy). Indian Statistical Institute, Calcutta, pp. 436-442.

Prakash, I. (Ed.) 1983d. Wildlife Protection in the Desert. (Wildlife Protection Society, Jodhpur), Bulletin 1:1 45.

Prakash, I., 1983e. Faunal resources of the great Indian desert: Problems and prospects. Proc. Indo-US Workshop on Arid Zone Research, pp. 263-266. DST, New Delhi.

Prakash, I., 1984a. Vertebrate pest problems in India. *The Organizational and Practices of Vertebrate Pest Control*, pp. 29-35. Hampshire, UK.

Prakash, I., 1984b. Combating desertification. *Yojana* (New Delhi), 28(12): 28-29.

Prakash, I., 1985a. Efficacy of conspecific urine in masking shyness behaviour in two gerbils. *Proç. Second Symp. On Rodent Advances in Rodent Control* (Eds. A.H.M. Mohammad Chal), pp. 131-140. Ministry of Public Health, Kuwait.

Prakash, I., 1985b. Minimising crop losses due to rodent pests in irrigated agriculture in arid area. *Proc. Development and Management Training Course on Irrigated Agriculture in Arid Areas* (Ed. S.D. Singh). Human Resources Development Group. Govt. of India (WAPCOS), New Delhi 609-704.

Prrakash, I., 1985c. Asian predators of livestock. Chapter 23 In: *Predators, Pests and parasites* (Eds. Gaafar, S.M., W.E. Howard & R.E. Marsh), pp. 405-410 Elsevier Scientific Publishing Company, Amsterdam..

Prakash, I., 1985d. Rodentology, an emerging discipline. In *Recent Advances in Zoology* (Eds. C.B.L. Srivastava and Suresh C. Goel) pp. 123-126, Rastogi and Co., Meerut.

Prakash, I., 1985e. Biotic factors in dry zone afforestation. *Proc. FAO/DANIDA Training Course on Sand Dune Stabilization, Shelterbelt and Afforestation in Dry Zones*. FAO Conservation Guide No. 10, pp. 193-202, FAO, Rome.

Prakash, I., 1986a. Faunal diversity in the arid tracts. *Proc. Indian Acad. Sci.*(B): 45-59.

Prakash, I., 1986b. Some aspects of the ecology of Indian rodents. In *Wildlife Wealth of India* (Ed. T.C. Majpuria), pp. 528-532. Craftsman Press, Bangkok.

Prakash, I., 1986c. Wildlife resources and its management in desert environment. In "*Conservation and Management*" (Eds. K.A. Shankarnarayan and Vinod Shankar), Monograph No. 25, pp. 19-22. CAZRI, Jodhpur.

Prakash, I., 1987a. Strategies for vertebrate pest management in India. In *Plant Protection in Field Crops* (Eds. Rao, M.V. and Sithanathan, S.) pp. 39-45 Plant Protection Association of India, Hyderabad,

Prakash, I., 1987b. Vanishing wildlife in western Rajasthan. In *Environmental Degradation in Western Rajasthan* (Eds. S.M. Mehnot & M.M. Bhandari), *Dainik Pratinidhi*, Jodhpur.

Prakash, I. and Idris, M., 1987. Olfactory communication among desert rodents. In *Recent Trends in Ethology in India* (Eds. M. Balakrishnan and K.M. Alexander), pp. 107-113. The Ethological Society of India, Bangalore.

Prakash, I., Idris, M. and Soni, G.R., 1987. Scent marking behaiour of wild and laboratory bred Indian desert gerbils, *Meriones hurrianae. Proc. Indian Nat. Sci. Acad.* 54: 31-34.

Prakash, I. and Mathur, R.P., 1987. *Management of Rodent Pests ICAR,* New Delhi, 1-133.

Prakash, I., 1988a. Rodent problems in India. *FAO Plant Prot. Bull.,* 36(3): 145-148.

Prakash, I., 1988b. Bait shyness and poison aversion. Chapter 18 in *Rodent Pest Management pp,*321-329. CRC Press, Boca Raton, USA, 1-480.

Prakash, I., 1988c. Changing patterns of rodent populations in India. Chapter 9 in *Rodent Pest Management,* pp. 179-190. CRC Press, Boca Raton, USA, 1-480.

Prakash, I., 1988d. Faunal resources of degraded land in northwestern plains. In: *Wastelands Development and Their Utilization* (Ed. K.A. Shankarnarayan) pp 112-117, Scientific Publishers, Jodhpur:

Prakash, I., 1988e. Wildlife-human-animal interactions, and conservation in the Rajasthan desert. In *Desert Ecology* (Ed Ishwar Prakash) pp. 169-178 Scientific Publishers, Jodhpur.

Prakash, I. (Ed.) 1988f. *Desert Ecology.* Scientific Publishers, Jodhpur, pp. 1-313.

Prakash, Ishwar (Ed.) 1988g. *Rodent Pest Management.* CRC Press, Boca Raton, USA, 1-480.

Prakash, I. and Idris, M., 1988. Olfactory communication among desert rodents. *Proc. All India Symposium of Animal Behaviour,* ZSI, Jodhpur, pp. 1-12.

Prakash, I. and Mathur, R.P., 1988. Rodent problems in Asia. Chapter 5 in *Rodent Pest Management* (Ed. Ishwar Prakash):pp 67-84, CRC Press, Boca Raton, USA, 1-480.

Prakash, I. and Rao, A.M.K.M., 1988. Vertebrate pest management. In : *Forty years of Plant Protection in India 1946-1986.* (Ed. D. Bap Reddy), pp. 46-56, Plant Protection Association of India, Hyderabad.

Prakash, I., 1989a. Degradation of desert rangelands due to grazing pressure. In: *Perspective in Ecology* (Eds. J.S. Singh & Brij Gopal) pp. 1181-1191, Jagmandir Book Agency, New Delhi.

Prakash, I., 1989b. Wildlife management: A factor in controlling soil erosion and desertification. In "*Workshop on Desertification, Drought, Soil Erosion and Related Problems*", CAZRI, Jodhpur.

Prakash, I. 1989c. Changing rodent pest situation in India. In *National Seminar on Changing Pest Situation in the Current Agricutlure Senario of India,* pp. 1-8, IARI, New Delhi.

Prakash, I., 1989d. Rodent conservation in the Indian sub-continent. In Rode*nts: A World Survey of Species of Conservation Concern*. (Ed. W.C. Lidicker). Occ. Paper IUCN, Switzerland, No. 4: 48-50.

Prakash, I., 1990. The need for research (key note address). *Proc. Group Meeting on Rodent and Rice*. International Rice Research Institute, Manila: 1-10.

Prakash, I., 1991a. Wildlife and its social aspects. In *Development of under developed regions- A case of Arid Region* (Eds. S.K. Lal and U.K. Nahar). Mittal Publications, Meerut: 76-87.

Prakash, I., 1991b. Rodent communities, epidemiological and economic considerations. In: *Indira Gandhi Canal Project* (Ed. I.P. Abrol), pp. 110-115. Indian Council of Agricultural Research, New Delhi.

Prakash, I., 1991c. Ecological role of snakes and their economic impact on human society. (In: Snake and human welfare. Ed.: Director, ZSI, Calcutta), pp. 1-7.

Prakash, I., 1991d. Ecology of artiodactyles in Thar desert: Their conservation in the desert biosphere reserve. In: Mammal in the Palearctic Desert. (Eds. A. McNeely and V.M. Neronov). The Russian Academy of Sciences & Russian Committee for the UNESCO MAB: 243-250.

Prakash, I., 1992a. Recent communities, epidemiological and ecological considerations. In *Prospects of IGP Project*. (Eds. I.P. Abrol and J. Venkateswarlu) pp. 110-115, Indian Council of Agricultural Research, New Delhi.

Prakash, I., 1992b. Bagwani phaslon ko jeenvon se bachaiye *Phalphool,* ICAR, New Delhi, 15 (July): 10-12 (In Hindi).

Prakash, I., 1992c. Future lines of research, In *Rodents in Indian Agriculture,* (Eds. Ishwar Prakash and P.K. Ghosh). Vol. 1, pp. 685-692, Scientific Publishers, Jodhpur.

Prakash, I. and Ghosh, P.K., 1992. *Rodents in Indian Agriculture*. Scientific Publishers, Jodhpur. Vol. 1. State of Art: 1-707.

Prakash, I. and Idris, M., 1992a. Scent marking behaviour. *Ibid*: 1: 445-460.

Prakash, I. and Idris, M., 1992b. The desert gerbil, *Meriones hurrianae*. *Ibid*. 1: 49-62.

Prakash, I., Kashyap, N. Mathur, N., 1992. The five-striped squirrel, *Funambulus pennanti*. *Ibid*. 1: 17-24.

Prakash, I. and Mathur, R.P. 1992. Acute rodenticides. *Ibid*. 1: 497-516.

Prakash, I, Saxena, S.K., Ghosh, P.K. and Bohra, H.C., 1992. Vegetation and wildlife in the Thar. In *Perspectives of the Thar and the Karakum*. (Eds. Kar, A., R.K., Abichandani, K. Anantharaman and D.C. Joshi) pp. 98-109. Department of Sci. & Tech. Govt. of India, New Delhi: 9.

Prakash, I., 1992. The *Euphorbia niche. Cactus Aurentules,* France, 13: 17-18.

Prakash, I. and Shyamsunder Rao, 1992. Management of vertebrate pests. In: *Agricultural Technologies,* ICAR, New Delhi.

Prakash, Ishwar., Mathur, R.P., Mathur, M., Parveen, F. and Patel, N., 1992. *Rodents in Indian Agriculture*. Scientific Publishers (India), Jodhpur. Vol. 2. Selected Bibliography: 1-246.

Prakash, I., 1993a. Wildlife management in the Thar desert-A factor in controlling soil erosion and desertification. In: *Desertification and its Control in the Thar, Sahara and Sahel Regions.* (Eds. A.K. Sen and A. Kar) pp. 353-362, Scientific Publishers, Jodhpur.

Prakash, I., 1993b. Food, feeding and shyness behaviour of the gerbils of the Thar desert. *Proc. Indian Natl. Sci. Acad.* (B) 59: 31-39.

Prakash, I., 1993c. The changing ecology of the Thar desert. *Proc. Indian Natl. Sci. Acad.* (B)59: 465-470.

Prakash, I., Rana, B.D. and Jain, A.P., 1993. Reproduction in the Cutch Rock-rat, *Rattus cutchicus cutchicus* in the Indian desert. *Mammalia* 37: 457-467.

Prakash, I., 1994a. Indira Gandhi Nahar Pariyojna: Kya hoga Jaivik Vividhta ka Paridrayashya. *Rajasthan Patrika,* 19 June 1994, p. 18. Paryavaran (In Hindi): 6(4): 39-42.

Prakash, I., 1994b. Biodiversity conservation in the Thar desert. *Indian Forester,* 120 (10): 873-879.

Prakash, I., 1994c. *Ecology and consevation of the Indian desert.* In: Tropical Ecosystems (Eds. M. Balakrishnan, W. Bie-stein and Reidar Borgstram) pp. 167-192.

Prakash, I., 1994d. Rodent management in India: Problems and new perspectives. In: *Behavioural Aspects sof Feeding.* (Eds. B.G. Galef Jr., M. Mainardi and P. Valesechi), pp. 369-385 Hardwood Academic Publishers, Chur (Switzerland).

Prakash, I., 1994e. Rukshshetra ki Samasyayen; Rokna hoga badte registana ko. *Rajasthan Patrika,* 11 Dec. 1994, p.9 (In Hindi).

Prakash, I., 1994f. The haunting desert. *Hornbill,* 2: 26-30.

Prakash, I., 1994h. *Mammals of the Thar Desert.* Scientific Publishers (India), Jodhpur, 1-114.

Prakash, I. and Agrwal, V.C., 1994. An Action Plan for Rodent Conservation in the Indian sub-continent. IUCN/SSC: 1-48.

Prakash, I., 1995a. Rodent pests the threat to the Indian subcontinent (Guest Editorial). *Phytoparasitica,* Israel 23(1): 3-6.

Prakash, I., 1995b. Invasion of peninsular small mammals towards the Aravalli ranges and the Thar desert. *Intl. J. Ecol. and Environmental Sci.* 21(1): 17-24.

Prakash, I., Saravanan, A. and Singh, P., 1995a. Ecology and taxonomy of the field mice in the Aravalli ranges. *J. Bombay nat. Hist. Soc.* 92(3): 372-377.

Prakash, I., Saravanan, A. and Singh, P., 1995b. Changes in the abundance of small mammals on the Aravalli ranges due to irrigated agriculture. *Proc. Natl. Acad. Sci.* (B) 65(1): 47-52.

Prakash, I. and Singh, P., 1995. Some observations on the birds of Abu hill, Aravalli ranges. *Pavo* 33: 99-110.

Prakash, I., Singh, P. and Saravanan, A., 1995a. Small mammals of the Abu hill, Aravalli ranges, Rajasthan, India. A comprehensive taxonomical and ecological study. *Zoology,* 5(1): 55-64.

Prakash, I., Singh, P. and Saravanan, A., 1995b. Niche alteration by the Cutch Rock-rat, *Cremnomys cutchicus* in the Aravallis *J. Bombay nat. Hist. Soc.,* 92(2): 259.

Prakash, I., Singh, P. and Saravanan, A., 1995c. Ecological distribution of small mammals in the Aravalli ranges. *Proc. Indian natl. Sci. Acad.* (B) 61(2): 137-48.

Prakash, I., 1996. Rodent communities in Thar sand dune ecosystem. *Ann. Arid Zone,* 35(3): 241-247.

Prakash, I., Singh, P. and Saravanan, A., 1996. Small mammalian diversity on the Abu hill. *Proc. Indian natn. Sci. Acad.,* B 62(2): 95-104.

Prakash, I., 1997a. *Thar Rukshashetra ke Stani:* Scientific Publishes (India), Jodhpur, 1-108 (In Hindi).

Prakash, I., 1997b. Desrtification. In: *Encyclopedia Asiatica.* VIII.

Prakash, I., 1997c. Community ecology and behaviour of mammals in desert and montane ecosystems of Rajasthan. *Association,* Calcutta. 1-33.

Prakash, I., 1997d. Ecology of desert mammals. *Curr. Sci.* 72(1): 30-34.

Prakash, I., 1997e. The risk-prone impacts of Indira Gandhi canal irrigation in the Thar desert. In: Souvenir "Research and Development Priorities for Arid Zone Beyond 2000 AD" CAZRI, 1: 69-72.

Prakash, I., 1997f. Dilemma of wildlife research in India. *Vacham,* 8 (2): 30-36.

Prakash, I., 1997g. Child hood in Mount Abu. *Hornbill* (4): 24-28.

Prakash, I., 1997h. Grass is green. *Down to Earth* 6(5) : 54-55.

Prakash, I., 1997i. The amazing desert gerbil. *Resonance* 2(10): 54-61.

Prakash, I. and Singh, P., 1997a. Ecological distribution and reproduction of the shrew, *Suncus murinus sindensis* on the Aravalli mountain hill *J. Ecol. And Env. Sciences,* 23(1): 17-24.

Prakash, I. and Singh, P., 1997b. Habitat preference of the Indian bush rat, *Golunda ellioti gujerati* in the Aravalli montane ecosystem. *J. Bombay nat. Hist. Soc.,* 94: 559-561.

Prakash, I., 1998a. Biodiversity decline in the Thar desert during twentieth *Centuary.* In: *Management of Arid Ecosystems* (Eds. A.S. Praroda *et al.*) Scientific Publishers (India), Jodhpur.

Prakash, I., 1998b. Northward migration of certain Deccanean elements. In: *Deccan Heritage* (Eds. H.K. Gupta *et al.*), pp. 27-34 Indian National Science Academy, New Delhi.

Prakash, I., 1998c. Impact of Indira Gandhi Canal on the desert ecosystem. In: *Ecology Today An Anthology of Contemporary Ecological Research* (Eds. B. Gopal, P.S. Pathak, K.G. Saxena) pp. 323-331, Intl. Scientific Publishers, New Delhi.

Prakash, I., 1998d. Australia ke vichitra standhari. *Strata Eklavya* : 1-8 (In Hindi).

Prakash, I., 1998e. Changing rodent ecology in India and public health. *Health and Environment* (Ed. Anil Agarwal), pp. 1-52 New Delhi.

Prakash, I., 1998f. Vilakshan registani gerbil. *Strotra* 119 (December): 25-28.

Prakash, I., Idris M., Kumari, S., 1998. Scent marking behaviour of three desert rodents *PINSA* B 64 (5&6): 319-334.

Prakash, I. and Singh, P., 1998. Ecology and behaviour of murids of south Asia. In *Mammals of South Asia.* (Eds. A.J.T. John Singh and R. Sukumar). Oxford IBH Publishing Co., New Delhi. Pp: 1-50

Prakash, I., 1999a. Decline of the Aravallis. *Down to Earth,* 8(3): 56.

Prakash, I., 1999b. Biodiversity rich spots in Rajasthan – Ecological changes *Second Dr. M.L. Bhati Memorial Lecture, University of Delhi, Delhi* 1-32.

Prakash, I., 1999c. Thar, Nature's gift. In: Survey of the Environment. *The Hindu,* 1999, 73-78.

Prakash, I., 1999. Marusthali Jahak. *Strotra,* 131: 35-38 (In Hindi)

Prakash, I. and Singh, H., 1999a. Food of the shrew, *Suncus murinus* inhabiting hilly tracts of south and southeastern Rajasthan. *Proc. Natl. Acad. Sci.* India, 69(B): 245-250.

Prakash, I. and Singh, H., 1999b. Distribution of three rodent species in the hilly tracts of Rajasthan. *J. Bombay nat. Hist. Soc.* 96:415-467.

Prakash, I. and Singh, P., 1999c. Species composition, relative abundance and attitudinal distribution of small mammals in the Aravalli montane range. *Intl. J. Ecol. & Env. Sci.* 25: 37-49.

Prakash, I., 2000a. Mammalian diversity on the Aravalli mountain. Presidential Address. Ninth All India Congress of Zoology. Bhuvaneswar, 1:1-6.

Prakash, I., 2000b. The hedgehog. *Hornbill : 9-10.*

Prakash, I., 2000c. Badalti ecology aur Thar ki Jaiv vividhta. *Strotra,* 142: 11-14. (in Hindi).

Prakash, I. and Singh, H., 2000. Small mammal diversity and ecology of small mammals in the Aravalli montane ecosystem in southern Rajasthan. *Proc. Natl. Acad. Sci.* India,

Prakash, I., 2001a. Survival strategies of desert vertebrates. In : *Ecology of Desert Environments* (Ed. Ishwar Prakash). Scientific Publisher (India), Jodhpur, 459-471.

Prakash, I., 2001b. Biological invasion and loss of endemic biodiversity in the Thar desert. *RESONANCE,* 6: 76-85.

Prakash, I. and Singh, H., 2001. Ecological Study of small mammals in the hilly tracts of southeastern Rajasthan. *Tropical Ecology* 42: 25-33.

Purohit, K.G., Kametkar, L.R. and Prakash, I., 1968. Reproduction biology and post-natal development of the Indian palm squirrel, *F. pennanti. Mammalia* 30: 538-546.

Rana, B.D., Prakash, I. and Jain, A.P., 1970. Morphological variation in *Tatera indica* Hardicke. 1807, inhabiting two types of Indian desert habitats. *Acta Theriol.* 15: 459-464.

Rana, B.D., Prakash, I. and Jain, A.P., 1975. Bait shyness and poison aversion in the hairy-footed gerbil, *Gerbillus gleadowi* Murray. *Proc. All India Rodent Seminar,* pp. 658-660, Ahmedabad.

Rana, B.D., A.P. Jain and I. Prakash, 1975. Morphological variation in the field rodents inhabiting the Indian desert II. *Gerbillus nanus indus* (Thomas, 1920). *Acta Theriologica* 20: 123-132.

Rana, B.D., I. Prakash and A.P. Jain, 1975. Variation in the weight of the adrenal glands of the Cutch Rock-rat, *Rattus c. cutchicus* (Wroughten). *Mammalia* 39: 479-480.

Rana, B.D. and Prakash, I., 1979. Reproduction biology and population structure of house shrew, *Suncus murinus sindensis* in western Rajasthan. *Z. Saugetierk*, West Germany 44: 333-343.

Rana, B.D. and Prakash, I., 1980. The metad: A serious rodent pest of Indian agriculture. *Indian Farming* 29: 21-23.

Rana, B.D. and Prakash, I., 1982a. Population structure of the metad *Rattus meltada Pallidior* in the Thar desert. *Z. angew. Zool.* 69: 79-87.

Rana, B.D. and Prakash, I., 1982b. Rodenticides and their future trends. *Krishi Chaynika*. Jan-March issue.

Rana, B.D. and Prakash, I., 1984. Reproduction biology of *Rattus meltada pallidior* (Ryley 1914) in the Rajasthan desert. *J. Bombay Nat. Hist. Soc.* 81: 59-70.

Rao, A.M.K.M. and Prakash, I., 1980a. Baitshyness among the house mouse, *Mus musculus bactrianus* Blyth to zinc phosphide and RH-787. *Indian J. Expt. Biol.* 18: 1490-1492.

Rao, A.M.K.M. and Prakash, I., 1980b. Laboratory evaluation of bait bases for the control of the house mouse, *Mus musculus bactrianus* Blyth. *Bull. Grain Technol.* 18(2): 111-118.

Rao, A.M.K.M. and Prakash, I., 1981. Activity and hoarding behaviour of the desert gerbil, *Meriones hurrianae. Indian J. Expt. Biol.* 19: 792-793.

Rao, A.M.K.M. and Prakash, I., 1986. Rodenticidal evaluation of zinc phosphide against house mouse, *Mus musculus bactrianus. Indian J. Plant Prot.* 13: 61-63.

Reichman, O.J., Prakash, I. and Roig, V., 1979. Food selection and consumption. IBP *Arid Land Synthesis* 16: 681-716.

Saxena, S.K. and Prakash, I., 1990. Rehabilitation of arid grazing lands in the Thar desert. In: *Environmental Rehabilitation* (Ed., M.K. Wali). SPB Academic Publishing bv, The Hague, The Netherlands, 203-216.

Saxena, S.K. and Prakash, I., 1992. Rehabilitation of arid grazinglands in the Thar desert. In: *Ecosystem Rehabilitation.* (Ed. M.K. Wali) Vol. 2, pp. 37-50,. S.P.B. Academic Publishing b.v., The Hague.

Singh, P. and Prakash, I., 1997a. Ecology and behaviour of the Cutch rock-rat, *Cremnomys cutchicus medius* in the Aravallis. *Proc. Nat. Acad. Sci. India.* 67(B): 89-110.

Singh, P. and Prakash, I., 1967b. Food preference and calorific requirement of the Cutch rock-rat, *Cremnoys cutchicus medius. Ann. Arid Zone,* 36(1): 65-72.

Soni, B.K., Prakash, I. and Mathur, R.P., 1979. Baits for the control of bush rat. *Golunda ellioti gujerati* Thomas. *Indian J. Rodentology* 1(1&2): 1-9.

Soni, B.K. and Prakash, I., 1981. Laboratory evaluation of WBA 8119 (Brodifacoum) against *Gerbillus gleadowi* Murray. *Pestology* 5(9): 23-24.

Soni, B.K. and Prakash, I., 1982. Relative acceptance of three acute rodenticides by five desert rodents. *Pestology* VI (9): 13-16.

Soni, B.K. and Prakash, I., 1984. Efficacy of brodifacoum against the house mouse, *Mus musculus*. *Pesticides* XVIII (7): 15-17.

Soni, B.K. and Prakash, I., 1984. Evaluation of Brodifacoum for the control of *Rattus meltada* and *Golunda ellioti*. *Indian J. Plant Protection*. 12: 19-23.

Soni, B.K. and Prakash, I., 1985. Evaluation of liquid brodifacoum against gerbils and rats. *Pestology* 9: 21-24.

Soni, B.K., Prakash, I. and Mathur, R.P., 1980. Laboratory evaluation of poison base for the control of *Funambulus pennanti*. *Proc. Indian Acad. Sci.*, 89(3): 227-233.

Soni, G.R. and Prakash, I., 1987. Effect of conspecific urine odour on behaviour of soft-furred field-rat, *Rattus meltada*. *Behav. Processes* 14: 175-182.

Soni, G.R. and Prakash, I., 1988. Mitigation of poison shyness in desert gerbil by adding conspecific sebum of the ventral scent marking gland and urine in poison baits. *Indian J. Expt. Biol.* 26: 476-478.

Tripathi, R.S., Jain, A.P., Kashyap, N., Rana, B.D. and Prakash, I., 1992. Rodents of northwestern desert. In: Rodents in Indian Agriculture. (Eds.: Ishwar Prakash and P.K. Ghosh), Vol. 1, pp. 357-396.

Changing Faunal Ecology in Thar Desert, pp. 27-36, 2005.
Eds.: B.K. Tyagi and Q.H. Baqri

2

SOME NATURAL AND ANTHROPOGENIC SPECIFICITIES OF THE THAR

R.P. Dhir

Principal Scientist and Director, CAZRI (Retd.), 498 Defence Colony, Jodhpur 342009 (Raj.) India

INTRODUCTION

Professor Ishwar Prakash was a towering figure in the field of animal ecology in the country as well as abroad - an exemplary achievement in view of the fact that much of his investigational effort was confined to rodents and that too in a restricted ecological domain of the Thar. He has left behind a huge legacy in the form of original research papers and finely edited tomes and synthesis which bear ample testimony to his love of the Thar Desert, holistic view of nature, keen observational ability, meticu-lous self-organization and an unparalleled capacity to do hard work. So much was his dedication to the cause of science that despite his affliction he continued to work till his last breath with his usual calm and serenity. The institutions he worked at: the Central Arid Zone Research Institute, where he spent much of his time and rose to occupy the most coveted position of Professor of Eminence; and Desert Regional Station of the Zoological Survey of India, where he chose to settle after superannuating from CAZRI, will continue to be inspired and bear imprints of his charm and luster for decades to come. I had the privilege of a long colleagueship and friendly association all through and therefore, when I got an invitation to contribute an article dedicated to the memory of Prof. Prakash, I accepted it as an opportunity to pay my own tributes to this most recognized, distinguished arid zone scientist and a kind-hearted human being.

This article covers some aspects of ecological setting, growth of biotic pressure and man-land relations - subjects, which have held some scientific interest for me and have appeared in some form or the other in press. These do not directly fall within the ambit of changing faunal ecology, the central theme of this book. Yet, I hope that the material may provide a backdrop on the Thar and lead to better understanding and appreciation of the findings on the matter presented later in this compendium.

NATURAL ENDOWMENTS

Like most other deserts, the Thar experiences abundant solar radiation, a high temperature regime with considerable diurnal variation, strong dust-raising winds and an acute atmospheric aridity. However, its distinguishing feature is the monsoon pattern of rainfall. Eighty-five to 90 per cent of the entire yearly rainfall is received during the 75-90 day period of south-western monsoon. Even though the actual amount and distribution of rainfall vary greatly from year to year, high atmospheric humidity, calm winds and moderate temperatures provide near ideal situation for a rainfall-dependent regeneration and growth of vegetation. The accumulated biomass continues to sustain secondary and tertiary consumers long after the monsoons have withdrawn.

Nearly seventy per cent of the area has highly sandy soils in form of dunes and aeolian plains. These soils lack any aggregation and thus have a high basic erodibility. A large-scale soil erosion and movement of drift sands under a sparse vegetation cover and strong winds of pre-monsoon dry summer is an annual feature. The nutrient reserve in these soils is low but it does not come in the way of establishment and maintenance of a healthy vegetation cover such as is permitted by the limited natural precipitation. These have a rapid infiltration rate but a very low water retention capacity, *i.e.*, 40-75 mm per meter depth of soil, as compared to 100 to 200 mm in loamy soils. But a favourable attribute of these sandy soils is their ability to conserve the stored moisture in the face of a high moisture deficit of the atmosphere. Several observations have shown that bare sandy soils below 60 cm depth remain almost at field capacity through out the year, including the most dry and windy period of May and June (Gupta, 1979). This conservation ability is due to extremely low unsaturated state hydraulic conductivity of sandy soils. For example, at 500 kPa a sandy soil has a conductivity of only 8.1×10^{-8} cm per day as compared to 1.7×10^{5} cm per day for a sandy loam soil. In other words, a sandy soil is two hundred times less conductive at this moisture tension than a sandy loam soil (Dhir *et al.*, 1991). It is this remarkable quality which prevents evaporative loss of stored moisture and permits maximum use by the vegetation. This attribute of sandy soils is a major contributing factor for survival of perennial trees and shrubs in an otherwise hostile setting marked by nine months of rainless period. It ensures also congenial conditions in the burrows and tunnels of the desert animals.

GROWTH OF BIOTIC PRESSURE

With a density of 110 persons per square kilometer, the Thar is indeed by far the most densely populated amongst the world arid zones. Much of this growth has taken place during the last 7-8 decades only. Though, almost the entire arid zone, including its most desertic parts in Jaisalmer and Barmer had not only been settled but also politically organized towards the end of the first millennium AD, the population all through has remained thin. It has been shown from some sample villages that during the two hundred and fifty years to the beginning of the twentieth century, the populations just doubled only. This near stagnancy in growth of human population is evident also from the fact that the population of the erstwhile Marwar State had remained close to 2 million during the hundred year period from year 1820. Of course from the year 1881 there are regular census data in existence. It shows that population up to year 1921, though fluctuating, possessed hardly any growth. However, from then on started a period of continuous, phenomenal growth, with population doubling in the first thirty-five years and redoubling in the next two and a half decades. The growth rate all the time has been higher than that of the Rajasthan State or the country as a whole (Fig. 1).

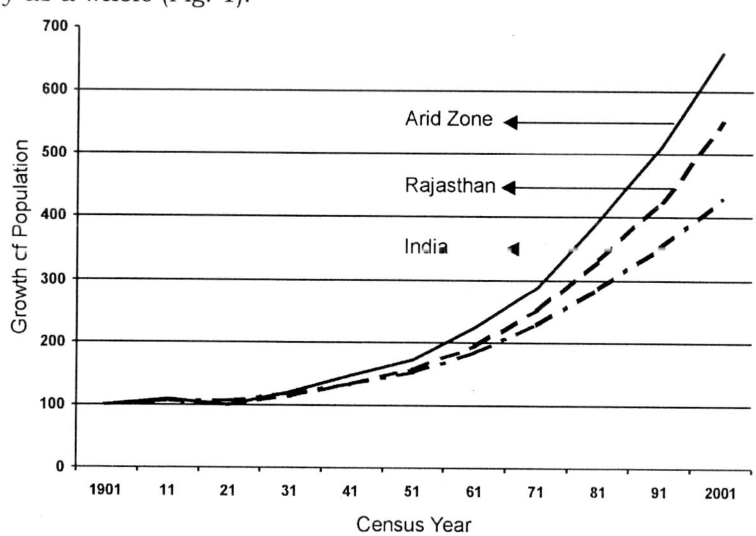

Fig. 1 Human population growth over the years (Census year 1901 = 100). The arid zone has recorded a far greater growth than the Rajasthan or country as a whole.

The increased population meant also more number of households. Since animal husbandry is an integral component of rural economy in the Thar, this meant parallel increase also in the number of livestock. The livestock population stood at 10 million in the year 1952 and steadily rose to a peak at 23.3 million in the census year 1982. It declined to 18.2 million following the droughts of the years 1987-88 but bounced back to an all time high at 28.5 million in the census year 1997. Besides, a profound change has occurred in composition of livestock. The cattle population which stood at 3.9 million in the year 1956, rose to just 4.6 million by year 1997 but that of sheep plus goat rose two and a half times during the corresponding period. Besides, the buffalo population has also steadily risen from 0.77 million in year 1956 to 3.2 million in the past four decades. In areas with substantial irrigated area, in fact, the buffaloe outnumber the cattle. The change in composition has some rationale behind it. The decline of cattle has been attributed to widespread deterioration of grazing lands. But a more significant reason seems to be greatly diminished cattle economy. The main value of cattle used to be the male calf but replacement of bullock power by tractor has taken away much of this value. The sheep and goat have a short gestation period and a ready, good market, are not so demanding in the type of fodder they need and are thus easy to maintain.

TRANSLATIONAL CHANGES IN LANDUSE

During medieval times right up to the close of the nineteenth century, animal breeding had remained a major pursuit with arable farming confined to the immediate vicinity of settlement. In favourable situations some irrigation was practiced also where crops like sugarcane, cotton, poppy, cucurbits were raised. However, these cultivated lands constituted no more than 5 to 20 per cent in different parts of arid zone. Situation started changing rapidly thereafter. Some census data, old maps and aerial photographs show that the period from 1930s to mid-fifties witnessed a major expansion of cultivation in the climatically favourable eastern half of arid zone of Rajasthan. Since this period is marked also by a rapid rise of human population, it is not difficult to correlate the two, though undoubtedly, the process of land reforms of the forties and fifties giving *khatedari* rights to actual tillers and ceiling on holding must have contributed also to the situation (Dhir, 1982). After this period, further expansion in farming of arid zone took place through: (a) reduced incidence of fallow farming in the established agricultural tract; (b) breaking of new lands considered marginal climate or soil wise, and (c) introduction of water from the Indira Gandhi Canal Project. Today, two-thirds of the arid Rajasthan is under cultivation, the rest is under village settlement and other utilities with usually two to four per cent area under community grazing lands in form of (i) *gauchars* (designated pasture lands in village records), (ii) *agor* (lands earmarked as catchment area of village ponds), and (iii) *orans* (lands in the name of a deity or temple). However, westwards incidence of cultivation declines and there is a sizable extent of open grazing lands, designated in revenue records variously as pasture lands or culturable waste lands. Present land use distribution is shown in Fig. 2. It shows also that the line separating the dominantly cultivated lands from the pasture lands has clearly shifted from the earlier 250 mm rainfall isohyet westwards to the 200 mm rainfall isohyet now.

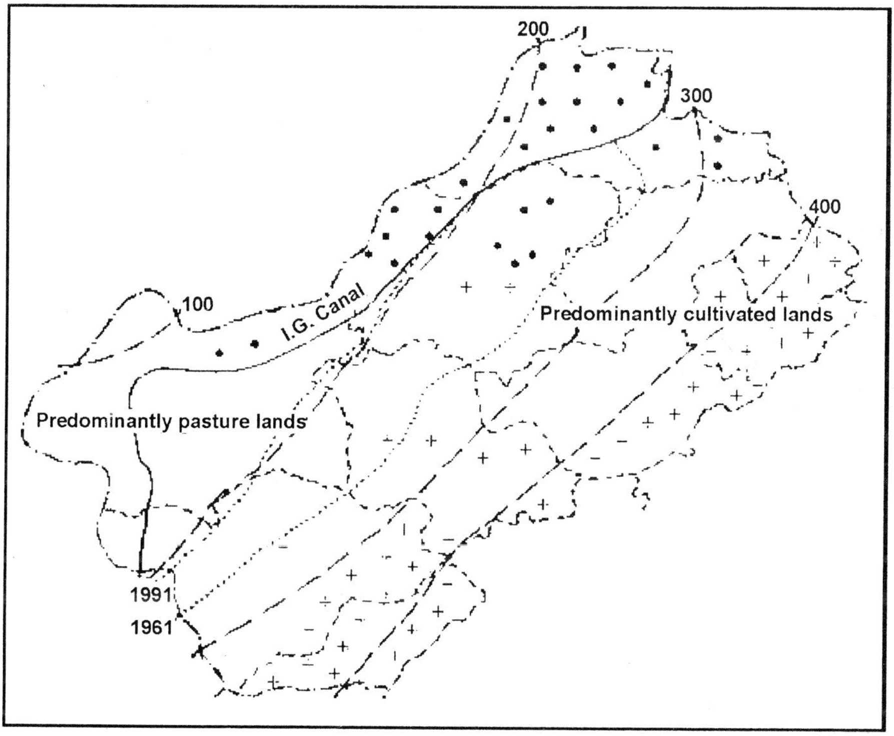

Figure 2. Present land use in the Thar. In the past thirty years the line of predominantly cultivated has shifted westwards to 250 mm isohyet

HIGHLIGHTS OF ARID LAND MANAGEMENT

Living in the desert calls for immense forbearance and endurance. Further to cope up with large uncertainity in land productivity the desert dwellers through their wisdom and experience over the generations have evolved several management styles and adaptation/adjustment mechanisms. Migration, both of human beings and livestock, has been a characteristic feature. The inter-state migration has come under considerable stress lately because of resistance from local communities. The other outstanding feature of arid land management is the rather efficient blend of animal husbandry and arable farming. Most farming households keep sizable herd of livestock that are sustained on crop residues, left over vegetation on crop lands, including the top-feed species. This mixed farming approach has several advantages (Jodha,

1985). It enhances the rural economy. In fact much of the cash in hands of a farmer comes from sale of livestock and their produce as the homegrown food grains go almost entirely for family consumption. Further, whereas crop activity is seasonal, animals provide a yearlong employment, particularly to the women. The system diversifies the economy basket and permits a risk coverage. In fact if one was to seek an explanation as to how an environmentally handicapped region like the Thar is able to sustain such a large population of both human beings and livestock, the answer will lie mainly in this efficient integration of animal husbandry and agriculture (Dhir, 2003).

Agroforestry, *i..e.*, conscientious promotion of trees on croplands is another example of indigenous wisdom. Most crop fields in 300-450 mm rainfall zone sustain 30 to 50 full-grown trees that give the lands the look of an open woodland. The trees provide protection to land against wind erosion but also top-feed, irrespective whether rain has occurred during a particular year or not. The tree does not compete with annual crops for soil moisture and in fact crops under the tree canopy are better because of improved soil fertility, a fact confirmed by scientific tools. Mixed cropping, scheduling of crops in relation to the date of onset, food and grain storage practices and rainwater harvesting and conservation structures such as naadis, taankas and khadins are other noteworthy examples in this genre.

IRRIGATION IN THE DESERT

Scarcity of water conditions life styles and arid land management and it is also a major constraint to all biological activity and development effort. Therefore, development of water resources has been a prime welfare measure and a philanthropic action for the nobility and other resourceful persons. Therefore, a number of tanks and dug wells for drinking and irrigation came into being in the last hundred years in the western foot-hill zone of the Aravalli Mountain. But such a development was restricted in the area to the west with deep-seated and often saline aquifers. A historic development, emulated from the irrigation canal system in the British territories in north-west India, was the construction in the year 1927 of the Gang Canal with a command area of 3.5 lac hectare by Maharaja Ganga Singh of Bikaner State. Since the resources of the State were far from adequate and riparian rights were not in its favour, it took lot of pleading, political maneuvering and borrowing of money to get the project going. The project was a grand success and set stage for further development and in post-independent early fifties of last century came the Bhakra Canal, which extended the irrigation by another 4 lac hectare. However by far a major stride has been the Indira Gandhi Canal Project (IGCP). This gigantic project started in late fifties aimed in its final form the construction of 650 km of main canal and a distribution network of over eight thousand kilometers and a command area of 1.8 million hectare. The project is in final stage of construction and it is already commanding nearly half of its target area and producing crops worth five billion of rupees. However, its operation has also thrown up problem of water logging and salinisation that necessitate remedial measures in form of drainage and efficient water management at field level.

Besides canal system, ground water based irrigation has expanded also. This has been made possible largely by rural electrification and an easy access to drilling technology. In fact of 2.45 million hectare net irrigated area, more than half is from ground water resource.

ENVIRONMENTAL CONCERNS OF DESERT DWELLERS

If the desert still supports a reasonable variety and numbers of wild-life, it is not due to the vast open lands and wilderness alone but also due to the care and concern of the people for life-forms. Bishnoi community is particularly reputed for its respect and religious ethos for protection and care of all living beings. Most of the 29 commandments are devoted to bio-diversity conservation and protection. Examples are not lacking of the members fighting for and even sacrificing their lives for protection of wild life. The most glorious example is of Khejarli village, where in the year 1780, 363 women, men and children sacrificed their lives in an attempt to protect khejri tree. Hence the common proverb:

"Sir katte, rungta bache, sasto sauda jaan"

(If a tree can be saved by sacrificing oneself, consider it a good bargain)

Though not with that intensity, other communities are kindly disposed towards plants and animals and observe care in not unnecessarily hurting these. Most villages maintain bird–feeding grounds and watering points on community basis. Guda Bishnoi, Kheechan Bird Sanctuary and Taal Chhapar Black Buck reserve are shining examples of peoples' initiative and support.

ENVIRONMENTAL IMPACT OF BIOTIC PRESSURE

Vegetation degradation

Despite the awareness and concern on the one hand and existence of meaningfully appropriate practices in a rather fragile ecosystem, on the other, the mundane considerations and an over-exploitative management have caused marked deterioration of natural assets. Vegetation degradation is alarming. Arable farming is more a transformed ecosystem where natural vegetation has been replaced by a cultured system. Though a conscious effort is made to protect and promote khejri tree, there is a obvious loss of biodiversity. *Calligonum polygonoides, Leptadenia pyrotechnica, Caparis aphylla* and several others have virtually disappeared from the more regularly cultivated eastern half. Even the pragmatically important shrub, *Zizyphus nummularia* and Khejri are under threat because of tractor cultivation.

The grazing lands constitute nearly one-fourth of the total area of the arid zone. A small part of this lies as individual parcels of lands, called *gauchar* and *oran*, in the immediate vicinity of the settlements in the agricultural region, the major chunk is in the western part, where rainfall is so low that crop husbandry is nearly impossible except in few favourable pockets (Fig. 2). These latter lands just four decades ago supported a fine stand of perennial grasses. Estimates of the early eighties showed that nearly two-thirds of these have become degraded to totally desertified (Dhir, 1988). The picture since then has worsened further and now patches with some respectable cover have become a rare feature. The resource base degradation is partly related to command area development and partly due to increased stress from over use and neglect. The *gauchars* had come to be degraded much before (Gupta & Saxena, 1971). The outcome is that the productivity of these lands has been reduced to one-fourth and less of the potential (Kumar, 1994).

The worst form of degradation has occurred on magra and rocky lands. The saprse cover of hardy species has gone and much of these have been turned into barren lands. The degradation is all the more grievous as it is well nigh impossible to rehabilitate these.

Wind erosion

Both the croplands and the pasture lands suffer from this problem. The deflation, loose sand sheets, piling of drift sands along field boundary and settlements, appearance of shifting sand dunes are the common manifestations of this process. The problem has got greatly accentuated due to extension of cultivation onto more vulnerable dunes and to climatically marginal lands. However, a major contributor is the tractorisation of cultivation. Their number was insignificant in early sixties and rose to 75,700 by year 1992 and is now over 1.15 lakh with tractor cultivation covering nearly eighty percent of the cropland. It's use removes the useful perennial vegetation and also makes the land easily erodible. Several studies have been made on the extent of soil loss due to wind erosion. For example Gupta *et al.* (1981) reported a loss of 325 to 615 t/ha for partially stable and unstable dunes. Dhir and Venkateswarlu (1996) reported for an above average windy year, *i.e.*, 1985, a mean loss of 207, 283, 932 and 2837 t/ha from long fallow, current fallow, tilled and disc ploughed lands respectively. The degraded grazing lands showed a mean soil loss of 472 t/ha with well-managed lands hardly any. Opinions differ as regards the impact of this problem. In authors considered opinion the loss of cropland productivity is of the order of 10-20 per cent. Further, the eroded lands are able to get back to their productivity in a couple of years.

Dwindling ground water resources

Rural electrification, subsidy on its supply together with loan facility and easy availability of drilling technology has brought in a great increase in groundwater use. This can be assessed from the irrigated area, the gross extent of which has increased from 0.25 mha in 1956 to 1.70 mha by 1999, a seven-fold increase. This has led to severe decline of ground water table. Most of the good aquifers except those with recharge from Aravalli drainage system are showing a decline of 1 to 2 meters per year. In fact it is apprehended that at this rate most of the aquifers will run dry in 25-50 years time. Though depletion of groundwater is a serious in other parts of the country, it is the most grievous in arid zone as recharge opportunities are very slim and also costly.

SUMMARY

The Thar has experienced profound changes in its complexion during the past few decades arising primarily from greatly intensified use of natural assets and an exploitative management, both dictated by greatly increased human pressure. The traditional, conservative management has lost ground to the expediency of making a living. Though people are by large having a better living condition that have been made possible by improved means of transport and communication, drinking water and public health schemes. There is also a governmental safety net during periods of drought for the human population and to an extent for the livestock, the land resources continue to degrade. There have been a centrally sponsored desert development programme in place for the past nearly three decades. Its important components are desert afforestation, pasture development, sand dune stabilization, surface water development and soil and water conservation. With an expenditure of nearly ten billion rupees there are some notable achievements, but the problem of natural resource degradation is far from solved. Though some of the processes are reversible, the weakened societal mechanisms for common good and collective functioning, on the one hand, and the continuous build up of biotic pressure and market economy, on the other, often call for greatly increased investments and societal organization.

REFERENCES

Dhir, R.P., 1982. The Human Factor in Ecological History- Arid Zone of Rajasthan. In "Desertification and Development: Dryland Ecology in Social Perspectives" B. Spooner and H.S. Mann Academic Press, London, pp. 311-332.

Dhir, R.P., 1988. Flux in Indian Arid Zone. In "Desert Ecology" I. Parkash (Ed.). Scientific Publishers, Jodhpur: 15-36.

Dhir, R.P., 2003. Thar desert in retrpspect and prospect. *Proc. Indian Natn Sci. Acad.* 69, A. No. 2, pp. 167-184.

Dhir, R.P., Chaudhary, M.R., Jagannath and Somani, L.L., 1991. Sandy Soils Bull. No. 15, Indian Soc. Soil Sci., New Delhi, pp. 58-69.

Dhir, R.P. and Venkateswarlu, J., 1996. Problem of Wind Erosion in relation to land use and management. In "Soil Management in Relation to Land Degradation and Environment" Indian Society of Soil science, New Delhi. Bull. 17: 56-65.

Gupta, J.P., 1979. Some observations on the periodic variation of moisture in stabilized and and unstabilised sand dunes. *J. Arid Hydrology* 41: 153-156.

Gupta, J.P. Agarwal, R.K. and Raikhy, N.P., 1981. Soil erosion by wind from bare sandy plains in the arid zone of western Rajasthan. *J. Arid Env.* 4: 15-20.

Gupta, R.K. and Saxena, S.K., 1971. Ecological studies on the overgrazed rangelands in the arid zone of western Rajasthan. *J. Indian Botanical Soc.* 50: 289-300.

Jodha, N.S., 1985. Report of the Institute of Development Studies, Jaipur.

Kumar, S., 1994. Ecology of degraded vegetation and its recovery in Indian arid zone, *JIRCAS Intern. series* 1: 27-37.

Changing Faunal Ecology in Thar Desert, pp. 37-44, 2005.
Eds.: B.K. Tyagi and Q.H. Baqri

3

IMPACT OF INTRODUCTION OF IGNP CANAL IRRIGATION ON MICRO- AND SECULAR CHANGES IN CLIMATE OF THE THAR DESERT REGION

A.S. Rao

Central Arid Zone Research Institute, Jodhpur 342 003 (Rajasthan), India

Human intervention in the form of introduction of irrigation from Indira Gandhi Nahar Parjyojana (IGNP) canal water in the Thar desert region in different phases, exposing different locations to different lengths of irrigation, has brought about changes in micro-climate, flora and fauna of the area. To know the impact of introduction of large-scale irrigation on micro-climate and climatic changes in the area as a result of the increased vegetation and replenishment of moisture to the rainfall systems, data pertaining to past one hundred years of annual rainfall and air temperatures (1901-2002) were analyzed from selected locations along the tracts of the canal. Though, Sri Ganganagar region has been imposed with canal irrigation since the past around seven decades, the increase in annual rainfall was apparent only during the last four decades (1961-2001). The long term rate of increase in the annual rainfall of Sri Ganganagar was 1.18 mm/year. However, in the areas where irrigation has been imposed during the past few decades, the rainfall decreased at a rate of 0.59 mm/year at Bikaner or no change in rainfall was observed at

Jaisalmer. In general, the decrease in air temperature was observed in the region at a rate of 0.008 °C/year at Sri Ganganagar, 0.005 °C/year at Bikaner and 0.006 °C/year at Jaisalmer.

INTRODUCTION

The Thar Desert region in western Rajasthan, which spreads in around 196,150 sq. km, is very fragile and subjected to excessive stresses due to frequent droughts and low rainfall. The arid phase of northwest India has a history of about 3000 years (Pant & Maliekel, 1987). Winstanly (1973 a,b) analyzed the rainfall at Bikaner and Jaisalmer and reported that the monsoon in the arid region of northwest India was favourable during 1700 to 1930 but unfavourable since 1970 onwards. The studies conducted on secular changes in rainfall and air temperatures of northwest India covering the meteorological sub-divisions of Punjab, Haryana, west Rajasthan and west Madhya Pradesh, showed that there was a marginal increase in the rainfall by 141 mm in the past 100 years (Pant & Hingane, 1988) and more so in irrigated belts of Sri Ganganagar region particularly during the past three decades (Rao, 1996). Studies on climatic changes over Jodhpur region showed that the rainfall and air temperatures were favourable, but the increase in human population (by 400%) and livestock (by 127%) during the twentieth century, resulting a major shift in land use pattern and tremendous pressure on surface and groundwater resources, were the main causative factors for drought and desertification conditions in the region (Rao & Miyazaki, 1997). Ramakrishna and Rao (1991) also observed an increase in the mean decadel rainfall of Sri Ganganagar which was under irrigation for a considerable period. In other irrigated arid areas of the Middle Asia also, an increase in humidity by 3 to 5%, while decrease in air temperatures by 2.5 to 2.0°C, resulting in decrease of the growing period by 200 to 500° days was reported (Orlovsky & Utina, 1977; Orlovsky & Durdyiev, 1978).

A study of droughts in the Thar region indicated that 47 to 62 per cent of the years experienced droughts of some intensity or the other, therefore, the Famine Commission (1890) and Irrigation Commission (1905) emphasized for an urgent need for irrigation in dry lands to bring stability and improve the agricultural productivity (Roy, 1983). The Gang Canal in 1927 and the Indira Gandhi Nahar Project (IGNP) in 1961 were initiated for diverting water from eastern rivers of Indus system, *viz.*, Ravi and Beas, to the arid western Rajasthan. The canal aims at utilizing 7.59 MAF of water from Ravi-Beas system to irrigate 11.43 lakh ha area (708,775 ha in Sri Ganganagar, 37,022 ha in Bikaner and 84 ha in Jaisalmer with net crop sown area of 78.02, 36.23 and 7.93%, respectively, in these districts initially envisaged under the IGNP), thereby aiming for additional food production of 31.00 lakh tones per annum (Roy, 1983). The stage-I of the project began in 1961, construction of 204 km feeder canal from Herike barrage to Hanumangarh and another 189 km long main canal extending up to Chattargarh. The work of stage-II on main canal of 256 km from Chattargarh to Mohangarh in Jaisalmer district began in 1970 (Chouhan, 1993).

The impact of irrigation in the arid Rajasthan is seen from vegetational and micro-climatic changes that occurred in the region. In the present review the trends in the annual rainfall and air temperatures at selected locations along the canal irrigated tracts of arid western Rajasthan, which are exposed for varying lengths of irrigation, are analytically presented and discussed.

MATERIAL AND METHODS

To study the possible impact of introduction of irrigation from IGNP canal, historical rainfall and air temperature data records (1901-2002) of Sri Ganganagar (29^0 55'N; 73^0 53'E; 177 m), Bikaner (28^0 00'N; 73^0 18'E; 224m) and Jaisalmer (26^0 54'N; 70^0 55'E; 242 m), all of which are located along the tracts of the canal, were selected for the study on long-term climatic changes. The meteorological droughts were based on departures from the aridity index of Thornthwaite in terms of standard deviation of aridity index (Subrahmanyam & Subramaniam, 1964). The long-term trends in respect of annual rainfall and mean annual air temperatures of the locations were analyzed using linear regression analysis with time as dependent variable and tested for significance. To support, the increase in annual rainfall over Sri Ganganagar region, rainfall trends of neighbouring stations, *viz.*, Anupgarh, Hanumangarh and Raisinghnagar, were also analyzed for the period.

RESULTS AND DISCUSSION

Climate and frequency of agricultural droughts

The IGNP canal region experiences an arid climate with a mean annual rainfall ranging from 100 mm at western parts of Jaisalmer to 300 mm at Sri Ganganagar with a high coefficient of variation. Compared to the available rainfall, the annual potential evapotranspiration values vary from 1600 mm at Sri Ganganagar to 1800 mm at Jaisalmer (Table 1). The water deficit thus results in aridity conditions at these locations. The frequency of agricultural droughts in the IGNP canal region based on aridity index is shown in Table 2. The Jaisalmer region experienced highest number of agricultural droughts (with 67 percent of the years), followed by Bikaner (with 45 percent of the years) and Sri Ganganagar (with 38 percent of the years). The periods 1900-10, 1911-20, 1961-70, 1981-90 and 1991-2000 recorded highest number of drought years in the region.

Table 1. Decade-wise percentage departure of from normal annual rainfall at some locations in the IGNP canal region

Decade	Sri Ganganagar	Bikaner	Jaisalmer
1921-30	+23	+04	-18
1931-40	-34	+03	00
1941-50	+09	+15	-09
1951-60	-06	+09	+30
1961-70	-17	-20	-08
1971-80	+12	-09	+08
1981-90	+22	-01	-22
1991-2000	+5	+7	+32

Prior to the introduction of Gang canal, about 20 million ha of land in Sri Ganganagar area was under grasslands and put under grazing for livestock. After the availability of canal water, traditional crops like pearl millet, greengram, clusterbean and sesame have been replaced in irrigated areas with crops like cotton, sugarcane, groundnut, rice and pegionpea during *kharif* and wheat, mustard, sugarbeet, chickpea and fodder crops in *rabi* season particularly in the Sri Ganganagar region. Thus, the human intervention of introducing canal water into the arid ecosystem brought about changes in the land-use a very great deal causing changes both in the micro- and macro-climates that also influenced the rainfall mechanisms directly or indirectly.

Table 2. Frequency of agricultural droughts in the IGNP canal region

Location	Moderate drought	Severe drought	Number of years with drought (in percentage)
Sri Ganganagar (1926-2002)	8	22	30 (38%)
Bikaner (1901-2002)	23	23	46 (45%)
Jaisalmer (1901-2002)	25	43	68 (67%)

Trends in annual rainfall

The normal annual rainfall of Sri Ganganagar is 283 mm with a coefficient of variation of 54%. Long-term annual rainfall at the location showed an overall increasing trend (Fig. 1a), at the rate of 1.18 mm/year, during the period 1926-2002 (Fig. 1a). Even in drought periods faced by western Rajasthan, Sri Ganganagar area received relatively higher rainfall, which is a factor attributed to the irrigation imposed for longer periods in the region directly by supply of moisture to rainfall systems and indirectly through energy balance from increased vegetation. The annual rainfall records of nearby stations to Sri Ganganagar, *viz.*, Anupgarh, Hanumangarh and Raisinghnagar, also showed a positive trend in annual rainfall during the period supporting the increase in rainfall of the region.

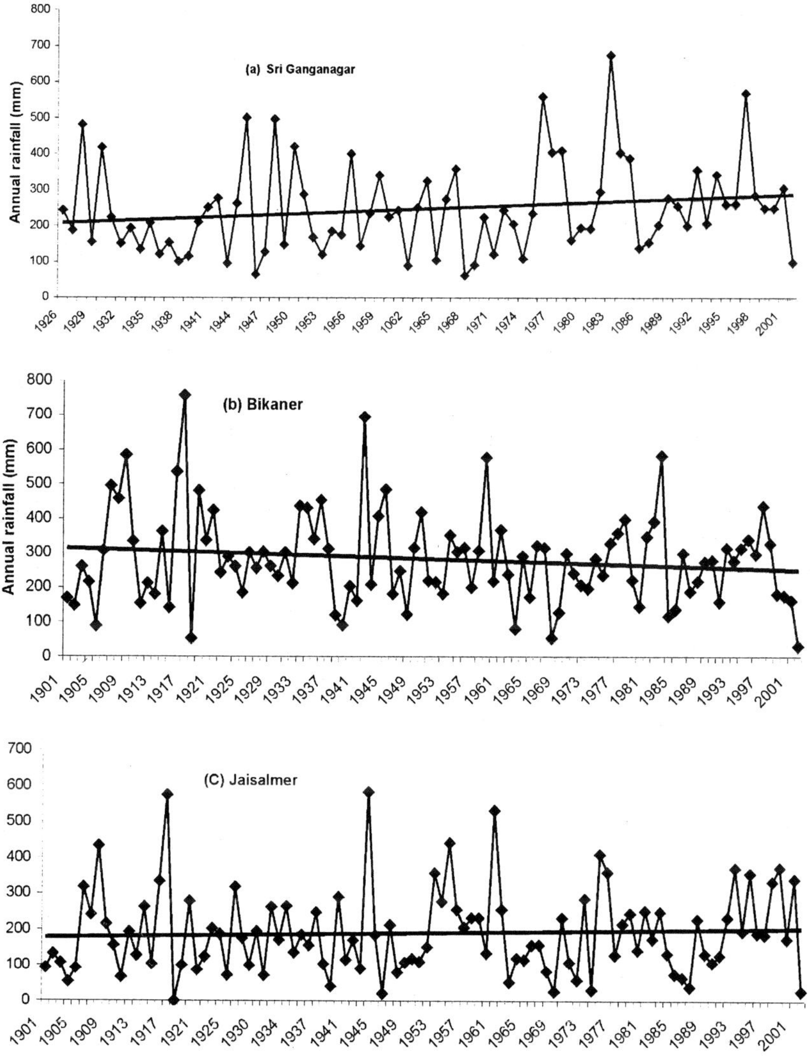

Fig. 1. Trends in annual rainfall (1901-2002) at (a) Sri Ganganagar, (b) Bikaner and (c) Jaisalmer

Bikaner receives an annual rainfall of 264 mm with a coefficient of variation of 48%. The annual rainfall trend at Bikaner showed an overall decreasing trend at a rate of 0.59 mm/year during 1901 to 2002 (Fig. 1b), the reduction in rainfall was due to droughts experienced after 1961.

Jaisalmer, which is located in the extreme western part of Rajasthan, receives a mean annual rainfall of 192 mm, with a coefficient of variation in annual rainfall of 64%. However, the overall trend in the annual rainfall at Jaisalmer (Fig. 1c) remained the same during 1901-2002.

Shifts in decade-wise annual rainfall

The decade-wise annual rainfall departures (Table 1) showed that Sri Ganganagar region received 5 to 22% higher rainfall from normal during the current decades 1971-1980, 1981-1990, and 1991-2000, the periods of intense irrigation given in the region. Bikaner region which had lesser exposure for irrigation, received above normal rainfall only during 1991-2000 showing length of irrigation might have played a role in vegetational changes associated with rainfall. Out of the seven decades commencing 1921-30 to 1991-2000, the annual rainfall was below the normal for a period of three decades at Sri Ganganagar and Bikaner, whereas for a period of four decades at Jaisalmer (Table 1). The cumulative departure values of rainfall from normal during 1921-30 to 1991-2000 decades were +14 at Sri Ganganagar, +8 at Bikaner and −13 at Jaisalmer indicating the lineament in the rainfall departure was more towards positive side at Sri Ganganagar, whereas, it was towards negative side at Bikaner and Jaisalmer. The decade-wise mean annual rainfall showed that the increase in rainfall was high during the last four decades beginning 1961, the periods of intense irrigation given in the region. In the recent decade, increased number of depressions originating from Arabian sea contributed for higher rainfall at Jaisalmer and the increase was not attributed to the canal irrigation.

Long-term trends in air temperatures

The air temperatures at all the three locations showed a general decreasing trend at a rate of 0.008°C/year at Sri Ganganagar, 0.005°C/year at Bikaner, and 0.006°C/year at Jaisalmer (Figs. 2a-c). The fall in air temperatures was due to rise in rainfall (Fig. 1a-c) at these locations.

Other environmental considerations

With the improper use of available canal water for irrigation, other environmental issues like water logging to an extent of 20,000 ha in the command area was reported. This changed the hydrological balance of the arid area with salinity build up from sub-surface soil layers. In arid regions, irrigated crops consume more water than the net energy available for evapotranspirational processes due to a phenomenon known as advection caused by dry surrounding areas with low humidity and high winds to transport the moisture. Therefore, a need has arisen for proper use of the canal based water for crop development with improved methods of irrigation like drip or sprinkler systems, ground water recharge and use of bio-drainage through tree plantations to arrest the water logging or through certain policies for proper utilization of precious water in the desert.

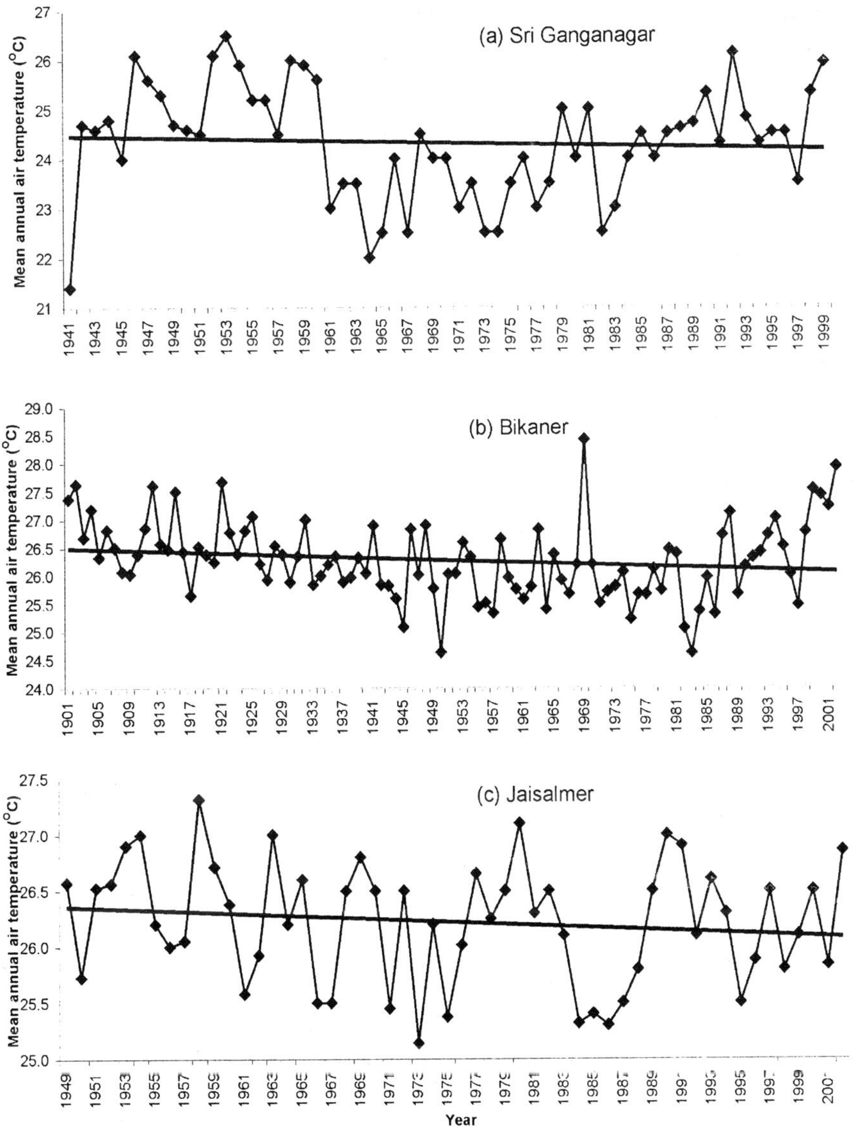

Fig. 2. Trend in mean annual air temperature in the IGNP canal region at (a) Sri Ganganagar, (b) Bikaner and (c) Jaisalmer.

Further, the humidity measurements over irrigated cropped surfaces were 5 to 10% higher than over the unirrigated surfaces, which indicate that the amount of canal water applied can influence atmospheric humidity, but it is not known how far this increase in humidity can enhance precipitation, which mainly depend upon the dynamical conditions of the atmosphere. Such human interventions favouring for improved rainfall patterns may even reverse the climate from arid phase to semi-arid phase over a period of time. However, the above study has clearly brought out that the irrigation imposed for longer periods in Sri Ganganagar region enhanced the precipitation processes in the Thar desert region.

ACKNOWLEDGEMENTS

The author is thankful to Dr. Pratap Narain, Director, Central Arid Zone Research Institute, Jodhpur for providing the facilities.

REFERENCES

Chouhan, T.S., 1993. Development of irrigation in Rajasthan. In: Natural and Human Resources of Rajasthan (Ed: T.S. Chouhan), *Scientific Publishers*, Jodhpur, pp. 125-145.

Orlovsky, N.S. and Durbyiev, A.M., 1978. Irrigation impact on thermal resources of Middle Asian deserts. *Prolemy Osvenine Pustya*. 4: 70-76.

Orlovsky, N.S. and Utma, Z.M.,1977. Micro-climate of desert areas irrigated and watered. *Proloany Obvpcia Pustyn*, 3:3-12.

Pant, G.B. and Hingane, L.S., 1988. Climatic changes in and around the Rajasthan desert during the 20th century. *Journal of Climatology* 8: 391-401.

Pant, G.B. and Maliekal, J.A., 1987. Holocene climatic changes over north-west India. An appraisal *Climatic Change* 10: 183-194.

Ramakrishna, Y.S. Rao, A.S., 1991. Climatic features of the Indira Gandhi Canal region. In: Prospects of Indira Gandhi Canal Project. (Eds: I.P. Abrol and J. Venkateswarlu), ICAR, New Delhi, pp. 1-10.

Rao, A.S., 1996. Climatic changes in the irrigated tracts of Indira Gandhi Canal Region of arid western Rajasthan, India. *Annals of Arid Zone* 38(2): 111-116.

Rao, A.S. and Miyazaki, T., 1997. Climatic changes and other causative factors influencing desertification in Osian (Jodhpur) region of the Indian arid zone. *Journal of Arid Land Studies* 7(1):1-11.

Roy, T.K., 1983. Impact of Rajasthan canal project on social, economic and environmental conditions. NCAER, New Delhi, 179 p.

Subrahmanyam, V.P. and Subramaniam, A.R., 1964. Applicability of water balance concepts for the climatic study in south India. *Annals of Arid Zone* 4(2): 46-56.

Winstanley, D.,1974a. Recent rainfall trends in Africa, the Middle East and India. *Nature* 243, 464-465.

Winstanley, D., 1974b. Rainfall patterns and general atmospheric circulation. *Nature* 245, 190-194.

Changing Faunal Ecology in Thar Desert, pp. 45-60, 2005.
Eds.: B.K. Tyagi and Q.H. Baqri

4

SOIL BIORESOURCES: A TECHNICAL MANAGEMENT IN ARID ENVIRONMENT

B.M. Sharma

Department of Zoology, J.N.V. University, Jodhpur-342 001, India

INTRODUCTION

One of the most important contributions of soil bioresources to ecosystem function is that of soil formation and turnover. They contribute to soil formation via both biochemical and biomechanical processes (Whitford, 2001). Soil biota constitutes the driving force of terrestrial ecosystems because it controls the rate of turnover and mineralization of organic substrates. In addition to plant and microbes, the microscopic animalcules such as protozoans are present in pore spaces where they ingest and process fine organic particles. Some protozoan species are found in gut of bigger soil dwelling fauna where they enhance the decomposition of cellulosic debris (Swift *et al.*, 1979). Among other invertebrates, soil arthropods are either saprophagous or carnivorous forms and include millipedes, mites, springtails *etc.* The saprophagous forms are of direct relevance to decomposition process. The significant effects of soil fauna on the nutrient dynamics of ecosystem have been documented (Butcher & Snider, 1971; Singh & Pillai, 1973; Moore *et al.*, 1988; Verhoeff & Brussaard, 1990; Tripathi *et al.*, 2003a). The underlying hypothesis is that soil fauna-associated decomposition improves physical, chemical and biological properties of soil, which in turn, increases protein synthesis and crop yield. It is a modification of proposition on the plant-soil ecosystem influenced by bio-remediated materials (Tomati & Galli, 2002). Soil fauna favour microbial activity, increase enzymatic activity, stimulate root development and maintain soil health.

There has been increased interest concerning the role of beneficial soil biota in enhancing soil productivity (Lavelle *et al.*, 1992). Soil contains dead organic materials in various stages of decomposition. In the majority of natural and agro-ecosystems all of these components are closely associated and combined to influence the physical, chemical and biological properties of soil. Soil biota impart in various activities to influence soil health. Soil biological health is used in a generic sense to describe the properties, processes and potentials of the soil system associated dead and living organic materials. Wood (1991) considered the biological quality of the soil in terms of the soil organismic populations or of the processes accomplished by the organisms. When plants die, the above ground plant parts are distributed within the soil by gravity, water movement and by soil organisms. Below ground root growth distributes carbon as does movement of dead plant parts by animals or water movement. In agro-ecosystems soil fauna play an important role in distribution and flow of carbon within the soil system by creating galleries, chambers, burrows, mounds and nests and producing faecal matters and plants have their root in the soil where there is an active interaction with soil biota including microbes and invertebrates. So, there is an urgent need to understand how to conserve the soil biota to maintain essential soil properties and how to access resilience of soil processes to environmental changes.

FAUNAL BIODIVERSITY

Biodiversity dictates a link among species, genetic and habitat diversities. It is acknowledged as foundation for sustainable development and food security. Commonly found soil protozoans are flagellates, amoebae and ciliates which occur in large numbers. Flagellates and amoebae occur thousands to millions in number and ciliates up to a thousand per gram of soil. Over 2000 species of soil inhabiting nematodes have been identified so far. They are ubiquitous soil fauna that interact in ecosystems directly as herbivorous on plants and indirectly as consumers of soil microfauna and flora, thus regulating decomposition and release of nutrients to plants (Coleman *et al.*, 1984; Sharma & Sharma, 1995). Most nematodes in natural and agricultural habitats are beneficial and contribute to soil process that enhance plant growth (Barker *et al.*, 1994). Entomopathogenic nematodes attack several agricultural insect pests and are safe to non-target organisms. Parasitic nematodes are also known as hidden enemies of farmers and their control is, therefore, an essential object of agriculture. Some 600 species of enchytraeids are known from different parts of the world. Dash (1983) studied diversity and biology of enchytraeids in temperate soil. He also demonstrated role of Enchytraeidae in functioning of agro-ecosystem (Dash, 1999). Recently, enchytraeids have been described as an important soil bioresources (Lal & Tripathi, 2003).

There are 20 families and more than 3000 species of earthworms (Reynolds, 1994). A total of 509 species belonging to 67 genera and 10 families have been recorded from Indian subcontinent, indicating a high degree of diversity (Julka, 1993). Among these, 385 species have been reported from India alone. About 68% of known species of earthworm in Indian subcontinent belong to 10 endemic genera. Remaining 57 genera are either monospecific or represented by less than 10 species. Earthworms have dynamic potentials and their applications have come up in the form of vermitechnology. They are very useful in pollution control and land reclamation, generation of growth stimulating and biopesticidal substances, soil improvement, organic waste management, protein and vitamin source, and drug production (Edwards *et al.*, 1985; Julka & Senapati, 1987; Kale, 1998; Tripathi & Singh, 2000; Tripathi *et al.*, 2003b). So, they are not only friends of farmers rather a real friend of both man and environment.

Singh (1991) reviewed progress of soil zoology in India and described different groups of soil fauna including spring-tails. Collembola are numerically one of the abundant groups among soil mesofauna. The most common soil forms are Isotomids, Podurids, Entomobryids and Onychiurids belonging to Arthropleona group. They play a prominent role in control of population of microorganisms and humification of soil organic matter. Protura and Diplura occur in all parts of the globe where sufficient moisture is available. They prefer humid soils rich in organic matter and feed on mycorrhiza (Singh & Pillai, 1981). Members of soil dwelling diplurans are omnivorous, but prefer animal prey (Pages, 1951). Singh and Mukherji (1971) listed different groups of Diplura from Indian fields. Soil zoologists often neglect proturans and diplurans because they are rarely collected in large numbers.

Mites constitute more than 90% of the total microarthropod population. About 45,000 species of mites have been described (Singh & Lal, 2001). The important orders of soil Acari are Astigmata, Prostigmata, Mesostigmata and Cryptostigmata. The soil-dwelling members of Astigamata (Acaridida) belong to the family Acaridae. Cryptostigmata (Orbatida) includes orbatid mites. They are one of the richest soil mites. Singh and Pillai (1975) recorded different species of orbatid mites including *Protoribates, Fosseremus, Galumna, Cryptocarus* and *Rhysotritia.* Sanyal (1996) presented a consolidated report on 10 species of soil mites of the Thar Desert including a record of six species of cryptostigmatid and one species of mesostigmatid mites for the first time from the Great Indian Thar desert.

There are more than 2000 species of termites reported from all over the world. Ahmad (1955) and Roonwal (1975, 1977, 1983) studied the distribution, biology and ecology of termites. A total of 46 species of termites belonging to 4 families spreading over 14 genera are known from the Thar region of Rajasthan, Gujrat, Haryana and Punjab States of India (Rathore, 1996). Coleoptera contributes about 4.4% of the world's beetle fauna in India (Ghosh, 1996). The superfamily Scarabaeiodea accounts for 8.5% beetles. Tenebrionid beetles (darkling beetles) are conspicuous insects of the Thar Desert. Only six species of tenebrionid beetles have been recorded from Indian desert namely *Adesmia* sp., *Blaps* sp., *Pimelia indica, Phytinota implita, Arthorodes* sp. and *Opatroides* sp. (Pruthi & Bhatia, 1952). A total 17 species of beetles belonging to 15 genera have been identified (Ghosh, 1996).

Pseudoscorpions are the smallest arachnids, which feed on small insects. Singh and Mukherji (1971), and Singh and Pillai (1981) described pseudoscropions from different fields. Like pseudoscropions, micro-whipscorpions are tiny animals with a long segmented tail. Singh and Mukherji (1971) also described *Koenenia* sp. from different fields. This is the first record of Palpigradi from India (Singh, 1991). Earwigs are mainly scavengers and occasionally herbivorous. Their unique forked abdominal tip can easily recognize them. Woodlice live under logs and stones, mostly in dark and damp places in order to keep their delicate gill-like appendages moist. They have the ability to conserve moisture due to the presence of their waterproof cuticle. Woodlice are largely omnivorous and average dwellers. They may help in identifying potential fertile land and water availability in arid region.

Myriapoda include puropods, symphylans, millipedes and centipedes. They are seldom found in sufficient number and are most agile and fragile group of soil animals. Millipedes and centipedes are found in arable land and garden soil. Millipedes enrich the soil system, acting as agents of decomposition and as accelerators in the nutrient release. Centipedes are carnivorous, feeding on aphids, flies, springtails, mites and other centipedes. Predatory associations of *Scutigera* sp. were found with termites. It helps to a certain extent in the biological control of crop pests.

SOIL BIOINDICATOR

Soil organisms serve as pedobiological monitors, indicating the quality of habitat and its degradation (Luff *et al.*, 1989). Below ground faunal biodiversity may be used as an indicator of soil quality due to its profound and usually beneficial effects on soil properties. They play an important role in influencing soil properties and processes that affect crop growth and environment regulatory functions. It can be said that higher the activity and species diversity of soil fauna, the better the soil quality. Soil biotic activity and species diversity depends on land-use, farming systems, and soil and crop management practices. Some land-use and management practices such as conservation tillage, residue mulch rotation, improved pastures and science-based agriculture enhance biodiversity of soil fauna. However, the soil management practices such as plough tillage and clean cultivation, mono-cropping, indiscriminate use of chemicals and biocides, and low-input or resource-based agriculture adversely affect soil biodiversity leading to nutrient depletion and soil degradation.

Soil faunal population influences soil biological processes, nutrient cycling and soil structure. Several properties or functions of soil fauna can be used to indicate soil quality such as the presence of specific organisms and their population or community analysis (functional groups and biodiversity) and biological processes including soil structure modification and decomposition rates (Linden *et al.*, 1994). Abundance of soil fauna may be linked to litter quality and nutrient cycling rates. Soil arthropods affect soil quality directly and indirectly both depending on their size and specific activity. Macroarthropods (millipedes, centipedes, insect larvae, termites and others) have the ability to modify soil structure by decreasing bulk density, increasing soil pore space, mixing soil horizons and improving aggregate structure (Abbott, 1989). Arthropods and earthworms can rearrange the soil profiles, mixing soil horizons through burrowing and nest-building activities.

Microarthropods (mites and collembolans) affect soil structure indirectly and nutrient cycling directly (Powers *et al.*, 1998). Many other studies have also demonstrated the importance of the soil fauna as a regulator in nutrient cycling (Anderson *et al.*, 1985; Persson, 1989). Knoepp *et al.* (2000) studied the biological indices of soil quality and compared with four common groups of soil biological indicators including soil micro-arthropods. Diverse soil fauna can reach tremendous species numbers in tropics which represents potential indicators of sustainability of rural activities (Erwin, 1997). Netruzhilin *et al.* (1999) assessed agricultural impact using and morphospecies as bioindicators in the Amazonian Savanna-forest ecotone.

IMPACT OF AGRICULTURAL PRACTICES

Arthropods indirectly affect increase of plant production both by their activity modifiers of soil fertility and water infiltration and by their effects as consumers (Whitford, 2001). Ground level arthropod community was collected from the two types of tea plots using pit-fall traps. Catches from these environments showed variation in the arthropod faunal structure with numerically and taxonomically greater abundance in the organic plot than conventional plot (Mukhopadhyay *et al.*, 2003). Increased biodiversity of soil animals, litter layer on the soil surface, soil-animal mediated nutrient cycling, water infiltration through biopores in the soil, and reduction in soil loss by erosion are common to undisturbed ecosystems and no-till cropping with stubble retention (House & Brust, 1989; Paul & Robertson, 1989; Robertson, 1990). There is ample evidence that some soil fauna increase decomposition of organic matter and availability of nutrients to crops (Kevan, 1968; Lee & Wood, 1971; Edwards & Lofty, 1972; Sayers *et al.*, 1974). Witkamp & Ausmus (1976) reported reduction in decomposition rate and nutrient release from the litter in absence of these fauna. More recently, Tian *et al.* (1993) found 41% and 177% increases in earthworm and termite populations, respectively, under mulched condition over a two-year period. It has been concluded that mulching generally provide low temperature and high moisture conditions which is conducive to the growth of the populations of earthworms, termites and other macrofauna. Badejo *et al.* (1995) showed that the densities of detrivorous and phytophagous microarthropods increase in mulch maize plots. Plots mulched with rice-stover support the highest numbers of microarthropods.

Specific crop types with differential nitrogen content may influence the dominance of different species of Collembola (Mitra, 1993). The mean population densities of Collembola (Isotomidae, Entomobryidae and Poduridae) are higher in the perennial lay treatments than in the tillage. However, the densities of araneae, pseudoscorpions, Diplopoda, Chilopoda, Symphyla, Protura and Diplura are low. Wardle (1995) reported reduction of Araneae population due to tillage when compared to perennial crop-covers. The densities of Coleoptera, Hymenoptera, Isoptera, Homoptera and Lepidoptera are higher in the perennial lay treatments than in the tillage (Reddy, 1999). Crops influence the density of termite populations through their annual crop-residue production and its palatability. The mean densities of workers of *Microcerotermes obesi* and *Odontotermes obesus* are higher in perennial population. Densities of *M. obesi* and the workers and soldiers of *O. obesus* are significantly affected by the perennial lay treatments between seasons (Reddy, 1999). It has been found that most of the soil surface arthropods were higher in perennial lay treatments than in annual tillage (Reddy *et al.*, 1995). These soil arthropods in perennial cropping systems experience less disturbances than under annual crops such as potatoes and barley (Braman & Penley, 1993). Abundance of soil-surface arthropods in the undisturbed natural vegetation area is equivalent to those of some of the perennial cover-crop treatments. Robertson *et al.* (1994) believe that perennial treatments favor the decomposer and predator population like a natural ecosystem.

Effect of Tillage

Tillage generally decreases the abundance of soil animals (Ghilarov, 1975; Stinner & Crosslay, 1980; Radford *et al.*, 1995). The indirect effect of tillage, such as drying of the uppermost part of the soil or removal of litter from the soil surface decreases the microarthropods densities (Gill, 1969; Wallwork, 1976; Andren & Largerlof, 1980; Lal & De Vleeschauwer, 1982). Whereas the total mite population remains higher under conventional tillage than under no-tillage (Boles & Oseto, 1987). Loring *et al.* (1981) also found that the Prostigmata are higher in number in tilled plots than in the no-tilled plots. These mites may adapt well to cultivated conditions and attain a considerable population size. The

Collembola population is significantly higher in the uncultivated undisturbed natural vegetation area than in the cultivated fields (Hazara & Choudhary, 1983). Loring *et al.* (1981) reported that there is virtually no difference in collembolan composition between different tilled treatments. Sminthuridae are higher in the tillage treatments than in the perennial lay treatments (Reddy, 1999). Nakamura (1988) reported that the Chilopoda are higher in the direct-drilling plots than in conventional tillage treatments in different soil management practices. Symphyla and Diplura populations are higher in zero-tillage treatment than in the deep-tillage and shallow-tillage treatments. It appears from the description that there is an urgent need to evolve fauna-specific agricultural practices for conservation of soil biota and agro-environmental sustainability.

Effect of Fertilizers

Application of inorganic fertilizers in agro-ecosystems usually brings reduction in the abundance of enchytraeids (Huhta *et al.*, 1967, 1969; Utrobina, 1976; Marshell, 1977; Lohm *et al.*, 1977; Nakamura, 1988). Standen (1982, 1984) found negative correlation between enchytraeid diversity and soil fertilization with nitrate and also negative correlation between their biomass and phosphorus fertilization. Fertilizer application has been reported to reduce enchytraeid diversity (Dash, 1999). Badejo and Akinyemiju (1989, 1993) reported that the populations of beneficial soil microarthropods are reduce tremendously in plots where herbicides are used to control weeds. These herbicides reduced the density by elimination of the microarthropods through death and migration to lower depths (Badejo, 1999). It has also been observed that sometimes the fertilizers practices do not influence the arthropod populations (Favretoo *et al.*, 1992). However, fertilizers may increase the population densities of Collembola but not much influence on acarine population is found. Mitra (1993) stated that chemical herbicides have positive influence on Collembola and Acari densities. Bhattacharya and Joy (1980) reported that nitrogen adversely affect oribatid density.

EFFECT OF PESTICIDES

The reduction in soil microarthropod population due to insecticide and herbicide treatments is well known (House *et al.*, 1987; Reddy, 1989; Reddy *et al.*, 1996). Carbofuran application has no residual effect on saprophagous microarthropods such as Prostigmata and Mesostigmata (Gamasida) (Mande *et al.*, 1993). Mitra (1993) reported greater susceptibility of Collembola to Thiodan and Furadan EC 35. The residual effect of Thiodan persists for more than 8 months and that of latter being less than two weeks in the field conditions. The Acarina abundance is more suppressed being susceptible to ecotoxicological effects of carbofuran because they are less agile, whereas the insecticide relatively affects less the collembolans and other microarthropods, as they move into the surrounding grassy margins rapidly (Reddy *et al.*, 1995). The population densities of soil-surface arthropods are adversely affected by insecticide applications (Braman & Pendley, 1993). Many authors reported that the Cryptostigmatid densities have increased over time in response to various agricultural practices, such as insecticides, herbicides or fertilizer applications, partly in response to an increase in decaying organic matter or to a decrease in predators, or both (Reddy, 1989; Turner *et al.*, 1987; Vavrek and Niemczyk, 1990; Braman & Pendley, 1993). Treatment of pesticides adversely affects the biochemical machinery of soil fauna. Tripathi and Sharma (2004) found reduction in aerobic activity of woodlice and suggested pesticide-induced impairment of metabolism in soil fauna.

EFFECT OF INDUSTRIAL POLLUTION

Industrial wastewater has been tried by several researchers for irrigation in developing and underdeveloped countries (Hansen *et al.*, 1980; Duncomb *et al.*, 1982; Somshekar *et al.*, 1984). The waste water although contains nutrients that enhance the crop growth also have toxic materials. Therefore, it is essential that the implications for the use of industrial wastes in the crop field and their effect on soil characteristics should be assessed before it is recommended for use in irrigation. It has been shown that heavy metals affect the abundance and the structure of the completely below-ground faunal community (Bengtsson and Tranvik, 1989). Species-specific distribution patterns in discontinuously metal contaminated soil may be due to a general tendency of organisms to disperse, adaptation to toxicants, interspecies interaction, and capability to perceive different concentrations of toxicants in their environment (Bengtsson *et al.*, 1994). The impact of chromium, which is an important toxic ingredient of tannery effluent, on soil organisms like earthworms, molluscs and other invertebrates has been studied (Kumar, 1989). The results presented are indicative of dangerous level of accumulation of chromium in various components of food chain and food products and their movement from soil and water to animals through agricultural products warranting cautious surveillance and safety measures. Martin and Coughtrey (1982) reported in their excellent review the use of terrestrial animals as monitors and indicators of environmental contamination by heavy metals form mining and sewage sludge.

CONCLUSION

Every species of below-ground soil animals silently and secretly contribute to the agro-environmental sustainability by improving soil chemical and biological health. They act as an indicator of soil quality and pedobiological monitor to a particular habitat. Increased biodiversity of soil fauna is a sign of a healthy and undisturbed ecosystem. Modern agricultural practices including use of fertilizers and pesticides have been proved to be detrimental to below-ground faunal resources. Industrial pollution has been demonstrated to be equally harmful to soil animals. Therefore, the conservation of beneficial soil fauna is suggested for improvement of soil biological health and environmental sustainability, particularly in a region like the Thar Desert which has largely remained uncultivated in the xeric environments and is rife with untapped reserves of bioresources.

ACKNOWLEDGEMENT

Author is grateful to Indian Council of Agricultural Research (ICAR), New Delhi for providing financial support and thankful to Dr. G. Tripathi, Dr. P.P. Singh and Dr. B.K. Tyagi for their valuable suggestions and help.

REFERENCES

Abbott, I., 1989. The influence of fauna on soil structure. In: *Animals in Primary Succession : The Role of Fauna in Reclaimed Lands.* (Ed., J.D. Major), pp. 39-50, Cambridge University Press, Cambridge.

Ahmad, M. , 1955. Termites of West Pakistan. *Biologia* 1 : 202-264.

Anderson, J.M., Leonard, M.A., Ineson, P. and Huish, S.A., 1985. Faunal biomass; a key component of a general model of N mineralization. *Soil. Biol. Biochem.* 17 : 735-737.

Andren, O and Largerlof, J., 1983. Soil fauna (micro-arthropods, enchytraeids, nematodes) in Swedish agricultural cropping systems. *Acta. Agric. Scand.* 33 : 33-52.

Andren, O. and Largerlof, J., 1980. The abundance of soil animals (Microarthropoda, Enchytraeidae, Nematoda) in a crop rotation dominated by ley and in a rotation with varied crops. In : *Soil Biology as Related to Landuse Practices, Proc. 7th Intl. Soil Zool. Colloq.,* (ed., D. Dindal), pp. 274-279, Washington, D.C.

Badejo, M.A., 1999. Soil management and beneficial soil biota in the humid tropics. In : *Management of Tropical Agroecosystems and the Beneficial Soil Biota* (Ed., M.V. Reddy), pp. 281-290, Oxford & IBH Publ., New Delhi.

Badejo, M.A. and Akinyemiju, O.A., 1989. Influence of atrazine on soil microarthropods fauna, and the performance of maize (*Zea mays* L.) in a humid tropical environment. *Ife J. of Agric.* 11 : 41-50

Badejo, M.A. and Akinyemiju, O.A., 1993. Response of soil mites to hexazinone application in Nigeria. *The Science of Total Environment,* pp. 1257-1263.

Badejo, M.A., Tian, G. and Brussaard, L., 1995. Effect of various mulches on soil microarthropods under a maize crop. *Biol. Fertil. Soil* 20 : 294-298.

Barker, K.R., Hussey, R.S., Krushberg, L.R., Bird, G.W., Dunn, R.A., Ferris, H., Ferris, V.R., Freckman, D.W., Gabriel, C.J., Grewal, P.S., MacGuidwin, A.E., Riddle, D.L., Roberts, P.A. and Schmitt, D.P., 1994. Plant and soil nematodes : societal impact and focus for the future. *J. Nematol.* 26 : 127-137.

Bengtsson, G. and Tranvik, L., 1989. Critical metal concentration for forest soil invertebrates. *Water, Air, Soil Pollution* 47: 381-417.

Bengtsson, G., Rundgren, S. and Sjogren, M., 1994. Modeling dispersal distances in a soil gradient : the influence of metal tolerance, competition and experience. *Oikos* 71: 13-23.

Bhattacharya, T. and Joy, V.C., 1980. Effects of two herbicides on soil intrabiting cryptostigmatids mites. In : *Soil Biology as Related to Land Use Practices, Proc. 7th Intl. Soil Zool. Colloq.* (Ed., D.L. Dindal), pp. 109-117, Washington, D.C.

Boles, M. and Oseto, C.Y., 1987. A survey of the microarthropod population under conventional tillage and no-tillage systems. *Agroecosystems* 8 : 247-253.

Braman, K.S. and Pendley, A., 1993. Relative and seasonal abundance of beneficial arthropods in Centipede grass as influenced by management practices. *J. Econ. Entomol.* 86 : 494-504.

Butcher, J.W. and Snider, J.R., 1971. Bioecology of edaphic Collembola and Acarina. *Ann. Rev. Entomol.* 16 : 249-288.

Coleman, D.C., Cole, C.V. and Elliot, E.T., 1984. Decomposition, organic matter turnover and nutrient dynamics in agroecosystems. In : *Agricultural Ecosystems : Unifying Concepts* (Eds., R. Lowrance, B.R. Stinner and G.J. House), pp. 84-104, Wiley, New York.

Dash, M.C., 1983. *Biology of Enchytraeidae (Oligochaeta)*. Ravandra Pal Singh Gehlot Publ., Dehradun, India.

Dash, M.C., 1999. Response of Enchytraeidae (Oligochaeta) to soil management in agricultural systems. In : *Management of Tropical Agroecosystems and Beneficial Soil Biota* (Ed., M.V. Reddy), pp. 213-221, Oxford & IBH Publ., New Delhi.

Duncomb, D.R., Larson, W.E., Clapp, C.E., Dowdy, R.H., Lindane, D.R. and Johnson, W.K., 1982. Effect of liquid waste water sludge application on crop yield and water quality. *J. WPCF* 54 : 1185-1192.

Edwards, C.A. and Lofty, J.R., 1972. *Biology of Earthworms*. Chapman and Hall Ltd., London.

Edwards, C.A., Burrows, I., Fletcher, K.E. and Jones, B.A., 1985. The use of earthworms for composting farm wastes. In : *Composting of Agricultural and Other Wastes* (Ed., J.K.R. Gasser), pp. 229-242, Elsvier, Amesterdam, The Netharlands.

Erwin, T., 1997. Biodiversity at its utmost : tropical forest beetles. In : *Biodiversity II* (Eds., M.L. Reaka-Kudla, D.E. Wilson and E.O. Wilson), pp. 27-40, J. Henry Press, Washington, D.C.

Favretoo, M.R., Paoletti, M.G., Caborali, F., Nannipieri, P., Onnis, A. and Tomei, P., 1992. Invertebrates and nutrients in a Mediterranean Vineyard mulched with subterranean clover (Trifolium subterranean L.), *Biol. Fertil. Soils* 14 : 151-158.

Ghilarov, M.S., 1975. General trends of changes in the soil animal population of arable land. In : *Progress in Soil Zoology, Proc. 5th Intl. Colloq. Soil Zool.* (Ed., J. Vanek), pp. 31-39, Academia, Prague.

Ghosh, A.K., 1996. The Thar desert ecosystem. In : *Faunal Diversity in the Thar Desert* (Eds., A.K. Ghosh, Q.H. Baqri and I. Prakash), pp. 1-18, Scientific Publishers, Jodhpur, Rajasthan.

Gill, R.W., 1969. Soil microarthropod abundance following oil-field litter litter manipulation. *Ecology* 50 : 805-816.

Hansen, E.A., Dawson, D.H. and Tolsted, D.N., 1980. Irrigation of intensively cultured plantations with paper mill effluent. *Toppi* 63 : 139-143.

Hazara, A.K. and Choudhary, D.K., 1983. A study of Collembola communities in cultivated and uncultivated sites of West Bengal in relation to three major soil factors. *Rev. Ecol. Biol. Sci.* 20 : 385-401.

House, G.J. and Brust, G.E., 1989. Ecology of low-input, no-tillage agroecosystems. *Agric. Ecosyst. Environ.* 27 : 331-345.

House, G.J., Worshman, A.D., Sheets, T.J. and Stinner, R.E., 1987. Herbicide effects on soil arthropod dynamics and wheat straw decomposition in a north Carolina no-tillage agroecosystem. *Biol. Fertil. Soils* 4 : 109-114.

Huhta, V., Karppinen, E. Nurminen, M, and Valpas, A., 1967. Effects of Silvicultural practices upon arthropod, annelid and nematode populations in coniferous forest soil. *Ann. Zool. Fenn.* 4 : 87-146.

Huhta, V., Nurminen, M. and Velpas, A., 1969. Further notes on the effect of silviculture practices upon the fauna of coniferous forest soil. *Ann. Zool. Fenn.* 6 : 327-334.

Julka, J.M., 1993. Earthworm resources of India and their utilization in vermiculture. In : *Earthworm Resources and Vermiculture,* pp. 27-31, Zoological Survey of India, Solan.

Julka, J.M. and Senapati, B.K., 1987. Records of the Zoological Survey of India. Miscellaneous Publication, Graphic Printall, Calcutta, India, 105p.

Kale, R.D., 1998. Earthworms : Nature's gift for utilization of organic wastes. In : *Earthworm Ecology* (Ed., C.A. Edwards), pp. 355-376, CRC Press, The Netherlands.

Kevan, D.K.M., 1968. *Soil Animals.* H.F. and G. Witherby Ltd., London

Knoepp, D.J., Coleman, D.C., Crossley Jr, D.A. and Clark, J.S., 2000. Biological indices of soil quality : an ecosystem case study of their use. *Forest Ecol. Manage.* 138 : 357-368.

Kumar, N.V.N., 1989. Tannery and chromate industrial effluent effect on soil, animals and plants. In : *Soil Pollution and Soil Organisms* (Ed., P.C. Mishra), pp. 149-170, Ashish Publishing House, New Delhi.

Lal, R. and De Vleeschauwer, D., 1982. Influence of tillage methods and fertilizer application on chemical properties of worm castings in a tropical soil. *Soil Till. Res.* 2 : 37-52.

Lal, V.B. and Tripathi, G., 2003. Enchytraeid Bioresources. In : *Vermiresource Technology* (Ed., G. Tripathi), pp. 173-187, Discovery Publishing House, New Delhi.

Lavelle, P., Spain, A.V., Blanchart, E., Martin, A. and Matin, S. (1992) Impact of soil fauna on the properties of soils in the humid tropics. In : *Myths and Science of Soils of the Tropics* (Eds., R. Lal and P.A. Sanchez), pp. 157-185, Madison, USA.

Lee, K.E. and Wood, T.G., 1971. Termites and soils. Academic Press, London, 251p.

Linden, D.R., Hendrix, P.F., Coleman, D.C. and Van Vliet, P.C.J., 1994. Faunal indicators of soil quality. In : *Defining Soil Quality for a Sustainable Environment.* (Eds., J.W. Doran, D.C. Coleman, D.F. Bezdicek and B.A. Stewart), pp. 91-106, *Soil Sci. Soc. A., Inc.*, Madison, WI.

Lohm, U., Lundkvist, H., Persson, T. and Wiren, A., 1977. Effects of nitrogen fertilization on the abundance of enchytraeides and microarthropods in Scots pine forests. *Studies Forestalia Suecia* 140 : 5-23.

Loring, S.J., Snider, R.J. and Robertson, L.S., 1981. The effect of three tillage practices on Collembola and Acarina populations. *Pedobiologia* 22 : 172-184.

Luff, M.L., Eyre, M.D. and Roushton, S.P., 1989. Classification and ordination of habitats of ground beetles (Coleoptera : Carabidae) in northeast England. *J. Biogeogr.* 16 : 121-130.

Mande, E.J.M., Raros, L.C., De-Pedrol, L.B., 1993. Effects of carbofuran treatments and Gliricidia mulch on the abundance of soil mites and collembolas in an upland rice farm. *24th Ann. Sci. Meeting of the Pest Manag. Conf. of the Philippines,* Inc. Coll., Laguna, 34p.

Marshell, V.G., 1977. Effect of manures and fertilizers on soil fauna : a review. *Commonwealth Bureu of Soils*, 79p.

Martin, M.H. and Coughtry, P.J., 1982. The use of terrestrial animals as monitors an indicators of environmental contamination by heavy metals. In : *Biological Monitoring of Heavy Metal Pollution* (Eds., M.H. Martin and P.J. Coughtry), pp. 221-310, Applied Sciences Publishers, London and New York.

Mitra, S.K., 1993. Effects of continuous and other agronomic practices on soil microarthropods-a unifying concept of agriculture and ecology for tropical agroecosystems. *Rec. Zool. Surv. India,* Occ. Paper No. 151, 90p.

Moore, J.C., Waiter, D.E. and Hunt, H.W., 1988. Arthropod regulation of micro and mesobiota in below-ground detrital foodwebs. *Ann. Rev. Entomol.*, 33 : 419-439.

Mukhopadhyay, A., Sherpa, P.W. and Pradhan, B., 2003. Diversity of ground arthropod community at organic and chemically intensive tea plantation of Darjeeling terai. *J. Environ. Biol.*, 24 (4) : 471-476.

Nakamura, Y., 1988. The effect of soil management on the soil fauna make-up of a cropped andosol in central Japan. *Soil Till. Res.* 12 : 177-186.

Nampoothiri, K.U.K., 2001. Organic farming : Its relevance to plantation crops. *J. Plantation Crops* 29 : 1-9.

Netuzhilin, I., Chacon, P., Cerda, H., Lopez-Hernandez, D., Torres, F. and Paoletti, M.G., 1999. Assessing agricultural impact using ant morphospecies as bioindicators in the Amazonian savanna-forest ecotone (Puerto Ayacucho, Amazon State), Venezuela. In : *Management of Tropical Agroecosystems and the Beneficial Soil Biota* (Ed., M.V. Reddy), pp. 291-352, Oxford & IBH Publ., New Delhi.

Pages, J., 1951. Contribution a la connaissance des diploures. *Bull. Sci. Bourgogne* 13, Suppl. 9, 97p.

Paul, E.A. and Robertson, G.P., 1989. Ecology and the agricultural sciences : a false dichotomy? *Ecology* 70 : 1594-1597.

Persson, T., 1989. Role of soil animals in C and N mineralization. *Plant Soil* 115: 241-245.

Powers, R.F., Tiarks, A.E. and Boyle, J.R., 1998. Assessing soil quality: practicable standards for sustainable forest productivity in the united states. In: *The Contribution of Soil Science to the Development of and Implantation of Criteria and Indicators of Sustainable Forest Management. Soil Sci. Soc.A. Special Pub.*, pp. 53-80, Madison, WI,

Pruthi, H.S. and Bhatia, D.R., 1952. Peculiarities of the insect fauna of the Rajasthan desert and the share of insects, in the maintenance of the desert. *Bull. Natnl. Inst. Sci. India* 1 : 241-245.

Radford, B.J., Key, A.J., Robertson, L.N. and Thomas, G.A. , 1995. Conservation tillage increases soil water storage, soil animal populations, grain yield and response to fertilizer in the semi-arid subtropics. *Aust. J. Exp. Agric.* 35 : 223-232.

Rathore, N.S., 1996. Termites (Isoptera of the Thar desert. In : *Faunal Diversity in the Thar Desert : Gaps in Research* (Eds., A.K. Ghosh, Q.H. Baqri, I. Prakash), pp. 1-18, Scientific Publ., Jodhpur, India.

Reddy, M.V., 1989. Soil pollution and soil arthropod population. In : *Soil Pollution and Soil Organisms* (Ed., P.C. Mishra), pp.209-253, Ashish Publishing House, New Delhi.

Reddy, M.V., 1999. Soil management and beneficial soil meso- and macrofauna with particulars reference to semi-arid tropical Alfisol. In : *Management of Tropical Agroecosystems and the Beneficial Soil Biota* (Ed., M.V. Reddy), pp. 223-271, Oxford & IBH Publ., New Delhi.

Reddy, M.V., Balashouri, P., Reddy, V.R., Kumar, P.V.K., Cogle, A.L., Yule, D.F. and Lee, K.K., 1995. Effect of soil management on seasonal relative abundance of soil-surface arthropods in a semi-arid tropical alfisol. *J. Soil Biol. Ecol.* 15 : 144-159.

Reddy, V.S., Reddy, M.V., Lee, K.K., Rao, K.P.C. and Srinivas, S.T., 1996. Response of some soil meso- and macro-fauna populations to soil management during crop and fallow periods on a semi-arid tropical alfisol (India). *Eur. J. Soil Biol.* 32 : 123-129.

Reynolds, J.W., 1994. The distribution of the earthworms (Oligochaeta) of Indiana:A case for the post-Quaternary introduction theory for megadrile migration in North America. *Megadrilogica* 5 : 13-22.

Robertson, L.N., 1990. *Interactions between agriculture and the Ecosystem in North Queensland* (Eds., K.L. Norman and A.L. Garside), pp. 56-69, Australian Institute of Agriculture Science Occasional Publication 51, Towonsville.

Robertson, L.N., Kettle, B.A. and Simpson, G.B., 1994. The influence of tillage practices on soil macrofauna in a semi-arid ecosystem in northeastern Australia. *Agril. Ecosys. Environ.* 48:149-156.

Roonwal, M.L., 1975.Thar desert termites. In : *Environmental Analysis of the Thar Desert.* (Eds., R.K. Gupta and I. Prakash), pp. 393-422, English Book Depot, Dehradoon.

Roonwal, M.L., 1977. Termite ecology in the Indian desert ecosystem : a review of recent work in desert eco-system and its improvement (Ed., H.S. Mann), pp. 323-328, *Monogr. No.1*, CAZRI, Jodhpur.

Roonwal, M.L., 1983. The ecology of termite swarming in the Indian desert. In : *Insect Inter-relations in Forest Agroecosystems* (Eds., P.K. Sen-Sarma, S.K. Sanyal and S.K. Kulshrestha), pp. 9-13, Jugal Kishore & Co., Dehradoon.

Sanyal, A.K., 1996. Soil arthropod population in two contrasting sites at Nadia, West Bengal. *Environ. Ecol.* 14: 346-350.

Sayers, J.K., Sharpley, A.N. and Keeney, D.R., 1974. Cycling of nitrogen by surface-casting earthworms in a pasture ecosystem. *Soil Boil and Biochem.* 11 : 181-185.

Sharma, R. and Sharma, S.B., 1995. Nematodes and tropical litter decomposition. In : *Soil organisms and litter decomposition in the tropics* (ed., M.V. Reddy), pp. 75-89, Oxford & IBH Publication, New Delhi.

Singh, J., 1991. Progress in Soil Zoology in India. In : *Advances in Management and Conservation of Soil Fauna* (Eds., G.K. Veeresh, D. Rajagopal and C.A. Viraktamath), pp. 127-139, Oxford and IBH Publishing Co. Pvt. Ltd., New Delhi.

Singh, J. and Lal, V.B.. 2001. *A Handbook of Soil Fauna.* Agricultural Research Station, Durgapura, Jaipur, Technical Bulletin No. 3, 44p.

Singh, J. and Mukherji, S.P.,1971. Qualitative composition of soil microarthropods in some fields at Varanasi (India), *Oriental Insects* 5 : 487-494.

Singh, J. and Pillai, K.S., 1973. The world beneath us. *Science Reporter*, pp. 318-321.

Singh, J. and Pillai, K.S., 1975. A study of soil micro-arthropods communities in some fields. *Rev. Ecol. Biol. Soil* 12 : 579-590.

Singh, J. and Pillai, K.S., 1981. Community structure and biocoenology of soil microarthropods in some tropical fields at Varanasi, India. *J. Soil. Biol. Ecol.* 1 : 5-20.

Skujins, J.J., 1967. Enzymes in the soil. In : *Biochemistry* (Eds., A.D. Mclaren and G.H. Peterson), pp. 371-414.

Somshekar, R.K., Gowda, M.T.G., Shettigar, S.L.N. and Srinath, K.P., 1984. Effect of industrial effluents on crop platns. *Ind. J. Environ. Hlth.* 26:136-146.

Standen, V., 1982. Associations of (Enchytraeidae, Oligochaeta) in experimentally fertilized grassland. *J. Anim. Ecol.* 51 : 501-522.

Standen, V., 1984. Production and diversity of enchytraeidae earthworms and plants in fertilized hay Neadon plots. *J. Appl. Ecol.* 21 : 293-312.

Stinner, B.R. and Crossley, D.A. Jr., 1980. Comparison of mineral element cycling under till and no-till practices : an experimental approach to agroecosystems analysis. In : *Soil Biology as Related to Landuse Practices* (Ed., D. Dindal), pp. 280-287, *Proc. 7th Intl. Soil Zool. Colloq.*, E.P.A. Washington, DC.

Swift, M.J., Heal, O.W. and Anderson, J.M., 1979. Decomposition in terrestrial ecosystems. *Studies in Ecology*, Vol. 5, Blackwell Scientific Publications, Oxford.

Tian, G., Kang, B.T. and Brussaard, L., 1993. Biological effects of plant residues with contrasting chemical compositions under humid tropical conditions : effects on soil fauna. *Soil Biol. Biochem.* 25 : 731-737.

Tomati, U. and Galli, E., 2002. Recycling of organic matter from wastes to restore soil fertility. In : *Bioresource and Environment* (Eds., G. Tripathi and Y.C. Tripathi), pp. 372-408, Campus Books International, New Delhi.

Tripathi, G. and Sharma, B.M., 2004. Impact of pesticides on aerobic capacity of *Porcellio laevis. Natl. Symp. Manag. Aqua. Resour. Biodi. Maint. Conser.*, JNVU, Jodhpur, p. 71.

Tripathi, G. and Singh, J., 2000. Earthworm biodiversity and vermitechnology in industrial waste management. In : *Industry Environment and Pollution* (Eds., A. Kumar and P.K. Goel), pp. 60-74, Technoscience, India.

Tripathi, G., Bhardwaj, P. and Lal, V.B., 2003a. Dynamic potentials of vermiresources. In : *Vermiresource Technology* (Ed., G. Tripathi), pp. 63-80, Discovery Publishing House, New Delhi.

Tripathi, G., Ram, S. and Sharma, B.M., 2003. Soil fauna in sustainable management of agroecosystem. In : *Potentials of Living Resources* (Eds., G. Tripathi and A. Kumar), pp. 584-617, DPH, New Delhi.

Turner, A.S., Bale, J.S. and Clements, R.O., 1987. The effect of a range of pesticides on non-target organisms in the grassland environment. In : *Proc. Crop Prot. in Northern Britain* (1987), pp. 290-295, Dundee Univ., Assoc. for Crop Prot. in Northern Britain.

Utrobina, N.M., 1976, Einflu B von Amoniak wasser auf den Bondentierbesatz under Kartonftein. *Pedobiologia* 16 : 206-218.

Vavrek, R.C. and Niemczyk, H.D., 1990. Effect of isofenphos on non-target invertebrates in turfgrass. *Environ. Entomol.* 19 : 1572-1577.

Verhoeff, H.A. and Brussaard, L., 1990. Decomposition and nitrogen mineralization in natural and agroecosystems : the contribution of soil animals. *Biogeochem.* 11: 175-211.

Wallwork, J.A., 1976. *The Distribution and Diversity of Soil Fauna.* Academic Press, London, U.K., 355p.

Wardle, D.A., 1995. Impacts of disturbance of detritus-food webs in agro-ecosystems of contrasting tillage and weed management practices. *Adv. Ecol. Res.* 26 : 105-185.

Whitford W.G., 2001. Invertebrates: their effects on the properties and processes of desert ecosystems. In : *Ecology of desert environments* (Ed., I. Prakash), pp. 333-356, Scientific Publishers (India), Jodhpur.

Witkamp, M. and Ausmus, B.S., 1976. Process in decomposition and nutrient transfer in forest ecosystems. In : *The Role of Terrestrial and Aquatic organisms in Decomposition Processes* (Eds., J.M. Anderson and A. Macfadyen), pp. 375-396, Blackwell Scientific Publications, Oxford.

Wood, M., 1991. Biological aspects of soil protection. *Soil Use Manag.* 7:130-136.

Changing Faunal Ecology in Thar Desert, pp. 61-83, 2005.
Eds.: B.K. Tyagi and Q.H. Baqri

5

IMPACT OF CHANGING ECOLOGY ON TERMITE FAUNA IN ARID AND SEMI-ARID REGIONS OF RAJASTHAN

N.S. Rathore and Madhumita Mandal

Desert Regional Station, Zoological Survey of India, Jhalamand Chauraha, New Pali Road, Jodhpur 342 005 (Raj.), India.

The Thar Desert is biogeographically the eastern most edge of the Sahara-Arabian Desert zone. Inspite of harsh climatic conditions it exhibits vivid and spectacular biodiversity. Vegetation consists mainly of xerophytes like *Prosopis cineraria, Capparis decidua, Calotropis procera, Salvadora oleoides, Laisurus sindicus* etc. Termite fauna of arid and semi-arid regions of Rajasthan is quite rich. A total of 34 species belonging to 4 families were recorded in this region. Man's continuous intervention in the fragile ecosystem of the most populated Thar Desert is the main cause for depletion of its rich biodiversity. Though water of Indira Gandhi canal is considered to be the lifeline of arid and semi-arid region of Rajasthan but it has changed the topography and ecology of about 11% of the Thar Desert. Moreover, due to perpetual availability of water, xeric species of termites are replaced by mesic species and also some minor pests may gain the status of major pest and extend their niche to crop fields.

INTRODUCTION

In the Indian subcontinent, the areas with arid and semi arid zone constitute 12% of the entire geographical area of the country and encompasses vast areas of western Rajasthan, western Gujarat and south western Punjab and Haryana states, covering together an extensive area of over 2,85,600 km^2. The focal centre for manifestation of aridity is Rajasthan (3, 42,274 km^2), which lies between 23^0 13' to 31^0 13' N latitude and 69^0 30' to 78^0 17' E longitude. It is of unusual ecological and biogeographical interests, as it occupies the major portion of the Thar Desert (61%). The Thar Desert in Rajasthan (lying between latitudes of 24^0 40' and 30^{0}12' N and longitude of 69^0 30' and 75^0 55' E) covers a total area of about 1, 96,150 km^2 and occupies nearly 2/3rd of the total area of the state. It is one of the most populated arid zones in the world (> 83 persons per square kilometer). The livestock population has also enhanced much beyond carrying capacity of the arid land. Overcrowding of the livestock animals has resulted into an inexorable deleterious vegetation composition (Mann and Prakash, 1983; Shankarnarayan, 1977).

Rajasthan state is administratively divided into 32 districts, with arid and semi arid zone spanning over 13 districts. The arid/semi arid zone has a well-defined boundary in the east marked by Aravalli ranges tending in the northeasterly direction. It is characterized by sparse and highly variable rainfall, high solar radiation, and extreme variation in diurnal and seasonal temperature with high summer temperature, low humidity of the atmosphere for the greater part of the year. Severe drought accompanied by high wind velocity, sandy and unretentive character of the soil, low water table, general soil and water salinity, evaporation far exceeding precipitation and too scanty to support any appreciable vegetation. The natural vegetation in this region consist of mainly xerophytes or short lived annuals having high degree of drought resistance because of the environmental constraint, *viz.*, limited moisture availability. Shrubs, herbs and grasses with few tree species which largely dominate vegetation in the arid and semi-arid region include like *Anogeissus pendula, Acacia catechu, A. senegal, A. tortilis Capparis decidua, Zizyphus nummularia, Z. mauritiana, Z. rotundifolia* (Ber), *Acacia nilotica, Prosopis cineraria, Calotropis procera, Ephedera foliata, Salvadora persica, Salvadora oleoides, Euphorbia caudicifolia, Cenchrus ciliaris, C. setigerus, Lasiurus sindicus* etc. A large portion of the land (63.9%) either remains fallow, uncultivated or not fit for cultivation (wasteland). The hostile climatic condition coupled with overgrazing, deforestation and overexploitation of the woody biomass adversely affect the regeneration. Added to it insects and rodents also play a significant role in the destruction of scanty vegetation. Among insects, termites cause considerable damage to agriculture, horticulture, stored products and plantation forestry particularly at the nursery level and thus threaten the very survival of the plantation under various afforestation programmes.

Inspite of the fact that, 2/3rd of the state is dry and arid, yet the termite fauna in arid and semi-arid region of Rajasthan is quite rich. Out of nearly 2,865 species of termites reported from all over the world (Myles 1998, unpublished data), 348 species are known from the Indian region, of which 44 species are so far recorded from Rajasthan state including 34 species from the desert region. A total of 15 species are endemic to the Thar Desert, while the remaining species are much versatile in their distribution. Recently some faunal studies were undertaken in the Thar Desert by various authors (Roonwal, 1982; Rathore, 1984; Kapoor 1985; Rahamani, 1989; Prakash *et. al.* 1992), but very little is known about the impact of changing ecology on termite fauna in arid and semi-arid region of Rajasthan. Zoogeographically Rajasthan lies virtually at the western end of the oriental region where it grades into Palearctic through Pakistan. The Aravalli range is very ancient in age (Archean to Cambrian) while the rest of the Rajasthan is recent one (Roonwal, 1976). The present day termite fauna of Rajasthan is overwhelmingly oriental in appearance, while the sole Palearctic element is *Anacathotermes macrocephalus*.

Termites are social and polymorphic insects comprising the order Isoptera. They live in small to large communities composed of reproductive forms together with numerous apterous sterile soldiers and workers. They exhibit divison of labour, performing different biological function. They are commonly known as 'white ants'; the latter expression, however, is unfortunate since these insects are only very distantly related to the true ants or Formicidae. In fact, termites are structurally more closely related to the cockroaches (Blattidae), the primitive genus *Mastotermes* of Australia possesses a number of characters similar to cockroaches especially the wing (both of them possess folded anal lobe in the hind wing). The two groups, that is "white ants' and 'ants', however, offer certain striking analogies of habits and structure, hence the name. Their remarkable social life and the presence of specialized reproductive individuals and sterile castes are cases in point.

A termite colony consists of functional reproductives, workers, soilders and immature individuals. Immature individuals are collectively termed as 'nymphs'. The functional reproductives are of two kinds, Primary and Supplementary. The Primary reproductives, the king and queen, are highly sclerotized and pigmented. The supplementary reproductives are either slightly pigmented, with short wing pads (brachypterous) or very slightly pigmented without wing pad (apterous). Both forms of supplementary reproductives may be termed neotenic, since they achieve sexual maturity without first reaching the imaginal form. There is usually only one pair of primary reproductives in a colony, but when they die they may be replaced by a number of supplementary reproductives. Besides carrying out the fundamental function of reproduction, the reproductive caste helps in distribution of species by swarming or colonizing flight.

The sterile castes, the worker and soilders, are wingless and, in the majority of the species, blind. The sterile caste sometimes exhibit polymorphism; in some species there are two types of worker – major (large) and minor (small), and three types of soilder-major, intermidiate (medium) and minor. The worker caste is considered to be absent in Mastotermitidae, Kalotermitidae and in subfamily Termopsinae of the family Hodotermitidae. Immature individuals in these groups are known as pseudoworkers and they carry out the function of the workers.

The castes most commonly available in the fields are workers (or pseudoworkers) and soldiers; the latter however, constitute only a small proportion of a colony population. Imagoes (winged reproductives or alates) are produced in a colony once or twice a year, when they swarm out of nest to start fresh colonies. For identification soldier caste is essential although soldiers are very rare or absent in a few genera like *Speculitermes*, *Anoplotermes* etc.

Termites are found throughout the tropics of the world and in the warm temperate countries. The Ethiopian region appears to be the richest in genera as well as in species. It was probably the most important center in the evolution of termite genera, because it possesses the largest number and proportion of endemic genera and species.

All sorts of invertebrates (beetles, flies, silver fish, roaches and others) have been adapted to live in the nests of termites. They have become so adapted over the course of evolution to live in close association with termites; they cannot survive independently and have developed interesting structural and physiological modification. These organisms are referred to as termitophiles.

Ecology

Based on their habitat preference there are two kinds of termites, *viz.*, the wood dwellers and the ground dwellers.

1. ***Wood dwellers***: They are confined to wood throughout their life in which they make tunnels. The attacked wood may be dead or living. The wood dwellers may be further separated into two categories as follows:

   (i) Damp wood termites: They require wood with a constant high amount of moisture. Examples are some Kalotermitidae (*Neotermes, Kalotermes*)

   (ii) Dry wood termites: They can live in dry, sound and seasoned wood. Examples are some Kalotermitidae (*Cryptotermes* sp.) which eject small faecal pellets from the holes in the attacked wood; an accumulation of these pellets on the ground betrays the presence of dry wood termites.

2. ***Ground dwellers:*** They live either exclusively, or atleast partly, in ground soil, the ground connection is necessary for their normal life and breeding. From ground the workers and soldiers come out for foraging either in the open (*Anacanthotermes*) or directly in wood having ground connection or under earthen runways or shelter tubes made by the workers on the ground, wall surface or tree trunks, *e.g.*, members of Mastotermitidae, Hodotermitidae, Rhinotermitidae, Stylotermitidae, Serritermitidae, Termitidae and Indotermitidae. The ground dwellers can be further classified into three categories as follows:

(i) Subterranean termites: They live under ground and reach their food above ground under covered runways. Their nest is made either underground and is diffuse or in wood (*Coptotermes*).

(ii) Mound builders: They construct discrete earthen mounds above ground which may reach a height of about four meters (South Asia), or much higher (tropical Africa). The shape and size of the mounds are characteristic of each species group, e.g., *Odontotermes*, *Macrotermes*, *Hypotermes* etc. The entire colony live in the mound and the underground part of the soil immediately beneath it, the whole composite structure is then called a termitarium. The royal cell, which contains the breeding pair (King and Queen) is lodged inside the termitarium, usually below ground level.

(iii) Carton nest builders: They construct discrete, rounded honey combed nests of wood-carton. The nest may lie half buried in the ground (*Microcerotermes*), or may be attached to tree branches (*Nasutitermes*), or may lie inside hollowed trunks, logs or planks (*Coptotermes*).

Catalogue of termites from arid and semi-arid region in Rajasthan

A classified list of termites (Insecta : Isoptera) so far recorded and known from arid and semi arid region of Rajasthan state is given below. Altogether 44 species have been so far recorded from Rajasthan, of which 34 species occur arid and semi–arid zone of Rajasthan (though not necessarily confined to it). These are marked with single asterisk (*) while the exclusively arid zone species are marked with double asterisks (**).

FAMILY KALOTERMITIDAE

**1. *Incisitermes didwanensis* Roonwal & Verma, 1973

2. *Cryptotermes havilandi* Sjostedt, 1900

FAMILY HODOTERMITIDAE

Subfamily Hodotermitinae

**3. *Anacanthotermes macrocephalus* Desneux, 1906

FAMILY RHINOTERMITIDAE

Subfamily Psammotermitinae

**4. *Psammotermes rajasthanicus* Roonwal & Bose, 1960

Subfamily Coptotermitinae

*5. *Coptotermes heimi* Wasmann, 1902
**6. *C. kishori* Roonwal & Chhotani, 1960

Subfamily Heterotermitinae

7. *Heterotermes gertrudae* Roonwal, 1953
*8. *H. indicola* Wasmann, 1902

FAMILY TERMITIDAE

Subfamily Amitermitinae

9. *Eurytermes mohana* Rathore, 1995
10. *Speculitermes cyclops* Wasmann, 1902
11. *Synhamitermes quadriceps* Wasmann, 1902
*12. *Amitermes belli* Desneux, 1906
**13. *Eremotermes fletcheri* Holmgren & Holmgren, 1917
*14. *E. neoparadoxalis* Ahmad, 1955
*15. *E. paradoxalis* Holmgren, 1912
16. *Microcerotermes annandalei* Silvestri, 1923
**17. *M. baluchistanicus*, Ahmad, 1955
*18. *M. heimi* Wasmann, 1902
**19. *M. laxmi* Roonwal & Bose, 1964
*20. *M. palestinensis* Spaeth, 1962
*21. *M. raja* Roonwal & Bose, 1964
22. *M. sakesarensis* Ahmad, 1955
*23. *M. tenuignathus* Holmgren, 1913

Subfamily Termitinae

**24. *Angulitermes dehraensis* Gardner, 1945
**25. *A. jodhpurensis* Roonwal & Verma, 1977

Subfamily Macrotermitinae

*26. *Odontotermes bellahunisensis* Holmgren & Holmgren, 1917
27. *O. brunneus* Hagen, 1858
**28. *O. dehraduni* Snyder, 1934
**29. *O. distans* Holmgren & Holmgren, 1917
30. *O. feae* Wasmann, 1896
**31. *O. giriensis* Roonwal & Chhotani, 1962
*32. *O. guptai* Roonwal & Bose, 1961
*33. *O. gurdaspurensis* Holmgren & Holmgren, 1917
*34. *O. kushwahai* Roonwal & Bose, 1964
*35. *O. latiguloides* Roonwal & Verma, 1973
*36. *O. obesus* Rambur, 1842
**37. *O. parvidens* Holmgren & Holmgren, 1917
*38. *O. redemanni* Wasmann, 1893
**39. *O. wallonensis* Wasmann, 1902
40. *Microtermes bharatpurensis* Rathore, 1989
41. *M. mycophagus* Desneux, 1905
*42. *M. obesi* Holmgren, 1913
**43. *M. unicolor* Snyder, 1933

Subfamily Nasutitermitinae

*44. *Trinervitermes biformis* Wasmann, 1902

Key to the termites recorded from arid and semi-arid zone of Rajasthan State

Key to the families

Imago caste

1. Fontanelle absent 2
- Fontanelle present 3

2. Ocelli absent; right mandible with a small subsidiary tooth at the base of anterior cutting edge of first marginal tooth; left mandible with three marginal teeth; cerci long, with 3-8 articles .... Hodotermitidae
- Ocelli present; right mandibles always without any subsidiary tooth; left mandible with two marginal teeth; cerci short, with 2 articles .... Kalotermitidae

3. Fontanelle without plate; right mandible with a distinct subsidiary tooth at the base of anterior cutting edge of first marginal tooth; left mandible always with three marginal teeth; forewing scales large, overlapping hindwing wing scales; wing membrane strongly reticulate .... Rhinotermitidae

- Fontanelle with plate; right mandible with a minute subsidiary tooth; left mandible with 1-2 marginal teeth; Forewing scale small, not overlapping hindwing scales; wing membrane either not reticulate or with a weak reticulatation .... Termitidae

Soldier caste

1. Head without fontanelle and frontal gland; eyes present; mandible with Prominent marginal teeth 2
- Head with fontanelle and frontal gland; eyes if present, dot like; Mandible with or without marginal teeth 3
2. Cerci long, with 3-8 articles. Antennae with 22-23 articles ....Hodotermitidae
- Cerci short, with 2 articles, antennae with 10-18 articles ....Kalotermitidae
3. Pronotum flat. .... Rhinotermitidae
- Pronotum saddle-shaped ....Termitidae

Key to the genera

1(2) Femora swollen .... *Incisitermes*

2(1) Femora not swollen

3(4) Larger species (10-16 mm. long), eye spots prominent .... *Anacanthotermes*

4(3) Smaller species (7.4-8.2 mm. long), eye spots absent; mandible with several large to small well-developed teeth on inner margin; soldier trimorphic ....*Psammotermes*

5(6) Head-capsule oval-shaped; fontanelle large, circular and lying on antero-median margin ....*Coptotermes*

6(5) Head-capsule subrectangular; fontanelle small, lying at the middle of head dorsum. .... *Heterotermes*

7(8) Mandible with serrated inner margin .... *Microcerotermes*

8(7) Mandible without serrated inner margin

9(10) Head-capsule with frontal protuberance; mandible slender, weakly incurved apically..*Eremotermes*

10(9) Head-capsule without frontal protuberance; mandible strongly built

11(12) Clypeus bilobed, frontal projection absent .... *Amitermes*

12(11) Clypeus not bilobed, frontal projection present ....*Angulitermes*

13(14) Mandible short, thin, delicate; soldier smaller than workers, smaller species .... *Microtermes*

14(13) Mandible large and generally strong, soldiers usually larger than worker, larger species .... *Odontotermes*

15(16) Mandible strongly asymmetrical, antero-lateral corners of head formed into tubercle-like projections

16(15) Mandible degenerated, non-functional, head capsule produced into a nasute, soldier dimorphic .... *Trinervitermes*

Key to the species of the genus *Coptotermes*

Soldier

1(2) Waist of postmentum lying in the middle of line connecting the level of max. width and hind margin; postmentum with minimum width of 0.25-0.34 mm.. *C. heimi*

2(1) Waist of postmentum lying below the middle of line connecting the level of max. width and hind margin; postmentum with minimum width of 0.23-0.25mm .... *C. kishori*

Key to the species of the genus *Eremotermes*

Soldier

1(2) Mandible distinctly longer than head capsule (head-mandibular index 1.12 and above); mandibular tooth weakly developed or absent, if present placed below the middle of mandible.. *E. neoparadoxalis*

2(1) Mandible either subequal or shorter than head capsule (head-mandibular index usually 1.00 or below); mandibular tooth prominent and placed in the middle of mandible

3(4) Larger species (max. head width 0.80-0.85 mm.; max. pronotum width 0.50-0.55 mm.); mandiblular length sub-equal to head length (head-mandibular index 1.00-1.05) .... *E. fletcheri*

4(3) Smaller species (max. head width 0.68-0.75 mm.; max. pronotum width 0.40-0.45 mm.); mandible distinctly shorter than head length (head-mandibular index 0.78-0.98) ....*E. paradoxalis*

Key to the species of the genus *Microcerotermes*

Soldier

1(7) Inner margin of the mandible finely serrated

2(3) Mandible shorter; index mandible-length/ head-length 0.82-0.87 mm.; head capsule in comparison to length wider, index width/ length 0.77-0.86 mm.. *M. baluchistanicus*

3(4) Mandible larger, almost as long as head; index mandible-length/ head-length 0.98 mm.; pronotum strongly emarginate ....*M. laxmi*

4(5) Head-length smaller (1.0-1.27 mm); mandible shorter than head-length by 3/4 times; index mandible-length/ head-length 0.70-0.79 mm; pronotum weakly emarginate .... *M. tenuignathus*

5(6) Head generally larger, length 1.19-1.25 mm; mandibular apices weakly incurved; labrum sharply pointed at tip ....*M. heimi*

6(5) Labrum with broadly rounded tip; mandible relatively weakly hooked .... *M. palestinens*

7(8) Inner margin of mandible coarsely serrated

8(9) Larger species; mandible with less incurved apices

9(8) Smaller species; mandible coarsely serrated without large tooth; labrum tongue-shaped, pentagonal .... *M. raja*

Key to the species of the genus *Angulitermes*

Soldier

1(2) Head weakly incurved on medio-lateral side; antero-lateral corners of head sharply angulate; length of frontal projection of head 0.10-0.13 mm.. *A. dehraensis*

2(1) Head straight on medio-lateral side, narrowed posteriorly; length of frontal projection of head 0.26mm. .... *A. jodhpurensis*

Key to the species of the genus *Odontotermes*

Soldier

(*O. dehraduni* is known only from imago, hence not included in the key)

1(17) Antennae distinctly darker distally and paler basally

2(3) Tooth on the left mandible placed near the apices; tooth-index less than 0.28 mm, head capsule moderately pilose; head mandibular index more than 0.50-0.59mm .... *O.giriensis*

3(4) Tooth on left mandible placed near apical third; tooth-index always more than 0.28 mm

4(5) Mandible longer (0.79-0.89 mm.); head-mandibular index higher (0.75-0.79 mm.) .... *O. latiguloides*

5(4) Mandible shorter (0.74-0.78 mm.); head-mandibular index lower (0.66-0.76 mm.)

6(7) Head and body more hairy; antennae with 15-16 segments; mandible shorter and stouter; labrum broadly rounded anteriorly; post-mentum convex laterally .... *O. guptai*

7(8) Head and body less hairy; antennae with 16-17 segments; mandible comparatively less stout and longer; labrum narrowing anteriorly; post-mentum laterally substraight .... *O. bellahunisensis*

8(9) Left mandibular tooth minute, rudimentary; mandible substraight, long, slender and very slightly bent inwards near distal tip; post-mentum laterally weakly arched

9(10) Head capsule subrectangular laterally, almost parallel upto the base of antennae .... *O. gurdaspurensis*

10(11) Head capsule oval, distinctly narrowed anteriorly

11(12) Mandible longer; outer margin strongly bent near the basal third; index mandible length/ head length to base of mandible 0.69-0.79 mm *O. redemanni*

12(13) Mandible short, slender, with weakly curved outer margin; mandible index length/ head length 0.59-0.68 mm; labrum short, broadly rounded anteriorly.. *O. obesus*

13(14) Head capsule subrectangular; teeth on left mandible placed near the distant end of the middle one-third portion

14(15) Head capsule with sides convex in the middle; inner margin of left mandible beyond the tooth sickle-shaped; index mandible length/ head length to base of mandible 0.61-0.67 mm.; tooth of left mandible not much forwardly placed (tooth distance 0.43-0.50 mm.); mandible tooth index 0.41-0.44 mm.. *O. kushwahai*

15(16) Tooth distance from the tip of left mandible 0.34-0.38 mm.; tooth index less (0.34-0.39 mm.); antennae 16-segmented

16(17) Tooth distance from the tip of left mandible 0.43-0.53 mm.; tooth index high (0.43-0.53 mm.); antennae 16-17 segmented .... *O. wallonensis*

17(18) Antennae uniformly coloured, not darker distally

18(19) Tooth on left mandible placed a little below the middle; mandible toth index 0.51-0.55 mm; smaller species

19(20)Tooth on left mandible rudimentary and placed near the basal third; mandible tooth index 0.58-0.66; cmparartively larger species; antennae 17-segmented .... *O. distans*

20(21) Tooth on left mandible minute and placed at basal third of mandible; antennae 16-17 segmented .... *O. parvidens*

21(20) Tooth on left mandible placed near the proximal end of mdiddle one-third

Key to the species of the genus *Microtermes*

1(2) Antennae with 15 segments .... *M .mycophagus*

2(1) Antennae with 14 segments; larger species, 3.2 to 5.0 mm

3(4) Head-capsule moderately hairy; second segment of antennae equal to the combined length of third and fourth segments; labrum reaching upto 2/3rd of mandible ....*M. obesi*

4(3) Head-capsule densely hairy; second segment of antennae shorter than combined length of third and fourth segments; labrum reaching upto 3/4th of mandible .... *M. unicolor*

MATERIAL AND METHODS

Subterranean termites and the mound building termites were collected with the help of shovel pushing it into the soil and then spreading the material on a tray and ultimately picking them up with forceps. Wood eating termites were also collected with forceps after cutting the wood with the help of hand axe. The collected termites were preserved in 70% alcohol.

The termite species were identified with the help of standard keys of systematic references provided in the Fauna of Isoptera Part I and II (Roonwal and Chottani, 1989; Chottani, 1997) and measurement of different morphological character (Roonwal, 1970) were taken with the help of ocular micrometer in the eyepiece of binocular.

OBSERVATIONS

I. Ecological relationship

(a) Relationship with rainfall and vegetation

With rainfall

Ecological zonation into three zones (arid, semi arid and wet zone) provides a useful framework for establishing relationships. Some species are restricted to one zone, some to another while others have wider tolerance. Few species occur almost exclusively in arid zone e.g. *Anacanthotermes macrocephalus*, *Psammotermes rajasthanicus*, *Microcerotermes laxmi* etc. Four species habitually inhabit dry wood, among them *Incisitermes didwanaensis* (Kalotermitidae) is an exclusively wood dweller. The remainder (Rhinotermitidae) have ground connection as well. *Heterotermes indicola* and *Coptotermes heimi* are wide spread in India and elsewhere and occur in variety of habitats; Psammotermes rajasthanicus is confined to arid zone in Rajasthan and Gujarat (Kutch), but also extends farther west into Pakistan where it inhabit semi arid areas. Some species of genera *Microcerotermes*, *Odontotermes* and *Microtermes* attack timber though not habitually and therefore cannot be classified as "wood inhabiting". *Eremotermes paradoxalis*, *Odontotermes obesus* and *Microtermes obesi* have wide tolerance.

With vegetation

Psammotermes rajasthanicus prefers sandy areas with little vegetation, but is also found in open scrub. *Microcerotermes sakeserensis*, *Angulitermes jodhpurensis* and *Microtermes unicolor* are restricted to open scrub. Two species, *Synhamitermes quadriceps* and *Odontotermes feae*, prefer thick forests, other species are found in several types of vegetation; some of them also occur in agricultural fields. Some species especially, those of the genera *Eremotermes*, *Odontotermes* and *Microtermes*, attack the roots of both wild and cultivated plants and derive food and moisture from them. As a result, the plants often wilt and die.

(b) Food gathering (Foraging and Non foraging species)

In Rajasthan all the termites, except the wood dwelling species *Incisitermes didwanaensis*, are soil dwellers. They live a subterranean life, except for the following situations:

1. For daily food gathering, the workers and soilders of the two truly foraging species, *Anacanthotermes macrocephalus* and *Trinervitermes biformis*, forage in the open. Some others (of the genera *Psammotermes*, *Heterotemes*, *Coptotermes*, *Odontotermes*, *Microtermes*) move under earthen shelter tubes; still others (of

the genera *Speculitermes*, *Amitermes* etc.) feed mostly underground or under dung or stones.

2. When the imagoes swarm out annually in the monsoon (June - September) to form new colonies.
3. *Psammotermes rajasthanicus*, *Heterotermes indicola* and *Coptotermes heimi*, while retaining their subterranean connections, also attack dry wood, e.g., wood work in buildings, lintels, etc.

(c) Relationship with soil cover

Soil preferences

Psammotermes rajasthanicus, the sand termite, occurs only in the more sandy part of the desert, usually in the non-cultivated areas, while *Anacanthotermes macrocephalus*, the harvester termite, is found in sandy and gravelly soil both near cultivated and uncultivated areas. *Microcerotermes sakeserensis*, *Microcerotermes laxmi* and *Angulitermes jodhpurensis* are confined to sandy desert soil with scruby vegetation. Other species are more adaptable and occur in several types of soil; they are *Eremotermes paradoxalis*, *Odontotermes obesus* and *Microtermes obesi*.

Adaptation to soil types

A number of species show adaptations which are probably related to soil types including their water retention capacities. *Anacanthotermes macrocephalus* builds extensive and fairly deep underground galleries in sandy or gravelly desert soil, both to avoid extreme surface temperatures and to trap sub-soil moisture. Similar but less extensive ramification is met within other species, e.g., *Angulitermes jodhpurensis*, *Microtermes* species, etc. *Odontotermes obesus* shows a most interesting adaptation as regards mound construction. In moist areas with heavy soil which retains water it builds earthen mounds, but in sandy desert soil which retains water it builds earthen mounds. In sandy desert soil with poor capacity for water retention no mounds are built as a rule except sometimes in the vicinity of subsoil moisture as near water reservoirs (Roonwal, 1976).

II. Pest status of Desert termite

The biotic pressure in arid and semi-arid region of Rajasthan is heavy due to large human population and the bovine stock. This situation has led to the indiscriminate destruction of whatever scanty vegetation was available in the arid region. Besides numerous biodeteriorating agents (rodents and other mammals, insect pests *etc.*) do considerable damage to young plantation. Among insects, termites have emerged as the *numero uno* notorious pest in the arid and semi-arid region and are considered as the single most important factor that is detrimental for plant growth, particularly in the drier season. Since they feed on cellulose, plants and/or plant-based products are mostly the most attacked ones.

According to the nature and intensity of damage, the desert termites have been classified as major pest, minor pest and the non-pest species.

(a) *Major pest*: Only five species, i.e., *Heterotermes indicola, Coptotermes heimi, Odontotermes obesus, Microtermes obesi* and *M. mycophagus,* are categorized as major pest causing immense damage to agricultural crops, cash crops, forest nurseries, fruit trees and wood works in buildings *etc.*

Heterotermes indicola is primarily a wood-infesting species, however it has also been reported to damage agricultural crops like germinating 'karela' (*Momordica charantia*) in Jodhpur (Roonwal, 1982) and fruit trees like 'ber' (*Zizyphus mauritiana*) in arid areas of Rajasthan (Roonwal and Bose, 1964, 1978). *Coptotermes heimi* is also a wood inhabiting species highly destructive to crops and fruit trees and cause serious damage to *Acacia senegal* and *Ficus* sp. *Odontotermes obesus* has been reported to damage bazra and chilli crops in the desert areas of Rajasthan (Roonwal and Bose, 1978). It also damages trees like *Prosopis cineraria, Delonix regia* etc. *Microtermes obesi* and *Microtermes mycophagus* are the very common species in the desert area that is evidenced by the swarm seen during the rainy season. It has been found damaging agricultural crops, vegetables, fruit trees and structures in houses.

(b) *Minor pest*: *Anacanthotermes macrocephalus* and *Trinervitermes biformis,* the harvester termites are responsible for denudation of the pasture land. The most affected grasses are *Cenchrus ciliaris, C. setigerus* and *Lasiurus sindicus* (Parihar, 1978, 1980; Thakur, 1985; Roonwal and Bose, 1978). In a way *Anacanthotermes macrocephalus* is a feed competitor of livestock. *Eremotermes paradoxalis* have also been reported feeding on grasses in arid areas of Rajasthan (Roonwal and Bose, 1964). *Microcerotermes baluchisthanicus* and *M. tenuignathus* are found damaging wheat crops in Rajasthan (Parihar, 1978, 1980). *Odontotermes bellahunisensis* damages plants at germinating, flowering and fruiting stage. *Amitermes belli* damages plant species like *Zizyphus mauritiana, Salvadora oleoides, Prosopis* sp. etc. *Psammotermes rajasthanicus* is a serious pest of *Tecomella undulata* trees.

(c) *Non-pest species*: Among desert termite only *Angulitermes jodhpurensis* are non-pest and soil dwelling species.

III. The Indira Gandhi Nahar Pariyojna and its likely impact on the desert ecosystem

A major event which made a profound influence on the ecosystem of northwestern parts of Rajasthan, *i.e.*, Thar Desert, is the construction of Indira Gandhi canal, which on completion will be one of the largest irrigation projects in Asia. The gross service area of IGNP will be over 2 million hectare of which the cultivable command area will be 1.3 million hectares.

The IGNP is a massive hydrological project with multiple objectives. It has been under process of completion over past four decades, and has invariably brought about enormous changes in the topography, physiography and ecology of the Thar Desert. The groundwater exploitation has altered the landuse pattern from grassland and open scrubs to irrigated land. As a result, the xeric flora and fauna is replaced by the mesic biota. A large number of plants, particularly weeds, are introduced along with canal waters that originate in from Punjab and enters into Rajasthan via Haryana. Cultivation of the land may result in elimination of characteristic desert plants. For example, *Lasiurus sindicus*, a perennial grass, endemic to the Thar Desert and adapted to low soil moisture content, is enhanced manifold with the inception of Indira Gandhi canal. This grass species is currently much threatened for its very survival. The natural vegetation in the canal command area now exists only in the wasteland and protected areas. Due to change in landuse pattern there has been a gradual increase of damage to various crops in the IGNP command area. Some minor insect species are gaining pest status and are expanding their niches into the crop fields. Insect pest are changing their host plants and some new insect pest are introduced in the desert zone (Prakash, 1998). It is quite probable that these ecological changes in the command area of IGNP may result the desertic termite species to be replaced by mesic species.

IV. Account of termite species of arid and semi-arid region of Rajasthan

Family Kalotermitidae

(1) *Incisitermes didwanaensis*

Ecology and biology: It is exclusively wood dwelling. Little else is known about it.

Family Hodotermitidae

(2) *Anacanthotermes macrocephalus*

Ecology and biology: this is the largest species in Rajasthan; length of Imagoes with wings 27.2 – 27.5 mm, without wings 12.4 – 14.6 mm; soldiers 10.0 – 16.7 mm; workers 5.9 – 8.8 mm. It lives in narrow but extensive underground galleries, and make small, subconical heaps of loose, granulated, excavated earth. Workers, and sometimes soldiers, come out daily for foraging; they gather semi-dry plant material which they take back to the galleries for storage in the storage chambers. The nest is built underground. Swarming occurs in July-August.

Family Rhinotermitidae

(3) Psammotermes rajasthanicus

Ecology and biology: It is a sand loving termite and confined to extreme arid portion. It prefers open scrub but occasionally it attacks wood work in houses and also some tree species like Tecomella undulata and *Prosopis cineraria*. The species is known from soldiers and workers. Nothing is known about its biology.

(4) Heterotermes indicola

Ecology and biology: It is a wood inhabiting termite and cause serious damage to dry wood; it nests underground. It also makes narrow earthen runways on the tree trunk. Swarming in the desert occurs in July and August.

(5) Coptotermes heimi

Ecology and biology: It is a serious pest of dry wood. The nest is built underground in soil or in dead logs. Swarming in the desert occurs in May to August.

(6) Coptotermes kishori

Ecology and biology: This species was collected from wood and no other observation is available from Rajasthan state.

Family Termitidae

(7) Amitermes belli

Ecology and biology: It is a soil dwelling species and is also common under bark and in dead wood lying on the ground. Swarming occurs from 7:45 to 8:30 pm in July month.

(8) Eremotermes fletcheri

Ecology and biology: not much is known about its ecology and biology.

(9) Eremotermes neoparadoxalis

Ecology and biology: It is a soil dwelling species found under dung, decaying wood, open as well as cultivated fields and other such situations. Swarming occurs from end May to July from small holes in the ground at 5:15 to 7:00 pm.

(10) Eremotermes paradoxalis

Ecology and biology: this is a soil dwelling species with small, whitish, sluggish soilders and workers. Swarming in Rajasthan occurs in May to August, the alates emerges from holes in the ground at 5:15 to 7:00 PM.

(11) Microcerotermes baluchistanicus

Ecology and biology: the species is evidently an arid zone species and endemic to Rajasthan. Little else is known about this species.

(12) Microcerotermes heimi

Ecology and biology: this species is basically a wet zone species but it strays into the arid zone of Rajasthan (Nagaur). Little else is known about it.

(13) Microcerotermes laxmi

Ecology and biology: It has been found in cylindrical earthen tubes at the base of desert trees, e.g.; *Prosopis julliflora*; only soilders and workers are known. The species is endemic to Rajasthan and evidently an arid zone species.

(14) Microcerotermes palestinensis

Ecology and biology: In Rajasthan it has been obtained from earthen runways on trunks of *Prosopis spicigera*, also from earthen mounds of *Odontotermes obesus* (Roonwal, 1976). It is basically an arid zone species but it is also reported from forest area of wet zone (Udaipur).

(15) Microcerotermes raja

Ecology and biology: it is basically soil dwelling species but it is also found under bark of trees (Palm stumps). Swarming occurs in June to August from holes in the ground; eggs are laid a few days after swarming. It is basically an arid and semi arid zone species, but its occurrence from udaipur area further extends its range in Aravallis.

(16) Microcerotermes tenuignathus

Ecology and biology: it is a soil dwelling species found not far from surface, under stones, cow dung, etc. sometimes occurs in association with *Odontotermes* sp. It is versatile species. Not much is known about its biology.

(17) Angulitermes dehraensis

Ecology and biology: Not much is known about its ecology and biology.

(18) Angulitermes jodhpurensis

Ecology and biology: It is a soil dwelling species. At Jodhpur it swarms in July to August from tiny holes in the ground; eggs are laid a few days later and hatches in about 18 days. It is an arid zone species and endemic to Rajasthan.

(19) Odontotermes bellahunisensis

Ecology and biology: Not much is known about its ecology and biology.

(20) Odontotermes dehraduni

Ecology and biology: Swarming occurs in July to September; not much is known abot its ecology.

(21) Odontotermes distans

Ecology and biology: It is a soil dwelling species. Swarming occurs in month of August in Rajasthan.

(22) Odontotermes giriensis

Ecology and biology: It is a soil dwelling species and is often found in rotten wood lying on the ground; occasionally builds mounds. Only soilders and workers are known.

(23) Odontotermes guptai

Ecology and biology: It is a soil dwelling species; only soilders and workers are known.

(24) Odontotermes gurdaspurensis

Ecology and biology: It is a soil dwelling species which builds complex underground nests. Swarming occurs in July and August. It mainly occurs in Wet and semi arid zone but now it is also known from arid zone.

(25) Odontotermes kushwahai

Ecology and biology: It builds low earthen mounds which rise to about 45 cm; only soilders and workers are known. No other details are known about this species.

(26) Odontotermes latiguloides

Ecology and biology: it is a soil dwelling specie; only soilders and workers are known. No other detail is known about this species.

(27) Odontotermes obesus

Ecology and biology: it is a soil dwelling species whose members come out for food gathering under the protection of earthen shelter tubes. Soilders and workers show a considerable size range within a colony and from colony to colony but they are not demarcated into major and minor forms and are connected by a series of intermediate forms. It is a mound building species but in Rajasthan, it either builds no mound (in arid zone); or the mounds are sub conical to dome shaped with the maximum height of a meter or so. Swarming occurs in July and August.

(28) Odontotermes parvidens

Ecology and biology: it is a soil dwelling species which often attacks dead wood lying on the ground, and sometimes even kills healthy trees. The nest lies underground; swarming occurs in February to April. In Rajasthan it is recorded from arid region only.

(29) Odontotermes redemanni

Ecology and biology: it is a soil dwelling species. Not much is known about its ecology and biology.

(30) Odontotermes wallonensis

Ecology and biology: it is a mound building species and builds a open mouthed chimneys. Swarming occurs in August in Rajasthan.

(31) Microtermes mycophagus

Ecology and biology: it is a small soil dwelling species which attacks living plants (cultivated crops). Swarming occurs during June to August at 8:00 PM to 9:00 PM from small holes surrounded by workers and soilders in the ground. Eggs are laid a few days after swarming and hatch in about 19 – 20 days.

(32) Microtermes obesi

Ecology and biology: it lives in small communities either underground or in earthen mounds of other termites, especially *Odontotermes obesus*. Swarming occurs during June to August but sometimes as early as April. The species is serious pest of agricultural crops.

(33) Microtermes unicolor

Ecology and biology: it is a soil dwelling species and nests underground. Swarming occurs at 8:00 PM to 9:00 PM during July to August, sometimes earlier.

(34) Trinrvitermes biformis

Ecology and biology: it is a foraging species, with nasutiform soilders which are dimorphic (major and minor forms). Nests are subterranean; underground galleries in Rajasthan go about 15 cm deep and may end in a blind basal chamber. It swarms in night. Little else is known about this species.

RESULTS AND DISCUSSION

Emerson (1952) postulated that during late Mesozoic, living termite families dispersed to major tropical areas of the world, but differentiation of subfamilies and genera took place later in Tertiary. Termites originating in South America dispersed to Australian, Oriental and Ethiopian regions and *vice versa* during late Mesozoic through a hypothetical land bridge. Several authors (Imms, 1937; Weidner, 1966, 1970; Krishna, 1970) have reconstructed the phylogeny of different termite taxa. Although a number of fossils dating back from as early as Tertiary or late cretaceous are available but such fossils are not clearly distinguishable from modern day Isopteran genera because of their fossilized nature. Above authors have opined that Problattoidea (roach like ancestor) which originated in the late Carboniferous or early Devonian served as the probable ancestor of Blattoidea and modern day Isoptera. Based on zoogeographical evidence Myles (1998) provided further evidence that modern day termites had evolved from roach like proto-termites prior to continental fragmentation.

Western Rajasthan was well-wooded in the late Pleiostocene, and sustained a richer, more varied and more tropical fauna than today. Its dessication is a recent feature, barely 7,000 year old (Roonwal, 1976), which resulted in the formation of the Indian Desert. No termite fossils have been found in Rajasthan, and geological evidence thus provides little guidance in zoogeographical matters. It may be presumed (Chottani and Bose, 1962) that the primitive elements, which evolved earlier, came from South America before the southern continents separated from one another in terms of Continental Drift Theoryof Wegener (1922). Those higher ones (Termitidae) which came from Africa, and were evolved later, probably also came later (Miocene or Pliocene?) through some ecologically suitable corridor between Africa and South Asia; that was the time when sea receded from the western Rajasthan. The desiccation probably resulted in: (i) elimination of inadaptable species, and (ii) a second series of intrusions of dry zone species may have occurred from the West, e.g., Iran, west Asia and even the Sahara; probable examples are *Eremotermes neoparadoxalis, Microcerotermes palestinensis* and *Microcerotermes tenuignathus* (Roonwal and Verma, 1977).

From ecological point of view Thar is dominantly a Sandy Desert. It was observed that it is a very good habitat for species like *Psammotermes rajasthanicus, Anacanthotermes macrocephalus, Microcerotermes laxmi* and *Angulitermes jodhpurensis.* The termite species, *P. rajasthanicus,* occurs in sandy and uncultivated soils, while another species, *A. macrocephalus,* prefers sandy and gravelly soil. In contrast, *M. laxmi* and *A. jodhpurensis* prefer sandy soil and scrubby vegetation. The later two species are endemic to the Thar Desert.

The Thar Desert of Rajasthan exhibit vivid and spectacular biodiversity but during the last few decades the productivity and the biodiversity it sustains have been diminishing as man has been exploiting the xeric environment indiscriminately. At this juncture Indira Gandhi canal has brought good quality water from Himalayas and hence, it is considered to be the life line of Thar Desert in Rajasthan, but it has changed the topography, ecology and physiography of about 11% of Thar Desert has changed. Rathore (1996) has reported, that with the inception of Indira Gandhi canal, soil moisture has enhanced manifold in the command area of IGNP and as a result xeric species will be replaced by mesic species, also some minor pest may gain the status of major pest by expanding their niche to crop fields.

It was observed that due to the above mentioned changes mesic species of termites are now recorded from extreme arid region of Rajasthan *viz. Amitermes belli,* a moisture loving species found in wet and semi arid areas of Rajasthan, but with the availability of moisture in the command area of IGNP, it has been now recorded from from extreme arid region of Jaisalmer and Barmer.

With the inception of Indira Gandhi canal the termite species adapted to severe arid conditions and dry arid soil, will not be able to survive in soil carrying perennial moisture. Due to ecological change, the command area of IGNP is no more suitable abode for the deserticulous termites. As a consequence they will either perish or migrate. Apart from the above result, further consideration is needed for ecological studies around the canal command area to know more about the impact of changing ecology on termite fauna of Rajasthan.

ACKNOWLEDGEMENT

Authors express their sincere gratitude to Dr. J.R.B. Alfered, Director, Zoological Survey of India, for his constant encouragement. The cooperation of Dr. Ram Sewak, Dr. Shalini Gaur and Mr. Alok K. Dwivedi is greatfully acknowledged. Grants received from the Ministry of Environment and Forests, Govt. of India, New Delhi for sponsoring the research project is also acknowledged.

REFERENCES

Chottani, O.B. 1997. Fauna of India – Isoptera (Termites). Zoological Survey of India, Calcutta Publication. Vol. II. xx + 800.

Emerson, A.E., 1952. The biogeography of termites. *Bull. Am. Mus. Nat. Hist.* 99: 217 – 225.

Imms, D. 1937. *Recent Advances in Entomology.* J. & A. Churchill Ltd., London. 2nd ed., x + 431 pp

Kapoor, J.P. 1985. Wild life management in Desert National Park. *Proc. Nat. Sympos. Eval. Environ.* (Spl. Vol. Geobios). Eds. S.D. Mishra, D.N. Sen and I Ahmad. 273 – 278 pp.

Krishna, K. 1970. Taxonomy, Phylogeny and distribution of termites. *In* : *Biology of Termites.* Vol. II (Eds. K. Krishna & F.M. Weesner), Academic Press, New York, : 127 – 152.

Mann, H.S. and I. Prakash 1983. *Halting the March: Eco-development in the Thar.* Ministry of Environment, Government of India and WWF-India, New Delhi. 1 – 36.

Myles, T.G. 1998. Proposed taxonomy of the order Isoptera. Internet material accessed on September 18, 2002. www.utoronto.ca/forest/termite/ taxon. htm.

Parihar, D.R. 1978. Field observation on the nature and extent of damage by Indian desert termites and their control. *Annals of Arid Zone* 17 (2): 192-199

Parihar, D.R. 1980. Termite problems in desert plantations. *Annals of Arid Zone* 19(3): 329 – 334.

Prakash, I., S.K. Saxena, P.K. Ghosh and H.C. Bohra 1992. Vegetation and Wildlife in the Thar. *In* : *Perspectives of the Thar and Karakum.* (Eds. Kar, A., Abihichandani, Anantharaman, K., Joshi, D. C.) Department Science and Technology, Govt. of India, New Delhi: pp. 98 – 109.

Prakash, I. 1998. Impact of Indira Gandhi canal on desert ecosystem. *In Ecology Today: An Anthology of Contemporary Ecological Research* (Ed. by: Gopal, B., Pathak, P.S. and Saxena, K.G.), pp. 323-331. (Published by International Scientific Publications, New Delhi).

Rahmani, A.R. 1989. The uncertain feature of the Desert National Park in Rajasthan, India. *Environmental Conservation* 16 (3): 237-244.

Rathore, N.S. 1984. Biological observation on Indian Desert Termite *Microtermes mycophagus* (Termitidae). *Oikoassay* 1 (1 & 2): 5 – 10.

Rathore, N.S. 1996. Termites (Isoptera) of the Thar Desert. *In : Faunal Diversity in Thar Desert : Gaps in Research* (Eds. Ghosh *et al.*), Scientific Publishers (India), Jodhpur : pp. 171 – 183.

Rathore, N.S. and Bhattacharya, A.K. 2004. Termite (Insecta : Isoptera) fauna of Gujrat and Rajasthan – present state of knowledge. *Rec. Zool. Surv. of India* Occ. Paper No. 223, 1 – 77.

Roonwal, M.L. 1970. Measurements of Termites (Isoptera) for Taxonomic purposes. *J. Zool. Soc. of India.* 21(1), pp. 9 – 66.

Roonwal, M.L. 1976. Field ecology and ecobiogeography of Rajasthan termites. A study in desert environment. *Zool. Jb. (Syst.)* 103: 455 – 504.

Roonwal, M.L. 1982. Fauna of Great Indian Desert (Past and present composition, zoogeography, ecology, biology, physiology and conservation). *In: Desert Resources and Technology,* Ed. Alam Singh, Scientific Publishers & Geo-tech Academy, Jodhpur, Vol. 1: 1 – 86.

Roonwal, M. L. and Chottani, O.B. 1989. *Fauna of India – Isoptera (Termites).* Volume I. viii + 672 + I pp. Zoological Survey of India, Calcutta publication.

Roonwal, M.L. and Bose, G. 1964. Termite fauna of Rajasthan, India. *Zoologica Stuttgart,* 40(3): (Heft. 113): vi + 58.

Roonwal, M.L. and Bose, G. 1978. Vegetational distribution of termites of Rajasthan (India) and their economic importance. *Proceedings of national Science Academy* (B) 44(5): 320 – 329.

Shankarnarayan, H.S. and Gupta, V.K. 1991. Soils of the region. *In Prospects of Indira Gandhi canal project* (Ed. By Venkateswarlu, J. and Abrol, I. P.), Indian council agriculture Research, New Delhi. pp. 19 – 35.

Thakur, R.K. 1985 Pest status of termite in Gujrat, India. *Indian Journal of Forestry-* 8(1): 37 – 40.

Changing Faunal Ecology in Thar Desert, pp. 85-133, 2005.
Eds.: B.K. Tyagi and Q.H. Baqri

6

INSECT DIVERSITY CONSERVATION STATUS WITH SPECIAL REFERENCE TO BUTTERFLIES (LEPIDOPTERA: RHOPALOCERA)

H.S. Rose

Lepidoptera Laboratory, Department of Zoology, Punjabi University, Patiala - 147 002, India

Like other invertebrates, insect diversity is one of the most important components of the biodiversity of a region. Though dominating yet no significance is being given to conserve them adequately. Our notion and approach of conserving vertebrates alone need to be changed. Huge insect diversity is yet to be explored in India besides updating of inventories. Even the *Rhopalocera (Lepidoptera)* which is considered an umbrella for other invertebrates is not thoroughly known, especially the studies on the larval-host plants and nectar food plants of different species of butterflies are miserably lacking. The studies on the immature stages of butterflies and their life histories are inadequate. In this respect, the position of *Heterocera* and other insect orders is also not very promising. In order to highlight these points, the life histories, larval host plants and nectar food plants of one representative species from each respective family has been dealt with in the present study, *i.e., Castalius rosimon* (Fabricius) (Lycaenidae), *Vanessa* (*Aglais*) *cashmiriensis* Kollar (Nymphalidae), *Acraea issoria* (Hubner) (Acraeidae), *Ypthima inica* Hewitson (Satyridae), *Euploea core* (Cramer) (Danaidae), *Colotis amata* (Fabricius) (Pieridae), *Papilio polytes* Linnaeus (Papilionidae) and *Pelopidas mathias mathias* (Fabricius) (Hesperiidae). Besides, the diagnosis of these species has been updated through studies on their genitalia. The need to look into different schedules of the Wildlife (Protection) Act, 1972 (as amended up to 1991) has also been emphasized by quoting suitable examples of certain butterfly species on the basis of recent field work undertaken in two different ICAR projects, first between 1991-1994, and another between 1995-1998. The conservation of insects is the conservation of biodiversity as a whole, and it has been highlighted in detail. Brief reference has also been made as to how the Ministry of Environment & Forests, Government of India has launched an All India Coordinating Project for Capacity Building in Taxonomy (AICOPTAX) to fulfill article 6 of Convention of Biological Diversity.

INTRODUCTION

Insect diversity in terms of number of species and populations is one of the most dominant components of biodiversity inhabiting any conceivable habitat except extreme polar regions and very high mountain peaks. There is no consensus as to how many species of insects actually dwell in different habitats/ecosystems world over. Though an estimate of insect diversity ranging between 2 to 30 million seems to be too vague, yet is close to be realistic. According to another estimate, more than four out of every five animal species are insects and more than two million of them have been taxonomically described from various parts of the globe as per International Code of Zoological Nomenclature (ICZN). Majority of the insects are terrestrial and only 2-3% of them are aquatic.

Insects possess a significant place in the environment due to their multifold activities. Apart from being associated with a variety of vegetation, they are also parasites and predators on many other organisms, including insects. Some of the species are detritus feeders as well. Amongst known species, it is being estimated that about 1% of the insect species are serious pests of various crops, forest trees and stored products. Many of the species also act as vectors of different diseases. The losses caused by major pest species are so enormous that huge estimates have been given from time to time, particularly in the tropical countries. In addition to this, the gravity of the situation gets further worsened due to damages caused by insects of medical and veterinary importance. To prevent such losses, many kinds of insecticides such as chlorinated hydrocarbons, organophosphates, carbamates and synthetic pyrethroid *etc.* have been used with limited success. It seems appropriate to record that as late as mid 1950s, an experienced entomologist counselled his colleagues, perhaps facetiously, to make collections of major pest species for deposition in museums since it would only be a matter of time before these pests would be eradicated by the new insecticides. Rather, the policy to eradicate the pests through the use of various insecticides and pesticides has proved fatal and counterproductive in many ways. The need to develop alternative methods for pest control have been initiated during the past 2-3 decades.

With this background, the question arises as to why do insects need to be conserved? The rate of animal extinctions, including insects, is on the higher side due to various factors such as deforestation, agro-industrialization, degradation of the habitats, urbanization and excessive use of pesticides *etc.* In fact, the habitat destruction by man to fulfill his insatiable greed has been ceaselessly resulting in a big loss to endemic fauna and flora. The national losses associated with many species, particularly those of the butterflies, have been well documented in the developed countries like the U.S.A and U.K. and run into billions of dollars per annum.

Further, barring negligible number of pest and vector species, insects are rather very valuable resources economically and ecologically owing to their manifold roles in sanitizing the environment. While competing with man, insects do perform an indispensable service to humanity simultaneously. In certain instances, their fundamental importance is likely to exceed than some other organisms, like mammals and birds - which are readily accepted as worthwhile conservation subjects do.

The importance of insects (particularly honeybees, moths and butterflies *etc.*) is largely proved by an act of pollination played by them for the enhancement of crop production. It is on record that many obligate pollinators play a significant role in conserving the gene pool of many rare plant species. So much so that even the most unwanted and dreaded mosquitoes also indulge in pollination of some scarce bog orchids. Further, the unwanted results of the pest control through the use of pesticides have resulted in the shifting of the idea towards the Integrated Pest Management (IPM). The highlighting of the role of biological control in this management programme, in fact, lends weight to the concept of their conservation. In order to keep a parasite alive, we need to protect some populations of the pest which is not only contrary to the earlier concept of eradication of the pest but, in other words, it also supports conservation for sustainable development to achieve long term environmental objectives.

The role of insect populations in the food chain/food web is highly significant for an overall betterment of a habitat. Many fish, amphibians, reptiles, birds and mammals predate upon the immature and adult stages of the insects. It may be noted that the Parulid Warblers are known to feed and entirely depend upon the arboreal arthropods mainly lepidopterous caterpillars and spiders. Insects also play a significant role as ecological indicators for depicting the health of a habitat/ecosystem. For instance, the butterflies are capable of delivering useful information on the changes taking place in a terrestrial ecosystem. They are greatly affected through industrial effluents and acid rain. The dragonflies are also very sensitive to pollution. Apart from this, many insects also help in recycling of the bulk of dead and decaying material.

Besides some major products like honey, lac, silk obtained from the insects, their eggs, caterpillars of Lepidoptera, grubs of the beetles, grasshoppers and adults of many other insects serve as a major food source for certain sectors of the society. Being aesthetic in appeal, many species of the insects are also used in commercial trade to enhance national income in some countries like Papua New Guinea and Taiwan *etc.* The trade in dead and live butterfly individuals gathers much publicity every now and then. The role of insects in impacting education in the biological discipline need hardly be emphasized. Had there been no research on *Drosophila*, there would have been scarcity of knowledge in genetics and evolutionary biology. The idea of conservation of insects also crops up from the fact that the excessive urbanization along water bodies, coast lines and bringing of more and more land under cultivation has deteriorated the habitats to an extent to cause enormous impact on the endemic insects.

Though the extinction and decline in population of certain exotic butterfly species is well known, yet we do not have much information on specific insect losses in different tropical habitats, particularly in India. The rate of excessive electrification in cities has resulted in an enormous loss to positively phototropic insects. The unproportional rate of conversion of land into agriculture, monoculture and introduction of exotic species have resulted in a direct loss of many host plants associated with a variety of insects. Some of the other factors such as overcollecting of insects for trade purpose, vegetation change pattern, global warming, forest cleaning, commercial exploitation of natural bioresources, increasing unplanned tourism in hilly areas, construction of multistoried buildings on the hill tops (*e.g.*, Simla, Manali, Dalhousie, Mussoorie, Nainital, Sikkim, Darjiling *etc.*), blasting of mountains for road construction, blasting of coal mines and pollution of water bodies *etc.* also result in decline in insect populations. This brings into discussion the aspect as to whether development without destruction is possible or not? Also, it awakens us to think as to how many individuals can actually live on the planet Earth, which is dotted with excessive population growth of humans in the Asian and African Continents. How many and how long can we continue to feed a newborn appearing every third second or so, at least in some underdeveloped as well as developing countries. In fact, these issues are highly sensitive and debatable, and look upon something that has a solution in its diversity *par excellence,* i.e., the insects .

Having given much attention to only about 1% major insect pest species, the remaining huge insect diversity, which has a general as well as a specific role to play in sustainable development, did not attract that attention which it deserves from conservation point of view. The decline in insect populations of a natural ecosystem definitely adds to the disruption in the natural processes, the integration of which is of utmost importance for a system to survive. The very basic concept to protect the environment is to know its biotic and abiotic components in all possible details. Obviously, this warrants more attention to that group which is dominating one, *i.e.*, the insects. We should not forget that the present era is an 'era of insects'!

The foregoing account clearly points out that insects are important bioresources and deserve long overdue attention for their conservation. The success of this work lies in scope of this work. The traditional approach of protecting mammals and birds need to be changed. Not only the insects, even other invertebrates have also not been minutely attended to from this angle. However, if some attention has been given in the past, it is too meager. The need to conserve many large species of vertebrates, particularly mammals and birds was duly recognized in early years of 20th century and it is since then that more and more thrust is being given with each passing day during the last 3-4 decades. Why the invertebrates got a stepmotherly treatment from this point of view is a soul-searching question ? Until recently, there was hardly any species of invertebrates on the International Wildlife Conservation Agenda and all resources of conservation agencies were directed towards large and conspicuous vertebrates (Pyle, 1978). Approximately 1.4 million species of invertebrates have been described and await the discovery of 2 to 10 million unknown species, especially from the tropics. The ecological and economic importance of the invertebrates in the complex web of living processes has a great bearing on sustainability. The mollusks and crustaceans form an important part of staple food of the coastal people. The earthworm maintains soil fertility.

Though, the issues of conservation of biodiversity and use of bioresources for sustainable development and safety of the planet Earth are being raised by many strata of people including politicians, scientists, bureaucrats and environmentalists after Earth Summits in Rio de Janeiro (1992) and New York (1997), yet invertebrates often come low on the list of priorities for conservation, mostly owing to lack of awareness about these fascinating creatures constituting nearly 90% of the entire faunal life (Cheesman & Brown, 1999).

In Indian context, out of 1,26,188 described species, as many as 50,717 (about 40%) belong to the class Insecta. Out of these, more than 7000 to 7500 species pertain to the Order Lepidoptera which is the second largest in the class Insecta (Hampson, 1893, 1894, 1895, 1896; Bingham, 1905, 1907; Evans, 1932; Fletcher, 1925a, 1925b, 1926, 1928, 1931; Bell & Scott, 1937; Talbot, 1939, 1947; Wynter-Blyth, 1957; Haribal, 1992; Varshney, 1994). It need hardly be emphasized that the Lepidoptera group serves as an umbrella for other arthropods as well. On one hand, the taxonomy and precise distribution of various Indian lepidopteran taxa of different ranks are far from complete while, on the other hand, studies on their immature stages of the moths and butterflies are extremely scanty. In the present communication, the latter aspect in butterflies is being taken up due to obvious reasons. On global basis, out of about 15,000 species of butterflies, 1400-1600 dwell in different habitats in India, which ranks sixth in species richness of all insects (Wynter-Blyth, *loc.cit.* 1994). As many as 451 Indian species are listed in different schedules of the Wildlife (Protection) Act, 1972 with a view to protect and conserve them adequately.

From an entomologist's point of view, the butterflies have spearheaded more interest in general insect conservation, owing to their charismatic aesthetic appeal. Though butterflies are insects yet inadvertently people make a distance by referring them as "butterflies and insects" as if these are different groups of animals (New, 1991). The critical scrutiny of literature reveals that we do not have ample information on the biology (life history, cheatotaxy, seasonal distribution of life history stages and ecology) of different species of butterflies in India (Bell, 1909-1914; Sevastopulo, 1933, 1935, 1940, 1941, 1946, 1959, 1973, 1975; Pant & Chatterjee, 1950; Wynter-Blyth, 1957; Brar, 1959; Mukerji, 1960; Rataul, 1960; Atwal & Joson, 1962; Atwal, 1963; Vaidya, 1970; Agarwala & Saha, 1984; Singh & Mavi, 1984; Bean, 1988; Radke & Kandalkar, 1989; Mavi, 1992; Haribal, 1990, 1992; Chaturvedi & Haribal, 1992; Jana & Ghosh, 1993, 1994; Arif *et al.*, 1994; Chandel *et al.*, 1995; Reddy & Reddi, 1996; Smetacek, 1997; Chaturvedi & Kehimkar, 2000). There are now no two opinions that for an effective and scientific conservation, appropriate authentic inventories besides studies on the life history and behavior of different species are of prime need. Accordingly, besides observations on two ICAR projects on butterflies from North-west India, some notes have also been added on the basis of field experience.

MATERIAL AND METHODS

Butterflies are holometabolous insects and pass through four stages, *i.e.*, the egg, the larvae, the pupa and the adult. In order to initiate life history studies, survey tours of varied duration were planned in different localities in the Shivalik adjoining Chandigarh/Punjab to procure any or all the immature stages along with the host plants. As per experience, it is obviously easier to locate ovipositing female in the field, flying around her host plant to collect the eggs rather than to search wild vegetation growing in an area to find out already deposited eggs on the host plants, especially the eggs which are cryptic and laid singly. The eggs and early larval instars brought to laboratory from the field were placed in small circular containers (diameter 10cm, depth 4.5 cm). Subsequently, the larval instars were shifted to relatively larger transparent containers (size 12x17 cm, 15x20cm, 18x23cm), along with fresh clippings of the food plants. The larvae nearing pupation were then shifted to wire-meshed eclosion cages (size 20x20x20cm and 30x30x30cm) for pupation and emergence of the adults. These rearing containers and cages were tightly closed and placed in a BOD in the laboratory which was maintained at constant temperature (28±1°C) and relative humidity (70± 5%) (Smart, 1985). All these rearing boxes were carefully examined for hatching, larval feeding, larval defense behavior, ecdysis, larval duration, pupation, pupal duration, eclosion *etc.* Due attention was given to the hygienic conditions as the rearing boxes were cleaned at regular intervals to remove fecal matter and dead insect stages. The fresh cuttings of the host plants were provided daily to the larvae for proper growth and to keep the mortality rate at the minimum. The gross morphology, colouration and measurements of different immature stages were regularly recorded. The freshly emerged adults were kept on nectar of plants or on artificial diet consisting of 20% sugar solution to record their longevity. The adults after death were pinned, stretched and preserved in air tight wooden boxes for identification. The genitalia were potashed in 10% KOH, dissected and dehydrated in different grades of alcohol. These were cleared in clove oil before writing the description of various constituent parts.

Table 1. List of larval host plants and study sites

S.No	Scientific Name	Common Name	Family	Larval host plant(s)	Study sites
1.	*Castalius rosimon* (Fabricius)	The Common pierrot	Lycaenidae	*Zizipus oenoplia* Mill. (Rhamnaceae)	Kansal forests, Dhanas forests, Sukhana lake (Chandigagh); Morni hills (Panchkula); Parwanoo, Nalagarh (Solan); Ropar.
2.	*Vanessa (Aglais) cashmiriensis* Kollar	The Indian Tortoise Shell	Nymphalidae	*Urtica dioica* Linn. (Urticaceae)	Solan Parwanoo, Nauni; Kandaghat, Sadhopull, Dharampur (Solan).
3.	*Acraea issoria* (Hubner)	The yellow Coster	Acraeidae	*Debregeasia hyploeuca*Wedd. (Urticaecae)	Parmwanoo, Dharmpur, Charbaghat, Solan, Kandaghat, Nauri, Sadhopull (Solan)
4.	*Ypthima inica* Hewitson	The Leser Three Ring	Satyridae	*Cynodon dactylon* (Linn.) (Graminaceae)	Dhanas forests, Sukhna lake, Kansal forest (Chandigarh); Nalagarh (Solan); Ropar
5.	*Euploea core* (Cramer)	The Common Indian Crow	Danaidae	*Ichnocarpus frutescens* R. Br. (Apocynaceae)	Sukhna lake, kansal forest, Napli forest (Chandigarh); Kalka, Morni hills (Panchkula); Parwanoo, Nalagath (Solan); Ropar.
6.	*Colotis amata* (Fabricius)	The small Salmon Arab	Pieridae	*Salvadora oleoides* Decaisne (Salvadoraceae)	Kansal forest, Napli forest (Chandigarh); Pakhrali (Ropar).
7.	*Papilio polytes* Linn.	The Common Mormon	Papilionidae	*Citrus auranti-folia* (Christrn.), *C. sinensis* (Linn.), *Limett-ioides tanaka, Aegle marmelos* (Linn.) (Rutaceae)	Botanical gardens, Bara-dari gardens, P.A.U. res. Stations, Bahadurgath (Patiala) and Balachur, Hoshiarpur, Bathinda
8.	*Pelopidas mathias mathias* (Fabricius)	The small Branded Swift	Hesperidae	*Oryza sativa* Linn. (Graminae).	Univ. Campus, P.A.U. res. Station, Bahadurgarh, (Patiala), Sahnewal, (Ludhiana), Rose gardens, Bathinda

Table 2. List of nectar food plants

S.No.	Scientific Name	Nectar food plant(s)
1.	*Castalius rosimon* (Fabricius)	*Lantana camara* Linn. (Verbenaceae), *Oxalis* sp. (Oxaliceaeae); *Ageratum conyzoides* Linn., *Tridax procumbans* Linn. *Tagetus erecta* (Linn.), *Vicoa* auriculata cass. (Compositae), *Sida acuta* Brum. (Malvaceae).
2.	*Vanessa (Aglais) cashmiriensis* Kollar	*Tagetus erecta* Linn. (Compositae); *Prunus bokharensis* Sahau (Rosaceae); *Lantana camara* Linn; *Verbena bonariensis* Linn. (Verbenaceae).
3.	*Acraea issoria* (Hubner)	*Erigeron* Sp. (Compositae), *Verbena bonariensis* Linn. (Verbanacae).
4.	*Ypthima inica* Hewitson	*Vemonia cinerea* (Linn.), *Cnicus arvensis* Hoffman (Compositae); *Momodica charantia* Linn. (Cruciferae); *Sida rhombifolia* Linn. (Malvaceae)
5.	*Euploea core* (Cramer)	*Ageratum conyxoides* Linn., *Tagetes erecta* (Compositae); *Alstonia scholaris* Linn.) (Apocynaceae).
6.	*Colotis amata* (Fabricius)	*Lantana camera* Linn. (Verbenaceae); *Tagetus erecta* Linn. (Compositae).
7.	*Papilio polytes* Linn.	*Lantana camara* Linn., *Stachytarpheta urticaefolia* (Salisb.), *Clerodondrum* inerme (Linn.) (Verbenaceae); *Murraya koenigii* (Linn.) (Rutaceae), *Cestrum diumum* Linn. (Solanaceae), *Phlox drummondii* Hook, (Polemoniaceae); *Acacia nilotica* Linn. (Mimosaceae); *Tagetes erecta* Linn. (Compositae); *Jasminum sambac* Linn. (Oleaceae); *Ipomoea batatas* Lilnn. (Convolvulaceae); *Hibiscus rosasinensis* Linn. (Malvaceae); *Ruellia brittoniana* Leonard (Acanthaceae), *Puinciana pulcherrima* Lilnn. (Caesalpiniaceae) and *Bongainvillea glabra* Choisy in DC. (Nytaginaceae).
8.	*Pelopidas mathias mathias* (Fabricius)	*Matthiola incana* R.Br. (Cruciferae), *Lantana camara* Linn. (Verbenaceae); *Parthenium hysterophorus* Linn. (Compositae).

OBSERVATIONS

Though, the life history studies of the species, viz., *Castalius rosimon* (Fabricius), *Chilades lajus lajus* (Cramer), *Euchrysops cnejus* (Fabricius), *Freyeria putli* (Kollar), *Lampides boeticus* (Linn.) *Pseudozizeeria maha* (Kollar), *Tarucus alteratus* Moore, *Zizeeria karsandra* (Moore) (Lycaenidae); *Ariadne merione* (Cramer), *Hypolimnas bolina* (Linn.), *Junonia almana* (Linn.), *J. hierta* (Fabricius), *J. Orithyia* (linn.), *Phalanta phalantha* (Drury), *Vanessa (Aglais) cashmiriensis* Kollar, *V. (Vanessa) indica* (Herbst), *V. (cynthia) cardui* (Linn.) (Nymphalidae); *Acraea issoria* (Hubner) and *A. terpsicore* (Linn.) (Acraeidae); *Ypthima inica* Hewitson (Satyridae), *Danaus (Anosia) chrysippus* (Linn.); *Euploea core* (Cramer) (Danaidae); *Anaphaeis aurota* (Fabricius), *Catopsilia pomona pomona* (Fabricius), *C. pyranthe* (Linn.), *Cepora nerissa* (Fabricius); *Colotis amata*

(Fabricius), *Eurema (Terias) hecabe* (Linn.), *Ixias marianne* (Cramer), *Pieris brassicae nepalensis* Gray, *P.* (*Artogeia*) *canidia* Sparaman (Pieridae), *Papilio* (*Princeps*) *demoleus* Linn., *Papilio polytes* Linn. (Papilionidae) and *Pelopidas mathias mathias* (Fabricius) (Hesperiidae) have been studied under two different ICAR projects but in the present communication only one representative from each family has been selected to highlight the significance of these studies. Accordingly, the names of the larval host plants and nectar food plants of eight species, presently dealt with, are given in Tables 1 and 2, respectively (*cf.* also Plates 1-8). Observations on the life history of these species are as follows:

FAMILY LYCAENIDAE

Castalius rosimon (Fabricius): The Common Pierrot, Fabricus, 1775, *Syst. Ent.*: 523 (*Papilio*) (Plate 1).

1. Egg: Incubation period 6.00±1.41 days; diameter 0.50±0.10mm, height 0.25±0.07; disc-like, flat; laid singly; chorionic surface sculptured with network of raised ridges, the latter thickened at angles, micropyle circular; light green in colour when freshly laid, turns white before hatching; affixed to leaf tightly.

2. Larva: Number of instars: 4

(i) First instar: Duration: 4.50±1.41 days.

Head: Width 0.16±0.00 mm; hypognathous, glabrous; light blackish brown; retractable into ventral recess on first thoracic segment, often hidden under prothorax, producible during feeding and walking; stemmata prominent, arranged in a semicircle laterally: primary setae conspicuous.

Body: Length 1.94±0.58 mm; width 0.35±0.04 mm; creamy white when newly emerged , later turns whitish green, well segmented, raised and rounded dorsally, flat ventrally, addorsal lines cream coloured; parallel; decorated with long; black setae dorsally, later white at base, the remaining setae; lenticels present, circular in shape; legs light creamy brown.

(ii) Second instar: Duration: 4.16±0.76 days.

Head: Width 0.35±0.04 mm; same as above, except the number of setae increases.

Body: Length 3.50±1.32 mm, width 0.83±0.28 mm; green, confluent with leaf colour of host plant; onisciform, broader in thoracic region, gradually narrows down towards anal end, addorsal lines cream colored, very distinct, become abscure on prothorax and anal segments, numerous secondary setae make their appearance; larva with seventh abdomen segment possessing a distinct median transverse opening of dorsal nectary organ, on eighth abdominal segment present a pair of dorsolateral eversible tentacular organs; lenticles increases in number; more dense around spiracles and dorsal nectary organs.

(iii) Third instar: duration: 3.33±0.70 days

Head: Width: 0.55±0.09 mm; same as above; head completely recurved under rounded prothorax, not visible from above.

Body: Length 5.33±1.52 mm, width 1.38±0.09 mm; dark green, addorsal white lines very clear, slightly get apart in thoracic region, broken at inter-segmental area; rest as in second instar.

(iv) Fourth instar: Duration: 7.00±1.00 days

Head: Width 1.02±0.04 mm; light green with mouth parts brownish; rest as above.

Body: Length: 8.66±2.08 mm, width 3.41±0.85 mm; dark green, with distinct addorsal light green lines, body appears whitish due to presence of numerous white glistening stellate chalazae bearing secondary setae, skin possess typical honey-comb like structure; myrmeophilous organs well developed, dorsal nectary organ further increases in size, consists of elongated large, transverse slit, made up of slightly sclerotized, elastic, unadorned cuticle, the pair of tentacular organs retractable each organ comprised of a cylindrical structure with elastic cuticle, possessing corona of knobbed setae at its apical end, lenticles further increase in number, more densely arranged around spiracles and anterior area of each segment.

Larval duration: 18.99±2.83 days.

3. Pupa: Duration: 8.88±0.57 days.

Length 8.33±0.57 mm, width 3.00±1.00 mm; formed on upper surface of leaf, lies parallel to it; firmly fixed by a large plate-like structure, the latter derived from last two abdominal segments, bearing hooked setae, proper cremaster absent, girdled along first abdominal segment dorsally; green colored; body free from protuberances, smooth; cephalic region rounded, prothorax, larger, mesothorax humped, very long; abdomen slightly yellowish dorsally, with obscure deep green mid-dorsal line; pupa nearly flattened ventrally; wing cases extending up to fifth abdominal segment, antennae touching termen of wings, lie adjacent to ventral meson; lenticles more concentrated around spiracles.

Adult longevity: 7.00±1.00 days.

5. Adult Diagnosis:

Male: Forewing upper surface white, with basal metallic blue scales, border dark brown, thicker at apex, narrower along costa, two outer discal spots in M_1 and M_2 present, two discal spots in Cu_{1a} and 1A+2A, Cilia along termen chequered; undersurface creamish, markings black, a basal streak from base of cell to costa, submarginal band discontinous. Hindwing upper surface white, with basal metallic blue scaling, border dark brown from Cu_{1b} to 3A. Cilia chequered; termen tailed at vein Cu_{1b} submarginal band and a row of marginal spots present; undersurface creamish, markings black, a streak from base of cell to costa, a row of marginal and sub marginal spots present, marginal spots in Cu_{1a}, Cu_{1b} and 1A+2A metallic.

Female: Wings above heavily suffused with dark scales, otherwise similar to male.

Male genitalia: Uncus bean-shaped, pilose; brachia long, stout, apices blunt; tuba analis with subscaphium extremely weak; tegument narrow, band-like; vinculum broad, with prominent, more sclerotized central fold; valva long, subtriangular, each valva with costa and sacculus narrow, the latter bears balloon-like, membranous process, ampulla and harpe well differentiated, slightly separated, former in the form of long, serrated spine; harpe small with blunt apex ampulla and harpe density, pilose; juxta U-shaped with a small depression at base, arms curtain-like; aedeagus short, thick, with apex sharp, zone in the middle, alulae present at zone, bulbus ejaculatorious entering cephalad.

Female genitalia: Lodix somewhat crescent-shaped; genital plate well sclerotized, semicircular, notched at middle of posterior margin; ductus seminalis attached near base of ductus bursae; the latter slender, portion after the entrance of ductus seminalis well sclerotized; corpus bursae very long, obovate, squeezed in middle; apophyses anterioris small; apophyses posterioris spiral; papilla analis trapezoidal, pilose.

Wing expanse: Half, Male: 15.00 mm; Half, Female: 15.00 mm

Material examined: 1♂, 20. X.1995, 1 ♂, 4.x.1997, Sukhana lake, Chandigarh; 3♂, 20.vi.1997, 1♂, 4.x.1997, Dhanas forests, Chandigarh; 2♂, 1♀ , 23.x.1996, 1♂, 8.viii. 1997, Kansal forest, Chandigarh; 1♂, 27.viii.1996; 1♂, 1♀, 25.x.1996, Napli forest, Chandigarh (Union Territory). 1♂, 22.x.1995; Skatari, Panchkula; 1♂, 15.x.1995, Kalka, Panchkula; 2♂, 4.viii.1996, 1♂, 1♀, 25.viii.1996, Morni hills, Panchkula; 3♂, 29.ix.1995, Pinjore, Panchkula (Harayana). 1♂, 26.viii.1995, 2♂, 1.x.1996, Nalagarh, Solan; 2♂, 24.x.1996, 1♀, 29.vi. 1996, Parwanoo, Solan; 1♀, 7.vii.1997, Sadhopull, Solan (Himachal Pradesh). 1♂, 28.xi.1995, Ropar (Punjab).

FAMILY NYMPHALIDAE

Vanessa (*Aglais*) *cashmiriensis* Kollar: The Indian Tortoise shell Kollar, 1844, *Hugel's Kaschmir* 4 (pt.2): 422 (*Vanessa*) (Plate 2).

1. Egg: Incubation period: 2.75±0.35 days height 0.72±0.33; width 0.57± 0.00; eggs laid in a single mass, number varies from 150-200, all eggs tightly packed, orderly arranged in 3-4 rows, with overlapping 2-3 piles above each row; each egg dark metallic green when freshly laid changes to greyish black before hatching; oval in shape, with 8-9 ridges and alternating grooves, ridges higher towards micropylar end; micropyle situated in a conspicuous depression at the centre on the top, encircled by inturning edges.

2. Larva: Number of instars: 5

(i) First instar: Duration: 3.35±0.35 days.

Head: Width 0.37±0.06mm; hypognathous, black with conspicuous epicraial and adfrontal sutures; primary setae prominent, black.

Body: Length 2.83±0.28 mm, width 0.48±0.05 mm; creamish black, 9th and 10th abdominal segments light grey, later changes to greenish black; body well segmented, uniformly widened, true legs on thorax black; abdominal legs greenish black; eversible prothoracic gland present midventrally in front of legs.

(ii) Second instar: Duration: 4.00±0.70 days.

Head: Width 0.66±0.00 mm, completely black, sutures black and prominent; same as above.

Body: Length 5.86±0.23 mm, width 0.78±0.04 mm; black dorsally and laterally, ventrally green except prothoracic segment, the latter black, middorsal line black , distinct, enflanked by yellow addorsal lines on either side, lateral yellow lines appear along spiracles; scoli black colored, secondary setae with bases yellow, body yellow spotted in appearance.

(iii) Third instar: Duration: 4.25±0.35 days.

Head: Width 1.18±0.06 mm; as that of second instar.

Body: Length: 9.00±1.00 mm, with 1.27±0.19 mm; same above except yellow spots increase in size; bases of scoli turn yellow; colour of prolegs change to greenish creamy.

(iv) Forth instar: Duration: 4.50±0.70 days.

Head: Width 1.80±0.17 mm; same as above, except vertex on which yellowish orange spots encircling bases of secondary setae make their appearance.

Body: Length: 15.00±1.00 mm; width 1.80±0.17 mm; same as except prolegs become green laterally, brownish from inner side.

(v) Fifth Instar: Duration: 5.00±1.41 days.

Head: Width 2.96±0.05 mm; black, appears orange dotted due to presence of numerous orange bases of secondary setae, more numerous on region, laterally mostly black; rest as above.

Body: Length: 30.66±1.15 mm, width 3.78±0.26 mm; black with yellow spots encircling the bases of secondary setae, middorsal line jet black, interrupted in the abdomen due to presence of median dorsal scoli; yellow addorsal lines present on either side of middorsal line, bearing addorsal scoli; two parallel supraspiracular and subspiracular yellow lines present on either side of body, broken at points of origin of supraspiracular and subspiracular scoli; all scoli black, covered with long black setae, metallic bluish at base, bases of the supraspiracular and subspiracular scoli with some color of orange subventral scoli orange yellow.

Larval duration: 21.00±3.51 days.

3. Pupa: Durtation: 8.00±1.00 days.

Length 22.00±1.41 mm, width 5.75±0.35 mm; hangs upside down, attached to substratum by anal cremaster, the latter with prominent hooks; freshly formed pupa green in colour, later turns to brown;

antenna and palps greenish grey; tubercles on dorsal surface golden, third to eighth abdominal segments with median row of tubercles smaller than addorsal tubercles, mesothorax with median dorsal prominence very distinct, thorax with lateral projections acute; head with cephalolateral projections very sharp, elevated well beyond edge of body, divergent.

4. Adult longevity: 6.00±1.00 days

Adult diagnosis

Male and female: Forewing narrower, apex produced at vein M_1, upper side tawny, with black bars in cell, at the end of cell and another beyond it, yellow bars present in between them, two black discal spots in spaces M_3 and Cu_{1a}, basal area brownish, marginal band dark brown, termen pale; underside brown, with tightly packed short, black, transverse striae, basal half of wing dark purplish brown, outlined by black, sinuous transverse line. Hindwing above tawny, with basal half dark brown, edged yellow outwardly, marginal blue spots inwardly black bordered, toothed at vein M_3; below brown, markings similar to upper side.

Male genitalia: Uncus undivided, short, straight, dorsoventrally compressed, slightly constricted in middle, apex squarish; tegument broad; vinculum band-like, somewhat wavy; saccus very long, tubular; valva somewhat wavy; a very a long pillar-like stout process arises from base, run parallel along axis of the valva, latter strongly bifid apically, ampulla somewhat bean shaped, minutely serrated near apex, tip sharp, bears a prominent hump-like process dorsally, harpe border, statulate, apex serrated; juxta U-shaped; aedeagus long, narrow, slender, slightly wavy, subzone much shorter than suprazone, wider, bulbus ejaculatorius cephalad.

Female genitalia: Genital plate poorly developed, elongated, strip-like; ductus bursae broad, indistinguishable from corpus bursae, apical portion membranous, otherwise straight, sclerotized, basal portion near ostium bursae highly sclerotized; ductus seminalis enters at a junction of membranous and sclerotized portion; corpus bursae very elongated, narrow; apophyses anterioris not developed; apophyses posteriorils rod-like, weakly sclerotized, apices swollen; papilla analis subovate, anterior half well sclerotized, posterior half nearly membranous, pilose.

Wing expanse: Half: Male : 23.00.25.00mm; Half: Female: 24.00-25.00mm.

Material examined: 1♂, 26. vii. 1996, Morni hills, Panchkula (Haryana). 3♂, 5.x.1996. Kandaghat, Solan; 1♂, 14.ix. 1995, 1♂,1♀ , 15.ix.1995, 1♂, 25.v.1996, 1♂,1♀ , 3.x. 1996, Nauni, Solan; 2♂, 8.x.1997, Solan; 4♂, 24.ix. 1995, Chail, Solan; 1♂, 25.v.1996, 2♂, 22.xi.1995, Dharampur, Solan (Himachal Pradesh).

Family Acraeidae

Acraea issoria (Hubner): The Yellow Coster; Hubner, 1818, *Zur.z. Samml. Exot. Schmett.*, 1: 27 (*Telchinia*) (Plate 3).

1. Egg: incubation period: 20.50 ± 1.53 days: height 0.86 ± 0.13 mm, width 0.62±0.06 mm; laid in a loose mass, number of eggs 150-175 in a cluster, eggs evenly arranged; each egg orange yellow, changes to dirty orange brown before hatching; oval, base broader than apex, chorion surface craved with shallow reticulate network of horizontal and longitudinal ribs.

2. Larva: Number of instars: 5.

(i) First instar: Duration 6.25 ± 1.06 days.

Head: Width 0.38 ± 0.00 mm; hypognathous, dark brown; epicranial suture black; stemmata prominent; primary setae present.

Body: Length 3.70 ± 1.02 mm, width 0.34 ± 0.03 mm; orange brown; beset with black colored primary setae; thoracic and abdominal legs blackish brown.

(ii) Second instar: Duration : 5.00 ± 0.70 days.

Head: Width 0.71 ± 0.06mm; brownish black; rest as above.

Body: Length 5.71 ± 0.58 mm, 0.66 ± 0.00 mm; brown; scoli make their appearance; rest as above.

(iii) Third instar: Duration: 5.50 ± 0.70 days.

Head: Width 1.66 ± 0.00 mm; jet black; epicranial suture black, adfomtal sutures whilte, base of clypeus and antennae white.

Body: Length 13.33 ± 152 mm, width 1.88 ± 0.01mm; blackish brown, ventral side lighter than dorsal; scoli black, prominent; a subspiracular cream coloured diffused line appears, thorax to ninth abdminal segment with a mid ventral line, the latter creamy in colour; thoracic and abodominal legs completely black from outer side and light brown from inner side; anal shield completely black.

(iv) Fourth instar: Duration: 5.50 ± 0.70 days.

Head: Width 2.13 ± 0.05 mm; black; epicrania; and adfrontal; sutures broader, whilte; parietals organe brown; rest as above.

Body: Length 22.66 ± 028 mm; the subspiracular line broken due to presence of subspsiracular scoli, bicoloured with organe bars present on dorsal edge at bases of scoli and creamy, V- shaped lines extend form one scolus to another; supraspiracular band with supraspiracular scoli also appears, off-white, broken.

(v) Fifth instar: Duration : 6.50± 0.70 days.

Head: Width 3. 76 ± 0.09 mm, epicranial and adfrontal sutures white, tinged with orange, frons black, area of parietals near ecdysial line narrowly or broadly (variable) black; otherwise head orange coloured.

Body: Length 35.00 ± 2.00mm, width 4.33 ± 0.33 mm; same as that of fourth instar; scoli as well as bands become more distinct and broader.

Larval Duration: 28.75 ± 3.86 days.

3. Pupa: Duration 8.66 ± 1.52 days .

Length 25.00 ± 0.00 mm, width 5.00 ± 0.70 mm: suspended by anal cremaster with head facing downwards: freshly formed pupa pinkish in colour, after 3-4 hours transforms to light pinkish brown, body quite elongated: cremaster broad, shorter, black, tenth segment black, addorsal black bands present from second to ninth abdominal segment: horizontal black band joins addorsal band form U-shaped, cup-like structure on each segment to accommodate turbercles: the latter sharp, orange coloured, with tip black: similar in size and not enflanked by black band: thorax with double lines of median dorsal black band, broadened in posterior part of mesothorx: cephalilc prominences poorly developed: mesothorax with sharp lateral prominences, median carina present on mesonotum: lateral abdominal band black, decorated with orange rounded spots on each segment: abdomen with a ventral median double black band present: ocular spots black: antennae black colour.

Adult longevity : 4.50 ± 0.70 days.

Adult diagnosis

Male: Forewing above yellow, veins blackened, a broad, curved black bar present at end cell, marginal band black, wider at apex, decreasing in width to tomus, transversed by a series of ground coloured spots; below yellow, paler towards apex, veins darker, dorsal black bar at end cell visible through undersurface. Hindwing upper side yellow, with nearly sagittate black submarginal line, coalescing with black slender marginal line along veins; underside pinkish white, veins prominently black, broad submarginal ochraceous lunular band present, limited both on inner and outer sides by black lines.

Female: Upper side of wings with ground color rather dull, black margins broader; otherwise as in male.

Male genitalia: Uncus undivided, slender, long, finger-like, downcurved with apex sharp, sparsely pilose; tegumen quite large, longer than broad somewhat triangular, hump-like; vinculum very broad in upper half, extremely narrowed down posteriorly, saccus well developed, long, tubular, dorso-venterally flattened, broad at base, apex rounded; valva reduced in size, basal half broad, distal half narrower, apex curved inwards, rounded, densely pilose; anellus sparsely hairy; aedeagus cylindrical, long, margins somewhat wavy, suprazone very longer that subzonal area, apical area curved downwards, apex obtuse, vesica wide, bulbus ejaculatorius dorsad, coecum short, rounded.

Female genitalia: Genital plate inverted U-shaped, medially bilobed, each lobe bean-shaped; sinus vagenalis well developed, consists of a pair of membranous elongated pouches and a large median weakly sclerotized sac, the latter minutely but densely spinose; bursa copulatrix reduced in size; ductus bursae extremely reduced; corpus bursae globular; ductus seminalis departing from base of corpus bursae; anterioris absent; apophyses posterioris short, very broad at base, apex dumble-shaped; papilla analis boat-shaped, minutely punctate, pilose in distal half.

Wing expanse: Half, Male: 26.00-34mm.; Half, Female: 32.00-34mm.

Materal examined: 3♂, 16. ix, 1995, kandhaghat, Solan; 1♂, 3. vi. 1995, Dharmpur, Solan; 1♀, 13.vi.1996, Solan; 1♂, 7♀. vii. 1997, Sadhopull, Solan (Himachal Pradesh).

Family Satyridae

Ypthima inica Hewitson: The lesser Three Ring, Hewitson, 1864, *Trans. Ent. Soc.* 2 (3): 264 (*Ypthima*) (Plate 4).

1. Egg: Incubation period; 4.75 ± 0.64 days; height 0.52 ± 0.03mm, width 0.57 ± 0.03 mm; spherical, surface texture smooth; deposited singly; shinning, light green in colour, prior to hatching turn light brown, firmly attached to leaf.

2. Larva: Number of instars: 4 to 5.

(i) Fist instar: Duration : 3.25 ± 0.05 days.

Head: width 0.50 ± 0.06mm; hypognathous; brownish yellow in freshly emerged larva, later tummy brown; large, broader that thorax; epicranial suture distant, primary setae brownish, verily long; vertex with a pair of small protuberances.

(ii) Body: Length 3.77 ± 1.59 mm, width 0.42 ± 0.09mm; creamy brown in newly hatched larva, turns green subsequently, last abdominal segment pinkish; legs light brown; studded with long primary setae, the latter white in color.

(ii) Second instar: Duration: 3.37 ± 0.75 days.

Head: Width -.96 ± 0.-05 mm; as above,

Body: Length 5.83 ± 0.28 mm, width 0.93 ± 0.11 mm; body grassy-green, mid-dorsal line dark green, all segments divided into distinct annuls, the latter light green, anal shield forked, caudae light brown; lateral white, obscure line appears; rest as above.

(iii) Third instar: Duration : 4.00 ± 1.2 days.

Head: Width 1.38 ± 0.04 mm; green, secondary setae numerous, greenish, each with base black; rest as above.

Body: Length 11.00 ± 1.73 mm, width 1.53 ± 0.05 mm; white lateral line becomes very clear, longitudinal light green addorsal line present; legs brownish green; otherwise tallies with second instar.

(iv) Fourth instar: Duration: 4.25 ± 0.95 days.

Head: Width 2.06 ± 0.11mm; same as above.

Body: Length 18.00 ± 3.60 mm, 2.36 +0.32 mm; same as explained above except the caudae turns brownish red.

(v) Fifth instar: Duration 4.33 ± 0.57 days.

Head: Width 2.20 ± 0.01 mm. Same as above.

Body: Length 21.66 ± 3.51mm, width 2.86 ± 0.25 mm; as explained above.

Larval duration: 19.20 ± 4.59 days.

3. Pupa: Duration: 7.00 ± 1.82 days.

Length 12.50 ± 0.70 mm, width 3.75 ± 0.35 mm, suspended vertically downward by cremasteral hook from silken network; newly formed pupa translucent green, transforms after 2-3 hours to brown with beige shade at thoracic region and on wing cases; long, somewhat spindle-shaped, tubercles absent; cephalic end with a transverse ridge, mesothorax with prominent median rounded elevation, extending caudal up to second abdominal segment, abdominal segments with dorsal indeterminate wavy transverse striations; wing venation very conspicuous, each vein ending in a black dot at termen.

Adult longevity: 6.33 ± 2.08 days.

Adult diagnosis: Male and Female: Forewing upperside uniformly dark brown, with apical black ocelus in Rs, bipupilled with yellow iris; underside pale grey, with numerous fine brown colored stirae, the latter evenly spaced, ocellus as on upperside. Hindwing dorsally with sub-marginal band ochraccous, broken, small ocellus present in Cu_{1a}, unipupilled; ventrally pale grey with numberous fine brown stirae, the latter evenlly placed, one black small apical ocellus present in Rs and two tornal ocelli present, apical and tornal ocelli not in line. Male with band on upperside of forewing. In dry season form, the ocelli on underside of wings obsolete or reduced to specks.

Male genitalia: Uncus simple, undivided, compressed dorsoven-trally, bulged in middle, squarish apical; tegumen narrow; vinculum band-like, with a projection on inner side near middle; saccus short, with apex rounded; valva quite elongated, narrow, fusiform broader apically, the apex bifid, with apex rounded, bears somewhat triangular inner flap; juxta U- shaped; aedeagus nearly slender, subzone much shorter than suprazone, alulae present at zone, apex wide, bulbus ejaculatorius enters cephalodorsal.

Female genitalia: Lamella antevagenalis nearly circular, pouch-like, encircle ostium bursae, lamella postvagenalis well sclerotized, flap-like apex bifid; ductus bur; epophyses anterioris absent; epophyses posterioris short, weak; papilla annals elongated, pilose distally.

Wing expanse: Half, Male: 12.00-16.00mm; Half, Female 16.00mm.

Material examined: 1♂, 29.iii. 1996, Sukhna lake; 1♀, 2.iv. 1996, kansal forest, Chandigarh (UT). 1♀, 31.iii. 1997, 1♀, 10.iv. 1996 2 ♂, 3.viii. 1996, Nalagarh, Solan (H.P.), 2♂, 13.iv. 1997 2♀, 12.iv. 1997, Kalka, Panchkula; 1♀, 4.ix. 1998, Sketari, Panchkula (Haryana).

Family Danaidae

Euploea core (Cramer); The Common Indian Crow. Cramer, 1780, *Pap. Exot.* 3: pl. 266, figs, EF. (*Papilio*). (Plate 5)

1. Egg: Incubation period: 2.16±0.76 days; height 1.50±0.14mm, width 1.00 ±0.14mm; laid singly; melon-shaped, broader in middle, narrow on either pole; off-white, turn greyish a day before hatching; chorion furnished with polygonal pits, ribs very low; egg firmly glued to substratum (like unopened flower buds of an inflorescence).

2. Larva: Number of instars: 5.

(i) First instar: Duration: 2.33 ± 0.57 days.

Head: Width 0.63±0.05mm; hypognathous, black; epicranial suture black, clear, primary setae present.

Body: Length 4.33±1.75 mm, width 0.86±0.11mm; newly hatched larvae creamy white, later changes to dirty yellow; well segmented, each segment further divided into annuli, but the latter poorly distinct; legs and anal segment blackish brown; mesothorax, metathorax, second and eighth abdominal segments with a pair of rudimentary filaments, the latter black and somewhat fleshy, present addorsally.

(ii) Second instar: Duration: 2.33 ± 0.57 days.

Head: Width 1.03 ± 0.05 mm; as explained above.

Body: Length 7.83 ±1.89 mm, width 1.93 ± 0.11 mm; yellowish brown, thoracic segments yellowish orange, abdominal segments with purplish tinge; four pairs of fleshy filaments explained above become longer; thoracic shield divided into two black lobes; white transverse bands representing annuli appear on each segment.

(iii) Third instar: Duration: 2.00 ± 1.00 days.

Head: Width 1.55 ± 0.13mm; black; vertex with a white band, the latter extending downwards, running through parietals, ending at genal area; antennae, clypeus and labrum white.

Body: Length 15.00 ± 1.00mm, width 2.53 ±0.05 mm; orange brown; each segment with five transverse bands, the later white in colour, spiracles and legs black; prothorax and last abdominal segment orange, anal shield black; a lateral white line appears on each side except prothorax and tenth abdominal segment; body below completely brown.

(iv) Fourth instar: Duration: 2.16 ± 0.28 days.

Head: Width 2.23 ± 0.05 mm; same as explained above.

Body: Length 22.33 ± 2.51 mm, width 3.33 ± 0.57 mm; brownish black, white transverse bands very prominent, alternating with black bands, prothorax and anal segment orangsh yellow, lateral bands yellow, distinct, decorated with prominent, black, oval spiracles, subspiracular band white, conspicuous, extends form mesothorax to ninth abdominal segment on either side; base of prolegs lined yellow fleshy filaments very large, brownish maroon, present in four pairs, mesothorax with longest pair, second abdominal segment with equalized pairs.

(v) Fifth instar: Duration: 5.83 ± 0.28 days.

Head: Width 3.45 ± 0.05 mm, same as above.

Body: Length 35.00 ± 5.00 mm, width 4.16 ± 0.28 mm; tallies with fourth instar.

Larval duration: 14.65 ±2.70. days.

3. Pupa: Duration: 7.33 ±1.52 days.

Length 16.50 ±0.70 mm, width 8.50 ± 0.07 mm; attached at anal end by cremaster, with head suspended downwards; somewhat pot-shaped, constricted in the metathoracic region; mesothorax and second abdominal segment very long; the abdominal segments caudal of wings tapering, form a hemispherical structure; cremaster black, narrow, broad at base; body shows a spectrum of dazzling and reflective colors. The freshly formed pupa very soft, greyish pink, flesh-like, after 2-3 hours becomes harder, glossy and glistening silver bands with golden tinge appears subsequently. A middorsal silver band extend from first to last abdominal segment but missing on second abdominal segment; subdrosal silver band present in mesothorax, and then from third abdominal segment to last segment, in mesothorax this band bears three black spots raised on tubercles; ocular region, base, Costa, termini and dorsum of wing cases silver-patched with reflective golden tinge. The colour of the pupa changes drastically and dramatically a day before eclosion; these abdominal segments become brownish, wooden-colored and non-reflective; the remaining pupil body blackish but reveals a mixture of variety of colors when rotated, thorax reflects green color, the ocular region blue, and the alar region violet and peacock shaded. The advanced pupa turn more blackish and spectrum of colours shrinks; it further reduces to small spots in middle of wing cases; before eclosion pupa becomes totally black except the abdominal segments, white spots on wings of imago visible through pupal case.

Adult Diagnosis

Male and Female: Forwing upperside glossy dark brown, termen broadly paler, with a row of submarginal and marginal white spots; below brown, with a spot at end cell, a small spot at Costa, two discal spots in M_3 and Cu_{1a} row of large submarginal spots and a row of small marginal spots, all spots white. Hindwing dorsally dark brown, with a row of white elongated submarginal spots, marginal spots smaller, whitish; ventral brown with obscurous white spots at end cell and a series of discal white spots, marginal and submarginal markings as on upperside.

Male genitalia: Uncus very weakly developed, extremely short membranous; tegumen narrow, directed backwards; vinculum narrow, extended anteriorly into a very long, tubular saccus; valva broad in distal half, proximal half much narrower, Costa lunar-shaped, apex squarish, lower margin bears thumb-like downcurved process, apex pilose; juxta longer than broad, bifid apical, apices triangular; aedeagus extremely large and stout, slender, arched, zone nearly in the middle, apex wide, bears single spine-like cornus nearer to zone, bulbus ejaculatorius dorsal, coech short, rounded.

Female genitalia: Genital plate somewhat squarish; a pair of ostium pouches present on either side of ostium burs, each pouch weakly sclerotized, with a knob-like process apically; ducts bursae quite long, slender, curved, well sclerotized, except basal one third portion membranous; ducts seminalis depravation ducts bursae near apex; corpus bursae elongated, rounded apical, lateral margins wavy, base elbow-shaped, well sclerotized except apical area; epophyses anterior wanting; epophyses posterioris narrow, rod-like, apices squarish; papilla annalis triangular, posterior half membranous, densely piles.

Wing expanse: Half, Male: 40.00-43. mm.; Half, Female : 39.00-50.00mm.

Material examined: 2 ♂, 22.×.1996, Sukhna lake, Chandigarth (UT). 1♂,27.viii. 1995, 2 ♀ , 4.iii. 1996, 1♂, 25.viii. 1996, 1♀, 23.×.1996 Morni hills, Panchkula, 1♂, 21.×. 1995, Pinjore, Panchkula; 2♂ 23.viii. 1995, 2 ♂, 24.×.1996, Kalka, Panchkula (Haryana). 1♀, 1.×. 1996 Nalagarh, Solan; 1♂, 15.× .1995, Nauni, Solan (H.P.)

Familly Pieridae

Colotis amata (Fabricius): The Small Salmon Arab. Fabricius, 1755, *Syst. Ent.*, 476 (*Papilio*). (Plate 6)

1. Egg: Incubation period: 1.66+0.57 days; height 0.77 ± 0.24, width 0.46 ± 0.02 mm; laid in loose clusters, each with 50-60 eggs, the latter evenly spaced within clutch, firmly glued to leaf surface; egg spindle-shaped, narrower at poles than middle; creamy when freshly laid, soon turns yellow on exposure to sunlight; vertical ribs prominent, extend on (egg) surface from micropyle to base, intersected by numerous diffused transverse ribs, resulting in somewhat rectangular pattern on chorin surface; micropyle shallow, narrow, encircled by high longitudinal ridges.

2. Larva: Number of instars: 5.

(i) First instar: Duration: 1. 16 ± 0.28 days.

Head: Width 0.27±0.00 mm; hypognathous; black, epicranial suture very clear; stemmata 2-5 present on dark papillae; primary setae black;

Body: Length 1.99 ± 0.44 mm, width 0.30 ±0.07mm; cream colored when freshly hatched, later becomes green, anal segment yellow green; well segmented; legs green.

(ii) Second instar: Duration: 1.33 ± 0/28 days.

Head: Width 0.43 ± 0.04 mm, same as above.

Body: Length 3.80 ± 0.29 mm, width 0.6o ± 0.00mm; pale green; segments divided into annulets; rest as explained above.

(iii) Third instar: Duration: 1.66 ± 0.57 days.

Head: Width 0.79 ± 0.06 mm; turns green; black colour confined to a pair of large, early circular spots on vertex, present adjacent to epicranial suture and extending unto epicranial notch; sutures green.

Body: Length 7.33 ± 1.52mm, width 1.02 ± 0.04 mm; mid-dorsal white band appears, distinct in thoracic and anterior abdominal segments, diffused on remaining body segments; lateral line white, obscurous; legs yellowish green; rest as above.

(iv) Fourth install: Duration: 3.00 ± 1.00 mm days.

Head: Width 1.27 ±0.09 mm; similar to third instar, except black patches on vertex reduced in size, more confined near epicranial notch, epicranial sutures highly clear, creamish green in color.

Body: Length 11.33 ± 2.30 mm, width 1.33 ± 0.27 mm; same as explained for third instar, except mid-dorsal and lateral white bands well defined, running from cephalic to anal end.

(v) Fifth instar: Duration: 3.00 ± 0.00 days.

Head: Width 2.00 ± 0.00 mm; same as above, except the black path further reduced in size.

Body: Length 18.33 ± 1.52 mm, width 3.50 ±0.s57 mm; as in fourth instar.

Larval duration: 10.15 ± 2.13 days.

3. Pupa: Duration 5.33 ± 0.57 days.

Length 12.50 +0.70 mm, width 3.50 + 0.70 mm; newly formed pupa green in colour, turns creamish brown after sometimes, suffused with minute black spots; formed on leaf of host plant, attached to substratum by cremaster and girdle; spindle-shaped, body with ventral line forming and obtuse prominent angle nearly in the middle, at the apex of costa of the wing sheath; thorax inconspicuously produced on dorsal surface, otherwise flat; cephalic projection somewhat triangular; cremaster very short, dorsoventrally flattened, cremastral hooks prominent, deep brown, covering the caudal end both dorsally and ventrally.

Adult longivity: 500 ± 0.00 days.

Adult diagnosis: Male and Female: Sexes similar. Forewing upper side salmon-pink brown bar at ent cell, joining with dark linear broader of costa, apex and termen dark brown, studded with ground coloured spots; underside greenish yellow, black markings of upper surface visible. Hindwing above salmon-pink, costa, termen and dorsum dark brown, marginal band bearing spots of ground colour; below greenish yellow, with black coloration of dorsal side otherwise visible through.

Male genitalia: Uncus finger-like, long, curved downwards, rounded apically; tegumen broad, triangular, vinculum narrow, straight; saccus extremely long, anteriorly swollen; valva narrow, triangular, costa highly oblique, apex obtuse, margins pilose; juxta U-shaped, very weakly sclerotized; aedeagus slender, sharply curved downwards at base, subzone extremely short, bulbus ejaculatorius dorsad, apex wide, beset with numerous comuti.

Female genitalia: Genital plate bilobed; each lobe bean-shaped, with a small projection in posterolateral side, both lobes join anteriomedially above ostium bursae to from small rectangular structure; ductus bursae with upper half very broad, flattened, sclerotized, lower half narrow, slender, membranous; ductus seminalils enters near middle of lower membranous half; corpus burse globular, imperceptibly goes into ductus bursae; appendix bursae somewhat spherical, attached to corpus bursae by moderately long, very narrow, slender, thread-like structure; apophyses anteriors wanting; apophyses posterioris very short; papilla analis triangular, pilose.

Wing expanse: Half, Male: 18.00-20.00 mm., Half, Female: 20.00 mm.

Material examined: 1♂, 1♀, 29.vi.1996, 2♂, 7.vi.1997. Pakhrali; 1♂, 29.vi.1996, Bari, Ropar (Punjab). 1♂, 1♀, 29.vi.1996, Kansal forest, Chandigarh (Union Territory).

Family Papilionidae

Papilio polytes Linnaeus: The common Mormon. Linnaeus, 1758, Syst. Nat. (Ed X) 1:460 (Papilio).

1. Egg: Incubation period: 4.00 ± 0.70 days. Laid singly, spherical, orange or pale yellow, mimetic dots present on egg sculpture when freshly laid, egg firmly attached to substratum; micropyle not distinct, later egg turns light brownish (dorsally); first instar visible from outside egg chorion.

2. Larva: Number of instars: 5.

(i) First instar: Duration: 3.80 ± 0.57 days.

Head: Width 0.74 ± 0.11 mm; hypognathous, distinct, light brownish, non- retractile inside prothorax, sutures conspicuous, primary setae present.

Body: Length 8.62 ± 0.30 mm, width 2.14 ± 0.04 mm; well segmented; light yellowish-brown, somewhat more brown dorsally and laterally; mid-dorsal stripe indistinct; a pair of tubercles (outgrowth) present on each body segment dorsolaterally, looks like bird-dropping, body furnished with primary setae; dirty whitish bands present on third to fifth and eighth to tenth abdominal segments dorsolaterallly; osmeterial gland present on prothorax.

(ii) Second instar: Duration: 4.00 ± 1.00 days.

Head: Width 1,98 ± 0.30 mm; same as above.

Body: Length 8.6 ± 0.65 mm, width 2,31 ± 0.24 mm; same as above, tubercles relatively more predominant on thoracic and eight, ninth and tenth abdominal segments; dirty whitish band comparatively more widened.

(iii) Third instar: Duration: 3.10 ± 1.14 days.

Head: Width 2.44 ± 0.37 mm; same as above.

Body: Length 14 ± 1.58 mm, width 3.4 ± 0.28 mm; same as above except body more or less cylindrical, greyish-brown; an additional dirty whitish thin streak appear on lateral sides of mesothorax and metathorax, and 1st abodominal segment, three pairs of black spots appear on first, second and eighth abdominal segments dorso-laterally.

(iv) Fourth instar: Duration: 3.50 ± 0.50 days.

Head: Width 2.69 ± 0.11 mm; same as above.

Body: Length 25 ± 3.60 mm, width 5.3 ± 0.83 mm; same as above.

(v) Fifth instar: Duration: 4.50 ± 0.55 days.

Head: Width 3.52 ± o.35 mm; same as above except light chocolate brown, primary and secondary setae present.

Body: Length 43.6 ± 4.04 mm, width 7.8 ± 1.30 mm; light greenish, well segmented, widest at thoracic region, mimicking with the leaf surface; mid-dorsal stripe indistinct; osmeterical gland well developed; spiracles well defined; whitish, broad, spongy, raised lateral stripe present above legs and prolegs, running throughout length of body.

Larval duration: 16.94 ± 3.71 days.

Pupa: Duration: 9.60 ± 1.14 days.

Length 22.2 ± 1.92 mm, width 4.40 ± 0.96 mm; light-greenish, protruded out and bent inwards mid ventrally and mid-dorsally, respectively, two light yellowish-green patches (one on either side) present mid-dorsally; additional whitish- yellow spots present on pupal case, scattered here and there, before eclosion turns blackish.

Adult diagnosis: Male: Forewing uppersurface dark brown/ black with terminal continuous row of white spot along termen; undersurface quite similar to uppersurface, lighter in colour, Hindwing upperside with small slightly spatulate tail, a complete discal band of white spots present, red marginal crescents absent or diminishingly present, undersurface blackish brown with a series of white marginal lunules and marginal white spots and an area of blue sealing in between submarginal and discal band.

Female: Polymorphic; either

ϖ form *cyrus*, similar to male with larger spatulate tail, marginal spots red, strongly marked, or

ϖ form *romulus*, tailed, black, mimics the crimson rose (*Papilio hector*); irregular /interrupted band on forewing from midcell to tornus, not well formed; hind wing upperslurface with four white medial spots sometimes accompanied by a smaller in cell with a submarginal row of reddish spots along termen and half of inner margin, with pale white to orange crescents on termen, under surface more of less similar to upper surface with radish spots apparently larger and semilunar, or ϖ form stichius, mimics the common rose (P.aristolochiae), tailed, black; forewing upperside with pale streaks between veins; hindwing upperside with 2 to 6 elongated white spots adjacent to cell, a white spot at the apex of cell, a series of red linules at margin, a series of narrow pale red spots on termen.

Male genitalia: Psedouncus with long, narrow, pointed, beak like projection, with apex curved downwards, fused up to middle of vinculum; uncus bilobed, each lobe long, narrow, with apex pointed, curved upwards, pilose in middle; tegumen reduced; vinculum narrow, curved, upper half slightly broader, weak; saccus very short, rounded. Valva base narrower, upper portion almost rectangular; costa narrow, oblique; sacculus well separated, with an inner blade-like fold; harpe arise as a saw-shaped long, arched fold, with dorsal and apical edge serrated; valva pilose fall over; juxta weakly sclerotized, apex narrower, base rounded. Aedeagus slightly curved, rounded; zone nearly in middle, apex opening widely; bulbus ejaculatorius entering broadly ventrocephalad.

Female genitalia: Genital plate covered with membranous, folded pouch of 7th sternum; lamella antevagenalis with medially a u-shaped, inwardly curved fold, present in less sclerotized area, laterally extended into elongated processes, each with inner surface straight, outersurface with two dents or one dent (variable), apex curved below ostium bursae, pointed, genital plate on either side with eighth tergite with scerotized fold of 8th pleuron; lamella postvagenalis circular, well scerotized plate; corpus bursae enormously large, oblong, globular, membranous; signum zipper-shaped, moderately developed; ductus bursae short membranous, slender; ductus seminalis origination near base; apophyses anterioris short knob-like; apophyses posterioris very thin, lean; papilla analis subovate, pilose.

Wing expanse: Half, Male: 44- 46 mm. , Half, Female: 90-108 mm.

Material examined: 3♂, 2♀, 26.iii. 1993, 4♂, 2♀, 19.iv.1994, 3♂ , 2♀. 29.Iii. 1995, 3♂, 10.ix.1995, 2♂,2♀, 30.iii.1996, 8 ♂,3♀, 15.iii.1997, University Campus, Patiala; 3♂, 28.v.1996, 2♂, 1♀, 17.viii. 1997, Shahpur Kandy, Gudaspur; 1♂, 1♀, 14.iii.1996, Mansa; 4♂, 3♀, 16.iv. 1996 Ropar; 2♂, 3♀, 29.iii.1997, Ludhiana; 10♂, 1♀, 21.iv.1996, Hoshiarpur; 1♂, 1♀, 13.iii. 1996, Muksar; 14♂, 8♀ (bred) (Punjab)

Family Hesperiidae

Pelopidas mathias mathias (Fabricius): The small branded Swift. Fabricius, 1798, *Suppl. Ent. Syst.* 433. (Plate 8)

1. Egg: Incubation period: 4.60 ± 0.54 days; height 3.62 ± 0.13 mm, with 5.18 ± 0.13 mm; freshly laid eggs shinning, hemispherical, light in colour, turn brown before hatching; each egg covered with transparent covering, the latter irregularly shaped, micropylar end not clear; colour changes to light yellowish-green at base, whitish near micropylar end movements of 1st instar become noticeable from outside egg chorion before hatching.

2. Larva: Number of instars:5.

(i) First instar: Duration: 5.80 ± 0.83 days.

Head: Width 0.91 ± 0.10 mm; Hypognathous, distinct, non-retractible, black; epicranial suture and other cephalic areas conspicuous; more wider as compared to body; ocelli distinctly visible laterally; primary setae present.

Body: Length 5.1 ± 0.89 mm, width 1 ± 0.07 mm; fusiform, well segmented, setose, light greenish when freshly emerged, turn greenish yellow later on; body uniformly wide except at prothorax, mesothorax, seventh and eighth abdominal segments; mid-dorsal stripe comparatively more greenish, indistinct; a blackad transverse spot present dorsally on prothorax; anal comb present on last abdominal segment.

(ii) Second instar: Duration: 7.60 ± 1.14 days.

Head: Width 0.96 ± 0.11 mm; same as above.

Body: Length 7.2 ± 0.83 mm, width 1 ± 0.07 mm; same as above.

(iii) Third instar: Duration: 7.50 ± 0.61 days.

Head: Width 1 ± 0.74 mm, width 1.42 ± 0.08 mm; same as above; mid- dorsal stripe turns dark- greenish and distinct; each body segment further get divided into distinct annuli.

(iv) Fourth instar: Duration: 8.10 ± 0.74 days.

Head: Width 1.46 ± 0.08 mm; hypognathous, obliquely placed, yellowish brown, reddish laterally, blackish-brown dorsally; epicranial suture, clypeal areas, ocelli and various clypeal setae distinct.

Body: Length 13.2 ± 1.30 mm, width 2.6 ± 0.1 mm; same as above, becomes stout and well segmented, annuli more distinct, setae arranged in transverse rows; width of larva relatively less in thoracic and last abdominal segments; dark green mid-dorsal stripe and two yellowish dorsolateral stripes (one on either side of mid- dorsal stripe) extend throughout length of body.

(v) Fifth instar: Duration: 13.00 ± 0.70 days.

Head: Width 2.66 ± 0.40 mm; same as above.

Body: Length 23.2 ± 1.30 mm, width 3.5 ± 0.70 mm, same as above.

Larval duration: 42. 00 ± 4.02 days.

3. Pupa: Duration: 7.60 ± 1.14 days.

Length 25.6 ± 1.14 mm, width 9.1 ± 0.74 mm; light greenish, continue to remain in the folded leaf blade, dorsal region of pupa visible, conical in shape, tapering anteriorly and caudally, abdominal segments distinct, covered with a jelly like covering, turns brownish after 4 to 5 days, wing region conspicuous, eyes and antennae wrapped around thoracic segments, wings visible through pupal case flutters last abdominal segments on disturbance.

4. Adult diagnosis

Male and Female: Forewing dull dark brown, with shinning whitish spots- two in dismal cell a postmedial series of six spots in male, eight and more prominent spots in female; undersurface light brown, decorated with light yellow scales in female, with various spots of uppersurface showing through in both sexes. Hindwing dark brown, covered with light yellow long scales and more densely in female, with two to three less conspicuous postmedial white spots; undersurface light brown, with a white spot in dismal cell and a series of postmedial white spots.

Wing expanse: Half, Male: 17.5-19.5 mm., Half, Female: 18.5-20.5 mm.

Material examined: 4♂, 4♀, 4.ii. 1993, 2♂, 20.iii.1994, 4♂, 19.iv. 1996, 5♂,2♀,, 3.iv.1997, 1♂, 1.xi. 1997, 4♂, 1♀,, 8.iv. 1998, 2♂,18.ix.1998, University Campus, Patiala; 4♂, 2♀,, 11.iv.1993, 3♂, 20.ix. 1997, Nabha, Patiala; 4♂, 1♀,, 10.ox. 1993, Dhuri, Sangrur; 13♂, 1♀,, 13.iv.1997, Sangrur; 2♂, 1♀,, 15.iv.1996, Gobindgarh, Khana; 4; ♂, 2♀,, 30.v.1993, 2♂, !.vi. 1997, Talwara, Hoshiarpur; 2 ♂, 1♀,, 10.iv.1993 Ropar; 3♂, 1♀,, 27.iv. 1996, Anandpur Saheb, Ropar; 4♂. 1♀,, 13/iv. 1995, Zira, Ferozpur; 4♂, 29.iii.1995, Sirhind, Fatehgarh Saheb; 6♂, 3♀, (bred); Pbi. Univ., Patiala (Punjab).

DISCUSSION

In butterflies, the oviposition behaviour seems to be highly evolved and complex. The female exercises a considerable amount of selectivity before actually depositing the eggs on a substratum/host plant. According to Thompson and Pellymer (1991), some of the major factors that influence the preference of egg laying by a female on a particular plant/plant part are behavioral, ecological and genetical. In fact, after obtaining cues, a female makes choices at three different levels while searching for an appropriate host. Firstly, she flies around to select a habitat in which to search the host plant; secondly, she chooses a particular plant on which to land, and finally a suitable substratum to deposit an egg (Chew and Robbins, 1984; Papaj and Probopy, 1989). While selecting a habitat or a particular host plant, she intends to use the sunlight factor hitting the plants (Grossmuller & Lederhouse, 1985) and shape of leaves (Gilbert, 1975; Young & Stein, 1976; Rausher, 1987; Stanton, 1982; Mackay & Jones, 1989), besides specific plant volatile (Feeny *et al.*, 1983). The sensory receptors involved are either present on the proboscis (olfactory) or the antennae (tactile) or the tarsi and the ovipositor (Renwick & Chew, 1994).

During the course of present studies, it has been observed that a female almost always patrols around and ultimately locates a host plant in a particular habitat before finally selecting an oviposition site on the host. The females of *Castalius rosimon* and *Ypthima inica* have been seen to examine the leaf/bud surface by repeated tapping with their abdominal tips before laying their eggs, whereas, those of *Acraea issoria, Vanessa (Aglais) cashmiriensis, Pelopidas mathias, Papilio Polytes* and *Colotis amata* quickly, though precisely, place their eggs on an appropriate substratum. While making a choice for oviposition, a female may sometimes even reject a potential host plant after landing, as has been observed by Young (1974a), and Brown and Oplar (1967). The species such as *Vanessa (Aglais) cashmiriensis, Castalius rosimon* and *Acraea issoria* have been observed to discard one or the other host plant after landing on the same but flying away without depositing an egg.

In the butterflies, the ovipositon generally takes place either in bright sunny weather during noon hours (Ganguli, 1961; Gibbs, 1970; Rutkowski, 1971; Young 1973b, 1974a, Kirton *et al.*, 1982; Damman & Feeny 1988; Fiedler & Schurian 1994; Schurian & Fiedler, 1996) or during morning and evening hours (Young, 1974c) or throughout the day (Young 1972c, 1973b,, Kitching, 1987). In all the presently studied species, the oviposition has been observed to takes place in bright sunny weather conditions between 9.00 am. to 3.00 pm.

While selecting a particular host plant or part of the host plant, it has been reported that a female makes a choice that ensures maximum larval growth and survival (Smiley, 1978; Rausher, 1981; Singer, 1983, 1984; Williams, 1983; Damman and Feeny, 1988). The present study reveals that the species, *Papilio polytes* and *Pelopidas mathias* choose fresh and young leaves for oviposition, whereas, *Euploea core* lays its eggs on young buds. The medium and large sized leaves are preferred by *Vanessa (Aglais) cashmiriensis, Acraea issoria* and *Colotis ammata*, who lay their eggs in clusters. Further, the workers like Atsatt (1981a, 1981b), Cotterell (1984), Henning (1983), Kitching (1987), Efgar & Price (1988), Fiedler and Maschwitz (1989a, 1989b, 1989c), Thompson and Pellymer (1991), and Wagner and Kurina (1997) have expressed that the ovipositing females in the families Lycaenidae and Riodinidae, where the larvae show mutualism with ants, prefer the host that harbours ants. However, in a presently observed Lycaenid species, *Castalius rosimon*, it has been noted that though the larvae show mutualism with ants yet the ovipositing female lays the eggs regardless of the presence of ants on the lost plant. This observation goes in accordance to the works by Thompson *et al.* (1990), Fiedler (1992a) and Fiedler *et al.* (1994), who have also worked on some other species of the family Lycaenidae.

In order to protect the eggs from predators, parasitoids and extreme hot weather conditions, it is indirectly inferred that most of the females opt to lay their eggs on under surface of the leaves (Jandu, 1942; Young, 1972b, 1974c 1979, 1984a, 1984b; Muyshondt *et al.*, 1973; Kirton *et al.*, 1982; Urich and Emmel, 1990, 1991a), though oviposition on upper surface of the leaf to has been recorded in certain species of butterflies (Young, 1973a; Otero and Aiello, 1996). In the present investigations, it has-been recorded that either side of the leaf has been used for egg laying by different species of the butterflies. For instance, *Vanessa (Aglais) cashmiriensis, Ypthima inica, Acraea issoria* and *Castalius rosimon* have been recorded to prefer under surface, whereas, *Colotis amata, Pelopidas mathias* and *Papilio polytes* opt the upper surface, of the leaf for the purpose of oviposition. A fertilized female may also make a further choose in selecting a particular site of the leaf for oviposition i.e., either near margin (Young, 1972a 1973a 1974b) or near midrib (Young, 1972b; Kirton *et al.*, 1982; Urich and Emmel, 1991a) or in the middle (Young, 1974c; Fiedler and Schurian, 1994) or at the apical area (Young and Muyshondt, 1973). Amongst the presently studied species, *Vanessa (Aglais) cashmiriensis, Acraea issoria, Castalius rosimon* deposit their eggs along midrib or harder veins, whereas, *Papillio polytes, Pelopidas methias, Ypthima inica* and *Colotis ammata* at the center of the leaf.

By and large, in the presently studied species it has been observed that the females mostly lay a single egg/cluster on each leaf, though very rarely two eggs may also be present on the same leaf. The ovipositng female often recognizes and avoids nonspecific egg load on the same host to avoid resource competition and cannibalism amongst the larvae (Thompson & Pellmyer, 1991). Prior to this, workers likes Young (1974c, 1983) and Soberon *et al.* (1988) Contrary to this, Baylis & Pierce (1991) have recorded that an ovipositing female may prefer and lay the eggs on the already loaded leaves.

According to Deither (1959) some species of the butterflies are not precise in egg laying and frequently commit errors. In different species of the genus *Pamassius* Latreille, Hauser *et al.* (1984) have pointed out such errors being frequently committed by the females, who rather laying the eggs on the larval host plant, do so far away from it on different objects on the ground. Reports on mistaken oviposition have also been published by some other workers as well (Deither, 1940, 1941, Young & Stein, 1976; Young, 1980). None of the presently studied species indulge in mistaken oviposition or oviposition error.

The study of relevant literature shows that the egg chorion is a potential source of food for newly hatched larva (Smart,1985) containing vital nutrients such as carotenoid pigments (Feltwell, 1982). However, Scoble (1995) has offered another explanation for the egg shell being eaten up by a cryptic larva is to escape perdation/parasitism, as an empty egg shell can easily be located. Whereas, on one hand, workers like Gibbs (1970), Young (1973c 1974a, 1974) Young & Moffett (1979), Young & Muyshondt (1985) have reported that the egg chorion forms first food of the newly hatched larva and, on the other hand, Muyshondt *et al.* (1973), Young & Muyshondt (1973), Fiedler and Maschwitz (1989b), Braby and Woodger (1994), and Schurian & Fielder (1996) have stated that the freshly emerged larva eats only the exit hole to come out of the egg shell. Upon hatching, the first instar larva devours the egg shell completely in the species, *Acraea terpsicore, Colotis amata* and *Ypthima inica,* consumes a part of it in *Papilio polytes* and *Acraea issoria,* and bites only an exit hole in *Pelopidas mathias* and *Castalius rosimon.*

However upon hatching in a cluster, the larvae of the species, *Vanessa (Aglais) cashmirensis* may consume the egg shell fully or partially.

Upon emergence, the first instar larvae have been seen feeding gregariously in *Vanessa (Agalais) cashmiriensis, Acraea issoria* and *Colotes ammata* which lay their eggs in clusters. This observation goes in accordance to Stamp (1980) Young (1983), Chew and Robbins (1984), who have categorically stated that all such aggregations are a consequence of egg clustering, at least in the first instar larvae. While making comments on the evolution of gregariousness in aposematic larvae, Sillen-Tullberg (1988) have inferred that unpalatability seems to be an important predisposing factor for the evolution of egg clustering and larval gregariousness in the butterflies.

The early instars may eat all the layers of the leaf except the vein network and the midrib (*Papilio polytes, Pelopodia mathias, Vanessa (Aglais) cashmiriensis, Ypthima inica*) or may scrab the upper or lower epidermis and chlorophyll, and accordingly leaving behind the lower or upper epidermis with vein network intact (*Castalius rosimon, Acraea issoria* and *Colotis amata*) or may feed on a flower/ bud (*Euploea core*). Further, it has been seen that the later instars (especially fourth and fifth ones) forage almost whole of the leaf/leaves, leaving behind the midrib and the harder veins (*Vanessa (Aalais) cashmiriensis, Papilio polytes, Pelopidas methias, Euploea core, Colotis amata*) or consume the harder veins or midrib, as well (*Acraea issoria*).

Amongst the presently studied species, the larvae of *Ypthima inica, Euploea core* and *Acraea issoria* are noctunal or crepuscular in their feeding activities and they rest during the day time in order to avoid direct sunlight and predators (Harrison, 1963; Gibbs, 1970; Young, 1972a, 1972b; Casagrande, 1979; Young & Muyshondt, 1985; Horvitz *et al.*, 1987; Fiedler, 1992a). In *Pelopidas* moths the larvae form tunnels and rest inside them. In *Colotis amata,* it has been noted that the first three instar larvae feed and rest gregariously remaining in a tight group in a very coordinated manner, with their heads facing in one direction. In different butterfly species, such a coordination in larval groups has also been observed by Barrett & Burns (1951), Fox (1967), Muyshondt et al. (1973), Ajmat & Teran (1970), Stamp (1977), Young & Moffett (1979), Braby & Douglas (1992).

The caterpillars have evolved a wide range of defense mechanisms against predators and parasitoids. Scoble (1995) has divided such mechanisms into two groups *viz.*, those having no direct adverse affect on the predators/parasitoids and those actively repelling the predators/ parasitoids. It may also be mentioned that the cryptic or protective resemblance falls in the first category (Silberglied, 1977). As per Mathews & Mathews (1978), crypsis is not only a matter of coloration, but behavior too. Except *Vanessa (Aglais) cashmiriensis* and *Acraea issoria*, the first instar larva of the remaining species shows crypsis with either leaf or bud colorations. In *Papilio polytes*, the earlier two instars resemble bird droppings, whereas, the later instars of this species and those of

Castalius rosimon have been found to completely camouflage with the leaf colour. Yet, in another case, the larvae of a Satyrid species, *Ypthima inica* exhibit excellent crypsis, in which the green colour punctuated with light green longitudinal parallel striations on the body completely match with venation of the blade of grass and the larvae are very hard to locate in the field. References to such cryptic coloration in the butterfly larvae is also available in the published literature (Young, 1972a,1972b 19712c, 1974a, 1974c; Aiello, 1980; Fiedler, 1992a; Otero & Aiello, 1996, Schurian & Fiedler, 1974c; Aiello, 1980; Braby, 1990; Fiedler, 1992a; Otero & Aiello, 1996, Schurian & Fiedler, 1996). The larvae of *Vanessa (Aglais) cashmiriensis* and *Acraea issoria* have been noticed to construct silken mats, which support them while feeding and resting. In *Pelopidas mathias*, the larvae from tunnels of silk and rest in them. This type of behaviour has also been discussed by Rainbow (1907), Common & Waterhouse (1972, 1981), Young (1972b, 1974a, 1974b, 1974c), Young & Moffett (1979), Oplar (1974), Rathcke & Poole (1975), and Chandal *et al.* (1995), in different butterfly species. Apart from this, the larva of *Castalius rosimon* shows a unique defecation behavior as a part of its protection. It does not allow the fecal pellets to remain on the surface of the leaf. In fact, the larva uplifts its rear end to expel the fecal pellet in such a way that they fall on the ground and thus not accumulate on the leaf, which otherwise would become conspicuous to attract predators. Callaghan (1991) has also reported this kind of defecation behaviour in a lycaenid, *viz.*, *Metacharis ptolomaeus* (Fabricius). Further, the larvae of *Castalius rosimon* are also attended to by the ants, *viz.*, *Pheido*l sp., *Compnotus* sp. and *Polyrhachis* sp. While giving due protection to the larvae from their enemies, the ants in turn harvest the secretions secreted by the lenticles and dorsal nectary organ of the caterpillar. This kind of mutuality is termed as myrmecophily and has been reported by a number of workers (Hinton, 1951; Malicky, 1969; Downey & Allyn, 1979; Eliot, 1980, 1992; Pierce & Mead, 1981; Pierce, 1984, 1985, 1987, 1989; Cottrell, 1984; Kitching & Luke, 1985; Pierce & Eastseal, 1986; Pierce & Young, 1986; Horvitz *et al.*, 1987; Pierce *et al.*, 1987; Fiedler, 1988a, 1988b, 1991, 1992a 1992b; Fiedler & Maschwitz, 1989a; Hill & Pierce, 1989; DeVries, 1977, 1991; Ballmer & Pratt, 1992; Peterson, 1993).

According to Kirton *et al.* (1982), Smart (1985) and Eliot (1992) as the larvae grow they entirely fill up their skin which consequently become very tight. Also, in order to grow and increase further in size, the old skin is regularly shed off from time to time, exposing a new and larger larva, which has formed beneath it. In the present studies, representatives of the families Nymphalidae, Danaidae and Peridae moult four times, *i.e.*, there are five instars, whereas, in the family Lycaenidae the larval ecdysis takes place three times, i.e., there are four instars. In the family Satyridae, it has been observed that the number of instars may be five or four in the pre- and post-monsoon populations. Such types of ontogenetical variability has also been reported in the taxa *Anartia* Hubner (Young & Stein, 1976; Silberglied *et al.*, 1979), *Dynastor darius* (Fabrius) (Aiello & Silberglied, 1978), *Dynastor macrosiris* (Linn.) (Urich & Emmel, 1991b), *Lycaena phlaeas hypophlaeas* (Boisduval) (Ballmer & Pratt, 1989), and *Papilio (Princeps) demoleus* Linn. (Papilionidae) (Badawi, 1991) by the respective workers.

Prior to pupation, the larva consumes enormous quantities of food in a relatively shorter period of time, Attaining complete maturity, it wanders restlessly in search of a suitable location for transforming to pupil stage. This period is called a nomadic or wandering period (Urquhart, 1960; Young, 1973b; Oplar, 1974; Feltwell, 1982). During this period or say before pupation, the length of the larva decreases considerably along with increase in width, besides change in ground coloration. After selecting a particular site, as presently observed, the larva starts forming a mat of silken threads slowly and steadily. Subsequently, it forms a thick cushion of threads in a rather smaller area and this is termed as button (Urquhart *loc. cit.*; Oplar *loc. cit.*; Young and Stein, 1976). After clamping on the button with its anal legs, the larva tends to change into pupa slowly and slowly. In the meantime, the larval skin splits along meson of the thorax and gradually it reaches up to caudal end, and thus liberating the enclosed pupa. The process of pupation has also been commented upon by various workers (Mosher, 1916; Urquhart, 1960; Ganguli, 1961; Young, 1972a, 1980; Aiello & Silberglied, 1978; Sillberglied *et al.*, 1979; Young & Moffett, 1979; Aiello, 1980; Kirton *et al.*, 1982; Horvitz *et al.*,1987; Fiedler, 1992a; Fiedler *et al.*, 1994; Bribe & Douglas, 1992).

In all the species studied presently, it has been observed that when the imago emerges out of the pupil case, a few drops of meconium (liquid of pupal excretory products, Smart, 1985) ooze out from the tip of the abdomen of the imago (Urquhart, 1960; Young, 1974c; Feltwell, 1982; Kirton *et al.*, 1982; Stehr, 1987; Shapiro, 1991).

According to Wynter-Blyth (1957) and Eliot (1992), the mating generally takes place immediately after emergence. The mating behavior has been recorded in four species, *viz.*, *Papilio polytes, Pelopidas mathias, Ypthima inica* and *Acraea terpsicore* in the field. After mating, it has been observed that a female would make herself busy in the concerned habitat for the location of a suitable host plant vis-à-vis substratum for ovipositon. The mating always takes place in and end-to-end position, and as and when, a pair in copula gets disturbed, it flies away with male as an active flight partner (Vane-Wright & Ackery, 1984; Wickman, 1988; Matsumoto, 1990; Schurian and Fiedler, 1996). As a part of courtship behaviour, the male of *Hypolimnas bolina* has been observed to show territorial behaviour, in which he chases away other butterflies entering in his domain (Fiedler & Schurian, 1994; Tennent, 1995).

Further, the rejection behaviour in courtship has been commented upon by Stride (1958) and Rutowaski (1978a, 1987b). This has been observed in two presently studied species, *i.e.*, *Pelopidas mathias* and *Papilio polytes.* In this behaviour, an unwilling female tends to lie flat on the ground and raises her abdomen upwards, as and when, approached by a male. However, David & Gardiner (1961) have interpreted this behavior as defensive and may also be adopted by a male, when attacked by another male as in the species *Pieris brassicae* Linn.

According to some reports the males of many Lepidoptera congregate of moist substrates along edges of mud puddles, streams, dung, carrion, perspiration-soaked clothings and mammalian eye to imbibe moisture (Collentte, 1934; Bauer, 1953; Sevastopulo, 1959; Downes, 1973; Young, 1973b, 1976; Young & Muyshondt, 1973; Buttiker, 1979; Pivnick & McNeil, 1987; Krizek & Oplar, 1989; Boggs *et al.*, 1981; Launer *et al.*, 1993; Tyler *et al.*, 1994; Viloria, 1997). Arms *et al.* (1974) have referred this activity as pudding. Though no specific comments can be made on the basis of present studies, yet it is evident from some works that sodium is a limiting source because many Lepidopterans feed primarily on nectar which contains very little sodium (Alder & Pearson, 1982; Hiebert & Calder, 1983; Pivnick & McNeil, *loc.cit.*) and accordingly, the pudding behaviour is a means to obtain sodium requirement (Weeks & Kirkpatrick, 1976). According to Alder & Pearson (*loc. cit.*), Greenfield (1982), Pivnick and McNeil (*loc.cit.*) the males transfer sodium to females during mating. During the course of present studies, the mud pudding has been observed in the species, *Euploea core, Vanessa (Aglais) cashmiriensis and Castalius rosimon.*

Most butterflies require an elevated body temperature (28° - 40°C) for active flight and the same is achieved during pre-flight by exposing the wings and body to solar radiations. This particular behaviour is termed as basking by Kingsolver (1985). According to Casey (1981), the pre-flight internal heat generation is rather common in larger moths but rare in the butterflies. The dorsal basking in which the butterfly completely spreads its wings has been reported to be common in the families Papilionidae, Danaidae and Pieridaae (Douglas, 1978; Rawlins, 1980). On the other hand, the lateral basking in which a butterfly completely folds its wings has been reported to be prevalent in the family Lycaenidae and subfamily Coliadinae of the family Pieridae (Watt, 1968). Kingsolver (*loc.cit.*) has also proposed another basking posture, termed as reflective basking, which involves the use of wings as solar reflectors. In this type of basking, the wings are partially stretched and reflect radiations into the body of the butterfly. During present studies lateral and reflective basking (*Pelopidas mathias*), dorsal, reflective and lateral basking (*Castalius rosimon, Vanessa (Aglais) cashmiriensis*) and reflective and dorsal basking (*Colotis amata, Ypthima inica*) have been noted in the respective species mentioned in parentheses. The sun basking behaviour has been dealt with by workers like Urquhart (1960), Gibbs (1970), Heinrich (1974), Young (1974a), Urquhart & Urquhart (1977), Douglas (*loc .cit.*), Turks & Brower (1978), May (1979), Rawlins (*loc. cit.*), Casey (*loc. cit*), Heath & Heath (1982), Kingsolver (*loc. cit*), Masters *et al.*(1988), Hagen & Eckursler (1998) in various butterflies.

Plate 1 : *Castalius rosimon* (Fabricius)

Adult (male)

Adult (female)

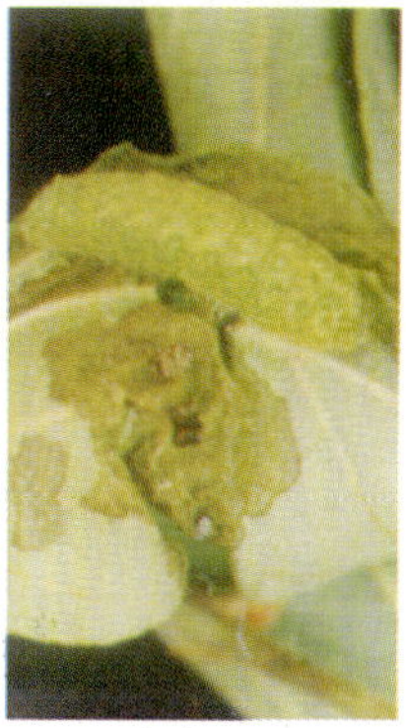

4th instar caterpillar

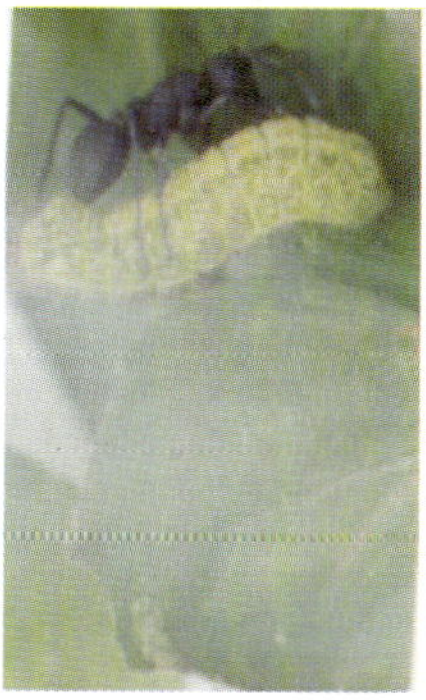

Myrmecophily-ant attending 4th instar

Pupa

Sun basking

Nectar sucking

Ovipositing female

Larval host plant *Ziziphus oenopilia*

Plate 2 : *Vanessa (Aglais) cashmiriensis* Kollar

Adult

Gregariously feeding
3rd instar caterpillar

5th instar caterpillar

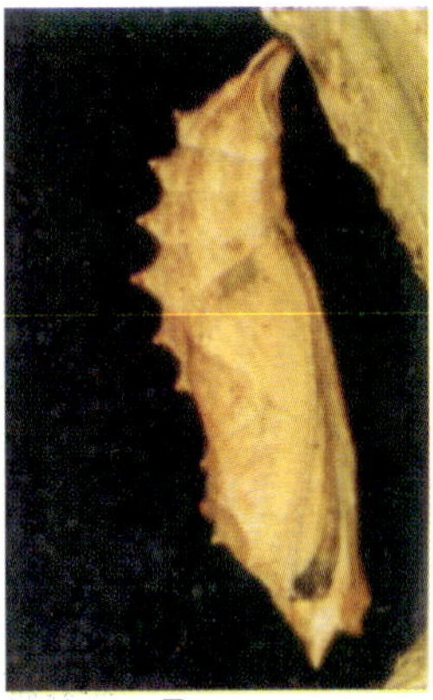

Pupa
(lateral view)

Nectar sucking

Larval host plant
Uritica dioica Linnaeus

Plate 3 : *Acraea issoria* (Hübner)

Adult (male)

Pupa (lateral view)

Adult (female)

Larval host plant
Debregeasia hypoleuca

5th instar caterpillar

Gregariously feeding
3rd instar caterpillar

Plate 4 : *Ypthima inica* Hewitson

Adult

5th instar caterpillar

Pupa (lateral view)

Larval host plant
Cynodon dactylon (Linnaeus)

Plate 5 : *Euploea core* (Cramer)

Adult

4th instar caterpillar undergoing moulting to change into 5th instar

5th instar caterpillar

Freshly formed pupa (lateral view)

Pupa nearing eclosion (lateral view)

Nectar sucking

Larval host plant *Ichnocarpus frustescens*

Plate 6 : *Colotis amata* (Fabricius)

Adult

Pupa (lateral view)

Gregariously feeding
5th instar caterpillar

Larval host plant
Salvadora oleoides

Plate 7 : *Papilio polytes* Linnaeus

Adult (female)

5th instar caterpillar

Pupa

Nectar sucking

Plate 8 : *Pelopidas mathias mathias* (Fabricius)

Adult

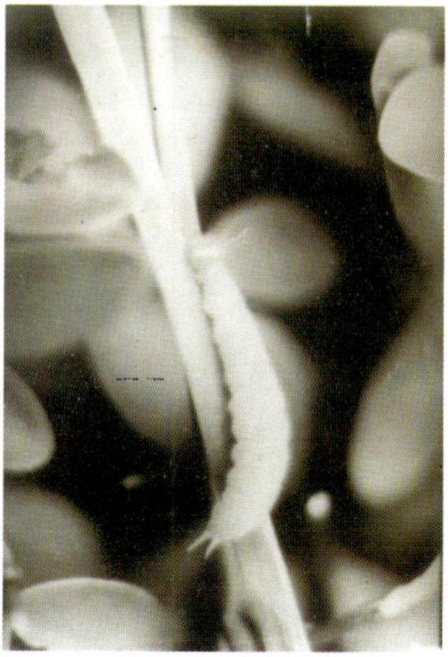

4th instar caterpillar

Eclosion (lateral view)

Pair in copula

The aspect of conservation of butterflies is mostly emphasized as a very sensitive issue (New, 1987, 1991, 1993). As a step towards their conservation, the above account dealing with various life history aspects, besides information on larval host and nectral food plants will hopefully be quite useful in devising strategies for ex-situ conservation. It seems also appropriate to mention here that the identification of the immature stages, particularly the feeding stage is very important from the conservation and phylogenetic point of view. This is possible through examination of the cheatotaxy, which though has been studied for all the above mentioned species for framing keys yet is purposely omitted here in the scope of this paper.

Till today, Rhopalocera group is being taken to serve as an umbrella for other invertebrates, including insects by almost all conservationists and environmentalists. Also there is a notion that this group of the order Lepidoptera is best known but the true picture is now being questioned/explored by many Lepidopterists. By and large, the adults have been attended to in a broader sense by some naturalists, amateurs, photoartists and taxonomists. In Indian context, except for serious pest species, the larval and nectar food plants of most of the Lepidopterans are unknown. It may also be categorically stated that though many species of butterflies have been dealt with by dwelling upon characters such as colour, maculation of wings, structure of eyes, antennae and venation (Bingham, 1905, 1907; Peile, 1937; Talbot, 1939, 1947; Wynter-Blyth, 1957; Cantlie, 1963) yet the genitalia, which are universally considered extremely important in delimiting various species' taxa and also solving various types of complexes, have not been attended to thoroughly. The early stages of different species of butterflies very rarely enter in analytical systematics, even on global basis. There are no keys for identification of immature stages, particularly the caterpillars of different species. To identify those, studies on their cheatotaxy is essentially needed. It will not only generate more field data but also the sound knowledge to understand inter-relationships amongst different butterfly taxa. In view of present author's acquaintance and experience in lepidopterology, it seems appropriate to make a specific statement that we do not have precise data on distribution, ecology, biology and habitat management available even for the umbrella group (Rhopalocera) and the situation is rather more deplorable in the case of the moths (Heterocera).

BUTTERFLY COLLECTIONS AND WILDLIFE (PROTECTION) ACT: A GLIMPSE

As many as, 451 species of butterflies have been included in different schedules of the Wildlife (Protection) Act 1972. As such, this legislation by the Government of India is meant to protect and conserve its fauna and flora. Since their inclusion not much has been done seriously to conserve them through scientific ways and means except through ban on their collection. Whether this is sufficient is debatable?

The inclusion of different species in the Act is based on insufficient collection/distribution data gathered randomly by Bingham (1905, 1907), Evans (1932), Talbot (1939, 1947), Wynte-Blyth (1957) and Cantilie (1963) *etc.* A scrutiny of the lists shows that the pest species *Lampides boeticus* (Linnaeus), which is widely distributed in India is also listed in Schedule II of the Act (Rose & Sidhu, 1994). The same is also true for *Euchrysops cnejus* (Fabricius), which is commonly called as gram blue butterfly. It is a polyphagous species with its larva infesting six different hosts belonging to the family Leguminosae. The Apollo butterflies (*Pamassius* Latreille) which are represented in the Himalaya at high elevations are the most sought after butterflies from the trade point of view. Almost all the species referable to the aforesaid genus are available beyond timberline and fly exclusively in high altitude above 10,000-12,000 ft. Out of 35 species, 12 have been listed from British India period but nothing concrete can be said about the number of their subspecies (Bingham, 1907, Evans, 1932; Piele, 1937; Talbot, 1939; Wynter-Blyth, 1957). According to Mani (1986) there are about 50 subspecies referable to 12 Indian species. Like other high altitude butterflies, our knowledge about Apollos in terms of their distribution and life histories is very inadequate. We are depending too much on old/outdated records. While dealing with the genus *Pamassius,* it has been seen that the species *P. charitonius charitonius* Gray has also been included in Schedule II, Part II of the Wildlife (Protection) Act, 1972 (as amended up to 1991). It is very interesting to record here author's own experiences, who in June, 1993, saw a swarm of this species in a specific habitat which is a rock prone area near Manimahesh (Himachal Pradesh), 3200m above sea level. With the sunshine disappearing and weather turning cloudy, within 12-15 mintes, all the flying individuals were seen hiding underneath boulders. The visual count shows that the species is abundant in that local area and its inclusion in the wildlife (Protection) Act 1972 needs reassessment. However, eleven other species, *i.e., Papilio (Princceps) alcmenor alcmenor* Felder & Felder, *P. (P.) arcturus arcturus* Westwood, *Chilasa agestor* (Gray), *Graphium (Graphium) doson axion* (Felder & Felder), *G.(G) agamemnon agamemnon* (Linnaeus), *G. (G) chiron* (Wallance), *G. (Pathysa) xenocles phrontis* (de Niceville), *G.(P) antiphates pimpilus* (Fabricius), *Lamproptera meges* (Zinken- Sommer) *Pammassius maharaja* Avinoff and *Atrophaneura (Atrophaneura) aidoneus* (Doubleday) of the family Papilionidae have been found to be rather.rare and, accordingly, it is being suggested that these be protected and necessary steps initiated for their conservation.

From the family Satyridae, whereas, two other species, *i.e., Maniola devendra devendra* (Moore) and *Callerebria annada* (Moore) included in the Act (schedule II, Part II) have been found to be abundant and, at the same time, *Maniola pulchella* (Felder), M. *pulchra pulchra* (Felder), *Lethe europa niladana* Fruhstorfer and *Pararge eversmanni cashmirensis* Moore which have previously been reported as common and distributed only in northwestern Himalaya have been found to be quite rare as per present surveys. With regard to the family Lycaenidae, six species, *viz., Oreolyce (Arletta) vardhana vardhana* (Moore), *Tarucus venosus* Moore, *Lycaena kasyapa* (Moore), *Chrysozephyrus birupa* (Moore), *Rapala selira* (Moore) and *Ancema ctesia* (Hewitson) which have previously been reported as common or not rare and distributed only in northwest India are found to be quite rare. Species such as *Dodona dipoea nostia* Fruhstorfer (Riodininae); *Euaspa ziha* (Hewitson) (wrongly mentioned as *Thecia ziba* in the Act) and *Spindasis nipalicus* (Moore) (Theclinae) enlisted in schedule II of the Act have been observed to be moderate in their distribution and are not so rare. Accordingly, a suggestion is being made for their transfer to Schedule III or Schedule IV. Some other species, *viz., Abisara fylla* (Westwood) (Riodininae), *Porita hewitsoni* Moore (Poritiinae), *Lycaena kasyapa* (Moore) (Lycaeninae), *Horaga albimacula viola* Moore, *Rapala pheretima* (Hewitson), *Arhopala paramuta* de Niceville (Theclinae) and *Azanus ubaldus* (Stoll) (Polyommatinae) have been observed to be rare/very rare. This is country to some earlier reports that these taxa are common/not rare. Besides initiating scientific ways and means for their conservation, these taxa need to be included in Wildlife (Protection) Act, 1972. Yet another observation based on the present field work reveals that two *Theclin* taxa, *i.e., Hypolycaena erylus himavantus* Fruhstorfer, *Zeltus amasa amasa* (Hewitson) and two Polyommatin, *i.e., Prosotas dubiosa indica* (Evans) and *Anthene emolus emolus* (Godart) though are common yet their natural populations are largely represented by male individuals alone. In fact, no females could be collected for these species during the recently conducted surveys. A proposal is being made for protection and conservation of these species, which is perhaps possible through overall habitat management. The Lycaenid taxa *Esakiozephryus mandara mandara* (Doherty), *Polyommatus icarus* (Rottemburg), *P. eros* (Ochsenheimer). *P. florenciae* (Tytler), *Albulina galathea* (Blanchard), *A. omphisa* (Moore) and *A. metallica metallica* (Felder & Felder) are exclusively confined to higher altitudes in the Greater Himalaya and their altitude range lies between 2680–4079 above msl.

The faunistic surveys undertaken between 1991-1998, have led to the collection of more than 400 species of the butterflies from various localities in North-West and North-East India. These are very well preserved in author's laboratory. Another inference drawn on the basis of these faunistic explorations points out that the butterfly collecting grounds at Bashisha (Assam), Sivoke (West Bengal), Rangpo (Sikkim), Mussoorie (Uttranchal), Runeka ji and Dalhousie (Himachal Pradesh) have been observed to be rich in butterfly diversity. Apart from above, it may also be mentioned that the nomenclature of many taxa mentioned in the Act need rectification as per recent developments in butterfly systematics (Rose & Sidhu, 1995). Though there are certain provisions in the Act to grant permission to collect material for scientific purposes yet these are insufficient from an entomologist point of view.

GENERAL REMARKS

The purpose of writing all above is to draw attention to the necessity of more vigorous and long term field work. Further, due to vastly unplanned as well as extensive urbanization, deforestation and environmental pollution, the butterfly diversity (being more sensitive), in particular, and insect diversity (invertebrates), in general, are facing a real threat to their survival. In fact, the successive demise of varied habitats is making many species threatened, rare and vulnerable before ultimately getting extinct. If Charles Darwin were alive today, his work would most likely focus not on origin of species but, rather, on obituaries of species! India, which harbours a wide diversity of invertebrate life, is expected to do much more in the current century. Though India has already 448 Sanctuaries spanning over 112,174,45 sq. km. and 85 National parks spread over 35,919, 03 sq. km. (From MoEF, 1998), yet it is a dubious attempt to resort to habitat protection in ignorance of its invertebrate biodiversity components. The concept of erecting "barbed wire" around the protected areas will not help till inventories and other biological, ecological and behavioural aspects of different species are not completed – hook, line and siker!. Are surveys of the most dominating group of biodiversity, *i.e.*, insects, valuable for impact assessment and national park designation? And, what taxonomic level is most useful requires to be determined. Insects belong to twenty eight orders and there is hardly anyone working on Apterygotes in India. The situation on Pterygote orders is also not very satisfactory. The position of insect taxonomy in our national museums is also not very promising at this time. This sorry state of affairs is due to the simple reason that insect taxonomy moved one step forward and two steps backward in India after independence. Even, it lost its charm in other counties, too.

Now with the coming of the catch word "Biodiversity" in 1980s and its frequent use and understanding after Earth Summit (1992), it is being felt that the wheel of alpha taxonomy has started moving again. India became a signatory to the Convention of Biological Diversity (CBD) in 1994 and the Ministry of Environment & Forests, New Delhi is a nodal ministry for CBD in India. The said ministry has prepared a National Report on implementation of article 6 of the CBD in India in 1998. Accordingly, the ministry has launched an All India Co-ordinated Project for Capacity Building in Taxonomy (AICOPTAX) sometime back. Under this programme, nine centers for research and two centers for training have been recognized throughout the country. The subject areas identified for establishing centers for research include non-pathogenic bacteria, viruses, fungi, lichens, bryophytes, palms, helminthes, nematodes and Microlepidoptera. This bold initiative of the Central Government to enhance interest in taxonomy is a first step towards conservation as a whole. This is to fulfill the implementation of article 6 of the Convention on Biological Diversity in India. Another exercise in conservation which is going on a war footing is the preparation of

National Biodiversity Strategy & Action Plan (NBSAP) and State Biodiveristy Strategy & Action Plans in India. Rightly implemented, there should be a great boost for invertebrate taxonomy, particularly insect taxonomy which will help exploring the vast insect diversity from different eco-regions in the country. The insect conservation and their role in conservation of the overall biodiversity will lead to interesting and meaningful results.

VISION STATEMANT

Insects are a unique gift of the mother nature to the mankind, in particular. they are unparalleled for their abundance and represent unrivalled diversity on the planet earth. They pose remarkable colours and have incredible powers of endurance and performance. Metamorphosis allows them temporal separation from potential competitors. Their wings increase their potential to spread on local as well as geographical scale. Their co-evolution with plants increases available niches. The short generation time permits rapid adaptation and finally evolution of social life permits some of them a considerable control over the environment. The unrecognized role they play for furthering human welfare and conservation need to be appreciated. The scope of the subject of diversity of insect fauna is wide and embraces such disciplines as ecology, biogeography, taxonomy, genetics and evolutionary biology. The real future of insect diversity lies much in keeping the habitats intact and allowing their exploration with open mind to arrive at suitable inferences meant for sustainable development. The conservation of insects has a great bearing on conservation of biodiversity of various different kinds of life forms, including welfare of man.

ACKNOWLEDGEMENTS

The author would like to thank Indian Council of Agricultural Research, New Delhi for funding two research projects on butterflies (1991-1994; 1995-1998). Thanks are also due to the staff working in the projects, particularly Ms. A.K. Sidhu, Mr. J. Sehgal and Ms. M. Kaur. The facilities provided by the Head, Department of Zoology, Punjabi University, Patiala are acknowledged gratefully.

REFERENCES

Adler, P.H. & Pearson, D.L., 1982. Why do male butterflies visit mud puddles? *Can. J. Zool.* 60: 322-325.

Agarwala, K. K. and Saha, J. L., 1984. Aphidophagous habits of the larvae of *Chilades lajus lajus* (Cramer)., *Tyo to Ga.* 34: 171-172.

Aiello, A., 1980. Life history of *Dismorphia amphiona beroea* (Lepidoptera: Pieridae: Dismorphinae) in Panama. *Psyche* 87 (3-4): 171-175.

Aiello, A. and Silberglied, R.E., 1978. Life history of *Dynastor darius* (Lepidodptera: Nymphalidae : Brassolinae) in Panama. *Psyche* 85 (4): 331-345.

Ajmat, Z.D. and Teran, H., 1970. Fauna del noroeste argentino-contribucion al conocimiento de los lepidodpterous argentinous, 1. *Mechanitis lysimia elisa* (Guerin, Meneville) (Rhopalocera- Ithomudae). *Acta Zool. Lilloana,* 26: 35-47.

Arif, M., Singh, H.B. & Joshi, M.C. 1994. *Pontia deplidice* L. (Lepidoptera: Pieridae)- a serious pest of cruciferous at high altitude in India. *Entomologist Record.* 106: 179-180.

Arms, K., Feeny, P. & Lederhouse, R. C. 1974. Sodium stimulus for puddling behaviour by tiger swallowtail butterflies, *Papilio glaucus. Science* 185: 372-374.

Atsatt, P.R.L., 1981a. Ant-dependent food plant selection by the Mistletor butterfly, *Ogyris amaryllis* (Lycaenide). *Oecologia* (Bert.) 48: 60-63.

Atsatt, P.R.L. 1981b. Lycaenidae butterflies and ants selection for enemy free space. *The American Naturalist* `18 (95): 638.

Atwal, A.S., 1963. Insect pests of citrus in Punjab. V. Biology and control of citrus caterpillar, *Papilio demoleus* L. (Lepidoptera: Papilionidae). *Punjab Horticultural J.*, 4: 40-44.

Atwal, A.S. & Josan, G.S., 1962. Citrus pests round the year. *Punjab Horticultural J.*, 2: 8-16.

Badawi, A., 1991. Studies on some aspects of the biology and ecology of the citrus butterfly, *Papilio demoleus* L. in Saudi Arabia (Papilionidae: Lepidoptera). *Sonderdruckaus, Bd.* 91 (3): 286-292.

Ballmer, G.R. & Pratt, G.F., 1989. Instar number and larval development in *Lycaena phlaeas hypophlaeas* (Boisduval) (Lycaenidae). *J. Lepid. Soc.* 43 (1): 59-65.

Ballmer, G.R. and Pratt, G.F., 1992. *Loranthomitoura,* a new genus of Eumaeini (Lepidoptera: Lycaenidae: Theclinae). *Trop. Lepid.* 3 (1): 37-46.

Barrett, C.L. & Bums, A.N., 1951. Butterflies of Australia and New Guinea. 187 pp. N.H. Seward, Malbourne.

Bauer, D.L. 1953. Butterflies at water holes in central Arizona *Lepid. News* 7: 146-147.

Baylis. M.& Pierce, N.E., 1991. The effect of host- plant quality on the survival of larvae and oviposition by adult of an ant-tended lycaenid butterfly *Jalmenus evagoras. Ecol. Ent.* 16: 1-9.

Bean, A.E., 1988. Observations on the occurrence and habits of *Nacaduba complex* of the Lycaenidae (Lepidoptera), mainly from Pune district, Western Ghats. *J. Bombay nat. Hist. Soc.* 85 (2) 332-363.

Bell, T.R., 1909. The common butterflies of the plains of India. *J. Bombay nat. Hist. Soc.* 19: 16-58, 438-74, pl. iii.

Bell, T.R. 1910. The common butterflies of the plains of India *J.Bombay nat Hist. Soc.*, 20: 279-330.

Bell, T.R. 1911. The common butterflies of the plains of India. *J. Bombay nat, Hist Soc.* 20: 1115-36.

Bell, T.R. 1912. The common butterflies of the plains of India, Pts xi, xii, *J. Bombay nat Hist. Soc.* 22: 92-100 : 320-44; 517-37.

Bell, T.R. 1914. The common butterflies of the plains of India. Pts xvi. *J. Bombay nat. Hist. Soc.* 23: 73- 103.

Bell, R.D. & Scott, F.B, 1937. Fauna of British India. Moths, 5. Sphingidae, Taylor & Francis Ltd., London. xiv+ 537 pp.

Bingham, C.T., 1905. The fauna of British India, including Ceylon and Burma, Lepidoptera, Butterflies. 1. (Nymphalidae, Nemeobidae). Taylor and Francis, London, 511 pp.

Bingham, C.T. 1907. The Fauna of British India, including Ceylon and Burma, Lepidoptera butterflies, II. (Papilionidae, Pieridae and Lycaenidae). Taylor and Francis, London, 480 pp.

Boggs, C.L. Smiley, J.T. & Gilbert, L.E. 1981. Patterns of pollen exploitation by *Heliconius* Butterflies. *Oecologia* (Bert), 48: 284-289.

Braby, M. F. 1990. The life history and biology of *Paralucia pyrodiscus lucida* Crosby (Lepidoptera: Lycaenidae). *J. Aust. Ent. Soc.*, 29: 41- 50.

Braby M.F. & Douglas, F., 1992. Observations on the biology of *Delias hyrpalyce* (Donovan) (Lcpidoptera: Pieridae) near mellbourne, Victora. *Aust. Ent. Maq.* 19 (1): 9-18.

Braby, M.F. & Woodger 1994. The life history of *Zizula hylax attenuata* (T.P. Lucas (Lepidodsptera: Lycaenidae). *Ausu. Ent.*, 21 (2): 39-42.

Brar, T.S. 1959. Bionomics and control of *Papilio demoleus* Linnaeus (Papilionidae). M. Sc.Thesis, P. U. Chandigarh.

Brown, R.M. & Oplar, P.A. 1967. Biological observations on *Callophrys viridis* (Lycaenidae). *J. Lepid. Soc.* 21: 113-114.

Buttiker, W., 1979. First records of eye-frequenting and anthrophilio lepidoptera from Saudi Arabia. *Fauna of Saudi Arabia* 1: 345-351.

Callaghan, C.J. 1991. Notes on the immature biology of two Riodinine butterflies: *Metacharis Ptolomaeus* and *Napaea orpheus* (Lycaenidae). *J. Res. Lepid.* 30: 221-224.

Cantlie, K. 1963. The Lycaenidae portion (except the *Arthopala* group) of Brigadier Evan's *The identification of Indian Butterflies* (India, Pakistan, Ceylon, Burma) *Bombay nat. Hist. Soc.*, Bombay: vi. + 156pp.

Casagrande. H.M. 1979. Sobre *Caligo beltrao* (Illiger), 1. Paxonomia biologia, morfologia das fases imaturase distribuicoes espocial e temporal (lepidoptera, Satyridae, Brassolinae). *Rev. Brasil. Biol.* 39: 173-193.

Casey, T.M. 1981. Behaviour mechanisms of thermoregulation. In insect thermoregulation. Wiley, S. and Sons, New Your, pp. 79-114.

Chandel, R.S., Kashyap, N.P. & Mehta, P.K. 1995. Bionomics of *Vanessa cashmiriensis* Kollar infesting hopes in Lahaul Valley of Himachal Pradesh., *Pest management and Economic Zoology* 3 (1): 41-43.

Chaturvedi, N. & Haribal, M., 1992. Cannibalism in butterfly larvae. *J. Bombay nat. Hist. Soc.* 89 (2): 261-262.

Chaturvedi, N. & Kehimar, I., 2000. Butterflies: their early stages. *Hornbill* (July-Sept.); p.14

Chew, F.S. & Robbins, R.K. 1984. Egg laying in butterflies. The Biology of butterflies, (eds.),Vane-Wright, R. and Ackery, P.R., 1984. Academic Press, New York, 65-79.

Cheesman, O.D. & Brown, V.K., 1999. Conservation of the wart biter bush cricket- flagship or pharmaceutical? *Antenna* 23 (2): 70-75.

Colelntte, C.L., 1934. On the sexes of some South American moths attracted to light, human perspiration and damp sand. *Entomologist* 67: 81-84.

Common, I.F.B. & Waterhouse, D.F., 1972. Butterflies of Australia. Angus and Robertson, Sydney: 498 pp.

Common, I.F.B. & Waterhouse, D.F., 1981. Butterflies of Australia (Angus and Robertson: Sydney).

Cottrell, C.B., 1984. Aphytophagy in butterflies: its relationship to myrmecophily. *Zool. J. Linn. Soc.* 80: 1-57.

Damman, H. & Feeny, P., 1988. Mechanisms and consequences of selective oviposition by the zebra swallowtail butterfly. *Anim. Behav.*, 36: 563- 573.

David, W.A.L. & Gardiner, B.O.C. 1961. The mating behaviour of *Pieris brassicae* in a Laboratory Culture. *Bull. Ent. Res.*, 52: 263-280.

Deither, V.C., 1940. Life histories of Cuban Lepidoptera. *Psyche* 47: 14-26.

Deither, V.G., 1941. Metamorphosis of Nymphalidae and Lycaenidae. *Psyche* 48: 70-78.

Deither, V.G. 1959. Egg-laying habits of Lepidoptera in relation to available food. *Can. Ent.,* 91: 554-561.

DeVries, P.J. 1977. *Eumaeus minyas,* an aposematic lycaenid butterfly. *Brenesia, 12: 269-270.*

DeVries, P.J. 1991. Call production by myrmecophilous Riodnid and Lycaenid butterfly caterpillars (Lepidoptera): Morphological, Acoustical, functional and evolutionary patterns. *American Museum Novitats,* 3025: 1-23.

Douglas, M.M. 1978. The behavioural and biophysical strategies of thermo-regulation in temperature butterflies. Dissertation, University of Kanas, Lawrence, Kanas.

Downes, J.A. 1973. Lepidoptera feeding at puddle margins, dungs and carrion. *J. Lepid. Soc.,* 27: 89-99.

Downey, J.C. and Allyn, A.C. 1979. Morphology and biology of the immature stages of *Leptotes cassius theonus* (Lucas) (Lepidoptera.: Lycaenidae). *Bull. Allyn Mus,* 55: 1-27.

Efgar, M.A. & Pierce, N.E. 1988. Mating success and fecundity in an attended Lycaenid butterfly. Insects 59-75. From Clutton-Brock TH; (ed.) Reproductive success: studies of selection and adaptation in contrasting breeding systems. The Univ. of Chicago Press, Chicago, pp. 59-75.

Eliot, J.N. 1980. New information on the butterflies of Malay Peninsula. *Malay. Nat. J.,* 33: 137-137.

Eliot, J.N. 1992. The butterflies of the Malaya Peninsula by A. Steven Corbet and Fl..M. Pendlebury. Fourth ed. *Malay. Nat. Soc.,* viii + pp. 1-595, pls. 1-69.

Evans, W.H. 1932. The identification of Indian butterflies (2nd ed., Revised, Madras). *Bombay nat. Hist. Soc., x* + 1 - 454, Pls. 1-2 figs. 1-9.

Feeny, P., Stadler, E., Ahman, I. & Cartr, M., 1983. Effects of plant odor on oviposition by the black swallowtail butterfly *Papilio polyxenes* (Lepidoptera: Papilionidae). *J. Insect. Behav.,* 2: 803-27.

Feltwell, J., 1982. Large white butterfly-the biology, biochemistry and physiology of *Pieris brassicae* (Linnaeus) (Pieridae). W. Junk Publishers, Hague-Baston-London, 535 pp.

Fiedler, K., 1988a. Die Beziehungen von Blaulings puppen (Lepidoptera: Lycaenidae) Zu Ameisen (Hymenoptera: Formicidae). Nchr. Ento. Ver. Apollo. Frankfurt, 9(1): 33-58.

Fiedler, K. 1988b. The preimaginal epidermal organ of *Lycaena tityrus* (Poda, 1761) and *Polymmatus coridon)* (Poda, 1761) (Lepidoptera: Lycaenidae) a comparison. *Nota lepid.,* 11(2): 100-116.

Fiedler, K. 1991. Systematic evolutionary and ecological implication of myrmecophily with in the Lycaenidae (Insecta: Lepidoptera: Papilionoidea). B. Zoologischo Forschungsinstith and Museum Alexander Kaenig Bom. 1.21 Opp.

Fiedler, K. 1992a. Notes on the biology of *Hypolycaena othona* (Lepidoptera: Lycaenidae) in West Malaysia. *Nachr. Ent. Ver. Apollo. Frankfurt.,* 13(2): 65-92.

Fiedler, K., 1992b. The life history of *Surendra florimel* Doherry; 1889 (Lepidoptera: Lycaenidae) in West Malaysia. *Nachr. Ent. Ver. Apollo. Frankfurt.,* 13(2): 107-135.

Fiedler, K. & Maschwitz, U., 1989a. The symbioses between the weaver ant, *Oecophylla smaragdina* and *Anthene emolus* and obligate myrmecophilous lycaenid butterfly. *J. nat. Hist.,* 23: 833-846.

Fiedler, K. & Maschwitz, U., 1989b. Functional analysis of myrmecophilous relationships between ants Hymenoptera: Formicidae) and Lycaenids (Lepidoptera: Lycaenidae). I. Release of food recruitment in ants by lycaenid larvae and pupae *Ethology,* 80: 71-80.

Fiedler, K. & Maschwitz, U., 1989c. Adult myrmecophily in butterflies the role of the ant *Anoplolepis longipes* in the feeding and oviposition behaviour of *Allotinus unicolour* (Lepidoptera: Lycaenidae). *Tyo to Ga.,* 40(4): 241-251.

Fiedler, K. & Schurian, K.G., 1994. Oviposition behaviour in *Lycaena theitis* Klug. (Lepidoptera: Lycaenidae). *Nota Lepid.,* 17(1/2): 25-29.

Fiedler, K., Schurian, K.G. & Hahn, M. 1994. The life history and myrmecophily of *Polyommatus candalus* (Herrich-Schaffer) from Turkeyfd (Lepidoptera: Lycaenidae). *Linneana Belgica.,* 14(6;): 315-322.

Fletcher, T.B. 1925a. Catalogue of Indian insects. 7. Lasiocampidae. Manager of Publications, Delhi.

Fletcher, T.B. 1925b. Catalogue of Indian insects. 8. Amatidae (=Syntomidae). Manager of Publications, Delhi.

Fletcher, T.B. 1925c. Catalogue of Indian Insect. 9. Zygaenidae, Manager of Publications, Delhi.

Fletcher, T.B. 1926. Catalogue of Indian insects. 16. Cosmopterygidae. Manager of Publications, Delhi.

Fletcher, T.B., 1928. Catalogue of Indian insects. 17. Yponomeutidae. Manager of Publications, Delhi.

Fletcher, T.B. 1931. Catalogue of Indian insects. 20. Alucitidae (= Pterophoridae). Manager of Publications, Delhi.

Fox, R.M. 1967. A monograph of the Ithomidae (Lepidoptera). Part III. The tribe Mechanitini. *Fox. Mem. Amer. Ent. Soc.*, 22: 1-190

Ganguli, R.N. 1961. Biology and control of *Eurema blanda silhetana* Wallaace, a pest of shade trees in tea gardens. *Ind, J. Ent.*, 23: 195-201.

Gibbs, G.W. 1970. The life history of the black mountain ringlet butterflly *Percnodaimen pluto* Fereday. *N.Z. Entomologist*, 4 (4): 12-18.

Gilbert, L.E. 19675. Ecological consequences of coevolved mutualism between butterflies and plants. In coevolution of animals and plants. Ed. L.E. Gilbert and Raven, P.H., pp. 210-240. Austin. TX: Univ. Texas Press.

Greenfield M.D. 1982. The question of paternal investment in Lepidoptera: male-contributed proteins in Plodia interpunctella. *Intemational Joumal of Invertebrate Reproduction* 5: 323-330.

Grossmuller, D. W. & Lederhouse, R.C. 1985. Oviposition site selection: an aid to rapid growth and development in the tiger swallowtail butterfly *Papilio glaucus. Oecologia*, 66: 68- 73.

Hampson, G.F. 1893. Fauna of British India. Moths, 1. Taylor & Francis Ltd., London.

Hampson, G.F. 1894. Fauna of British India. Moths, 2. Taylor & Francis Ltd., London.

Hampson, G.F. 1895. Fauna of British India. Moths, 3. Taylor & Francis Ltd., London.

Hampson, G.F. 1896. Fauna of British India. Moths, 4. Taylor &I Francis Ltd., London.

Haribal, M. 1990. Behaviour of the Indian tortoiseshell, butterfly *Aglais (Vanessa) cashmiriensis* (Kollar) in the Himalaya. *J. Bombay nat. Hist. Soc.*, 87 (1): 163-164.

Haribal, M. 1992. The butterflies of Sikkim Himalalya and their natural history. Published by Sikkim nat. Conser. Fund, pp 217.

Harrison, J.O. 1963. On the biology of three banana pests in Costa Rica. *Ann. Ent. Soc., Amer.*, 56: 87-94.

Hauser, C.L., Naumann, C.M. & Tremewan, W. G., 1984. On the biology of *Pamassius charllonius* Grey, 1852 (Lepidoptera: Papilionidaae). *Entomologist's Gazette*, 36: 5-13.

Heath, J.E. & Heath, M.S., 1982. Energetics of locomotion in endothermic insects. *Annual Review of Physiology,* 44: 133-143.

Heinrich, B., 1974. Thermoregulation in endothermic insects. *Science,* 185: 747-756.

Henning, S.F., 1983. Chemical communication between lycaenid larae (Lepidoptera: Lycaenidae) and ants (Hymenoptera: Formicidaae). *J. Ent. Soc. Sth. Afr.,* 46: 341-366.

Hiebert, S.M. & Calder, W.A., 1983. Sodium, potassium and chloride in floral nectars: energy free contribution to refractive index and salt balance. *Ecol.,* 64: 399-402.

Hill, C.J. & Pierce, N.E., 1989. The effect of adult diet on the bilology of butterflies. 1. The common imperial bule, *Jalmenus evagoras. Ecologia,* 81: 249-257.

Hinton, H.E. 1951. Myrmecophilous lycaenidae and othr lepidoptera - a summary. *Proc. Trans. S. Lond. Ent. Nat. Hist. Soc.,* 1949-50: 111-175.

Horvitz, C.C., Tumbull, C. & Harvey, D.J., 1987. Biology of immature *Euryvia elvina* (Lepidoptera: Riodinidae) a myrmecophilous matelmark butterfly. *Ann. Ent. Soc. Am.,* 80 (4): 513-519.

Jana, A.K. & Ghosh, M.R., 1993. Record of rice horned caterpillar *Melanitis leda* Linn. in West Bengal. *Agric. Sci. Direst.,* 13 (2): 59-62.

Jana, A.K. & Ghosh, M.R., 1994. Biology of horned caterpillar *Melanitis leda* Linn. on rice in West Bengal. *Crop. Res.,* 7 (2): 273-275.

Jandu, A.S., 1942. Biological notes on the butterflies of Delhi. *Ind. J. Ent.,* 4. (2): 201-205.

Kingsolver, J.G., 1985. Thermoregulatory significance of wing melanization in *Pieris* butterflies: Physics, posture and pattern. *Oecologia* (Berlin), 66: 546-553.

Kirton, L.G. Wah, T. & Kirton, C.G., 1982. The life histories of *Euploea crameri bremeri* and *Idea hypermnestra linteata* (Lepidoptera: Danaidae). *Malay. Nat. J.,* 36: 29-43.

Kitching, R.L., 1987. Aspects of the natural history of the lycaenid buterfly *Allotinus major* in Sulawesi. *J. Nat. Hist.,* 21 : 535-544.

Kitching R.L. & Luke, D., 1985. The myrmecophlious organs of the larvae of some British Lycaaenidae (Lepidoptera): a comparastive study. J. Nat. Hist., 19: 259 -276.

Krizek, G.O. & Oplar, P.A., 1989. Water pumping in *Lamproptera meges* (Papilionidae). *J. Res. Lepid.,* 217: 13.

Launer, A.E. Hurphy, D.D. Boggs, C. L. Baughnan, J. F., Weiss, S.B. & Ehrlich, P. R., 1993. Puddling behaviour by Bay checkerspot butterflies (*Euphydryas editha bayensis*). *J. Res. Lepid.,* 32: 45-52.

Mackay, D.A. & Jones, R.E., 1989. Leaf shape and the host finding behaviour of two-oviposition monophagous butterfly species. *Ecol.Ent.,* 14 (4): 423-431.

Malicky, H., 1969. Ubersicht uber praimaginalstadien, Bionomie and Okologie der mitteleuropaeschen Lycaeniden. *Mitt. Ent., Ges. Basel,* 19: 25-91.

Mani, M.S., 1986. Butterflies of the Himalaya. Oxford & IBH Publishilng Co. New Delhli, Bombay, Calcuta. pp. 181.

Masters, A.R. Malcolm, S. B. & Brower, L.P., 1988 Monarch butterfly (*Danaus plexippus*) thermoregulatory behaviour and adaptations for overwintering in Mexico. *Ecology*, 69: 458-467.

Mathews, R. W. & Mathews, J.R. 1978. Insect Behaviour: Wiley-inter science publication, N. York, 1-507 pp.

Matsumoto, K., 1990. Population dynamics of *Luchorfia japonica* Leech. (Lepidoptera: Papilionidae). II. Patterns of mortality in immatures in relation to egg cluster size. *Res. Popul. Ecol.*, 32 (1) 173-188.

Mavi, G. S., 1992. A critical review on the distribution and host-range of Pea Blue butterfly, *Lampides boeticus* (Linn.) *J. Insect. Sci.*, 5(2): 115-119.

May, M.L., 1979. Insect thermoregulation. *Annual Review of Entomology*, 24: 313-349.

Mosher, E., 1916. A classification of the Lepidoptera based on the characters of the pupa. *Bull Ill. St. Lab. Nat. Hist.*, 12: 17-159.

Mukerji, G.P. 1960. On the biology of 'Cabbage White', *Pieris brassicae. J. Zool. Soc. Ind.*, 13: 121-127.

Muyshondt, A., Young, A.M. & Muyshondt, A.R. 1973. The biology of the butterfly *Dione juno huascama (Nymphalidae: Helioconiinae) in EI Salvador. J.N.Y. Ent. Soc.*, 81 (3) : 137-151.

New, T.R., 1987. Insect conservation. By Dr. Junk W., Kluwer Academic Publishers group, Dordrecht: 184pp.

New, T.R., 1991. Butterfly conservation. Oxford Univ. Press, Australia.pp.224.

New, T.R., 1993. Conservation. By Dr. Junk WE., Kluwer Academic Publishers Group, Dordrecht: 184 pp.

Oplar, P.A., 1974. Studies on Nearctic *Euchloe*. Part 7. Comparative life histories, hosts and the morphology of unmature stages. *J. Res. Lepid.*, 13(1): 1-20.

Otero, L.D. & Aiello, A., 1996. Descriptions of the immature stages of *Adelpha alala* (Nymphalidae). *J. Lepid. Soc.*, 50(4): 329-36.

Pant, G.D. & Chatterjee, N.C., 1950. A list of described immature stages of Indian Lepidoptera. Part I. Rhopalocera. *Ind. For. Rec.*, 7(7): 213-255.

Papaj, D.R. & Prabopy, R.J., 1989. Ecological and evolutionary aspects of learning in phytophagous insects. *Ann. Rev. Ent.*, 4: 315-350.

Peile, H.D. 1937. A guide to colleccting butterflies in India. Staples Press, London: 523 pp. 72

Pierece, N.E. 1984. Amplified species diversity: A case study of an Australian lycaenid butterfly and its attendant ants from the biology of butterflies. Academic Press, London: 197-200.

Pierce, N.E. 1985. Lycaenid butterflies and ants: Sėlection for nitrogen fixing and other protein-rich food plants. *Am. Nat.*, 125: 889-895.

Pierce, N.E. 1987. The evolution and biogeography of associations between lycaenid butterflies and ants from Oxford surveys in evolutionary biology. Oxford University Press, 4: 89-116

Pierece, N.E. 1989. Butterflies, ant Matualisms. Book towards a more exact ecology. Blackwell Scientific Publications, Boston, Melbourne: 299-324.

Pierce, N.E. 1986. The selctive advantage of attendant ants for the larvae of a Lycaenid butterfly, *Glaucopsycne lygdomus. J. Anim. Ecol.*, 55: 451-462.

Pierce, N.E & Mead, P.S., 1981. Parasitoids as selective agents in the symbiosis between lycaenid butterfly larvae and ants. *Science,* 211: 1185-1187.

Pierce, N.E. & Young, W.R. 1986. Lycaenid butterflies and ants: two species stable equilibria in mutualistic, commensal and parsitic interactions. *Am. Nat.* 128(2): 216-227.

Pierce, N.E., Kitching, R.L., Buckley, R.C. Taylor, M.F. & Benbow, K.F. 1987. The costs and benefits of cooperation between the Australian Lycaenid butterfly *Jalmenus evagoras* and its attendants ants. *Behr. Ecol. Sociobiol.*, 21: 237-248.

Peterson, M.A., 1993. The nature of ant attendance and the survival of larvae *Icaricia acmon* (Lycaenidea). *J. Lepid. Soc.,* 47(1) : 8-16.

Pivnick, K.A. & McNeil, J.N., 1987. Puddling in butterflies. Sodium affects reproductive success in *Thymelicus lineola. Ecol. Ent.,* 12: 461-472.

Pyle, R.M., 1978. IUCN and Insect conservation. *Antenna,* 2(2): 36.

Radke, S.G. & Kandalkar, H.S.G., 1989. Observation on the biology of lemon butterfly, *Papilio demolus* Linnaeus (Lepidoptera: Papilionidae). *P.K.V. Res. J.*, 13 (2) 176-177.

Rainbow, W.J., 1907. A Guide to the study of Australian Buterflies. 272 pp. T. Lothian, T.C. Melbourne.

Rataul, H. S., 1960 . Studies on biology of *Pieris brassicae* Linnaeus (Pieridae). *Ind. J. Hort.*, 16: 255-265.

Rathcke, B. J. & Poole. R. W., 1975. Coevolutionary race continues: Butterfly larval adaptation to plant trichomes. *Science,* 187: 175- 176.

Rausher, M. D., 1978. Search image for leaf shape in butterfly. *Science*, 200: 1071- 1073.

Rausher, M. D. 1981. Host plant selection by *Battus philenor* butterflies: the roles of predation, nutrition and plant chemistry ecology. *Monogr.,* 51: 1- 20.

Rawlins, J. E., 1980. Thermoregulation by the black swallowtail butterfly, *Papilio polyxenes* (Lepidoptera: Papilionidae). *Ecol.*, 61. 345- 357.

Reddy, T.B. & Reddi, C.S., 1996. Pollination ecology of *Duranta repens.* (Verbenaceae). *J. Bombay nat. Hist. Soc.*, 93: 193- 201.

Renwick, J.A.A. & Chew's, F.S., 1994. Oviposition behaviour in lepidoptera. Ann. Rev. Ent. 39:377-400.

Rose, H.S: & Sidhu, A. K., 1994. Llycaenid butterfly diversity (Lepidoptera) of Mussoorie (Uttar Pradesh, now Uttranchal) alongwith comments on Wild life (Protection) Act, 1972. *Cheetal,* 33 (3-4): 10-16.

Rose, H.S. & Sidhu, A. K., 1995. Indian butterfly diversity versus Wildlife (Protection) Act, 1972. *Proc. XI Ann. Conf. NESA* pp. 37-44.

Rutkowski, F., 1971. Observations on *Papilio aristodemus ponceanus* (Papilionidae). *J. Lepid. Soc.*, 25 (2): 126-136.

Rutowaski, R. L., 1978a. The courtship behaviour of the small Sulphur butterjly, *Eurema lisa*. Anim. Behav., 26: 892- 903.

Rutkowski, R.L. 1978b. The form and function of ascending flights in *Colias* buterflies. *Behav. Ecol. Sociobiol.*, 3: 163-172.

Schurian, K. G. & Fiedler, K. 1996. Adult behaviour and early stages of *Lycaaena ochimus* (Herrich-Schaffer, 1851) (Lepidoptera: Lycaenidae). *Nachr. Ent. Ver. Apollo.*, 16 (4): 329- 343.

Scoble, M. J., 1995. The Lepidoptera form , function and diversity. *Nat. Hist. Mus. Associaation with Oxford Uni. Press*: 404 pp.

Sevastopulo, D. G., 1933. The early stages of Indian Lepidoptera. *Entomologist's record*: 148-149.

Sevastopulo, D.G. 1935. The early stages of Indian Lepidoptera. *Entomologist's record*: 18-19.

Sevastopulo, D.G. 1940. The early stages of Indian Lepidoptera. *J. Bombay nat. Hist. Soc.*, 40: 391-408.

Sevastopulo, D.G. 1941. The easrly stages of Indian Lepidoptera. *J. Bombay nat. Hist. Soc.*, 41: 311- 320.

Sevastopulo, D.G. 1946. The early stages of Indian Lepidoptera. *J. Bombay nat. Hist. Soc.*, 46: 253-269.

Sevastopulo, D.G. 1959. Attraction o Lepidoptera to moisture. *The Entomologist*, 92: 108- 109.

Sevastopulo, D.G. 1973. The food plants of Indian Rhopalocera. *KJ. Bombay nat. Hist. Soc.*, 70 (1): 156-163.

Sevastopulo, D.G. 1975. A. List of food plants of East African macrolepidoptera. I. Butterflies. *Bull. Amat. Ent. Soc.*, 34: 84s-92.

Shapiro, A.M. 1991. The biology and zoogeography of the legume-feeding *patagonian* Fuegian white butterfly *Tatochela theodice* (Lepidoptera: Pieridae). *J.N.Y. Ent. Soc.*, 99: (2): 251-260.

Silberglied, R., Aiello, & Lamas, G. 1979. Neotropical butterflies of the genus *Anartia*: Systematics, life histories and general biology. (Lepidoptera: Nymphalidae). *Psyche*, 86 (2-3): 219-260.

Silberglied, R.E. 1977. Communication in the Lepidoptera, pp. 362-402, in How Animals communicate, T. A. Sebeok, ed. Indiana Univ. Press, Bloomington.

Sillen-Tullberg, B. 1988. Evolution of gregariousness in aposematic butterfly larvae. A phylogenetic analysis. *Evolution*, 42 (2): 293-305.

Singer, M. C., 1984. Butterfly-host plant relationship: host quality, adult choice and larval success. In: Biology of butterflies: Vane-Wright, R. I. and Ackery, P.R. (eds.) London, Academic Press, London: 81-88 pp.

Singh J. & Mavi, G.S. 1984. A spider as predator of *Lampides boeticus* Linnaeus (Lepidoptera : Lycaenidae) from Punjab. *J. Bombay nat. Hist. Soc.*, 81 (2) 501.

Smart, P. 1985. The illustrated encyclopedia of the butterfly world. Salamander Books Ltd., England. 1-275 pp.

Smetacek, P. 1997. *Pieris brassicae* Linnaeus (Lepidoptera: Pieridaae) in Delhi. *J. Bomb. Nat. Hist. Soc.*, 94 (3): 584-585.

Smiley, J.T. 1978. Plant chemistry and the evolution of host specificity: new evidence from *Heliconius* and *Passiflora. Science*, 201: 745-747.

Soberon, J. Cordero, C.R. Benrey, B. Parlange P., Garcia-Saez. C. & Berges, G. 1988. Patterns of oviposition by *Sandia xami* (Lepidoptera, Lycaenidae) in relation to food plant apparency. *Ecol. Ent.*, 13: 71-769.

Stamp, M. 1977. Aggregaton behaviour of *Chlosyn lacinia* larvae (Nymphalidae). *J. Lep. Soc.*, 31: 35-40.

Stamp, N.E. 1980. Egg deposition patterns in butterflies. Why do some species cluster their eggs rather than deposit them singly? *Amer. Natur.*, 115: 367-380.

Staton, M.L., 1982. Searching in a patchy environment; foodplant selection by *Colias eriphyie* butterflies. *Ecol.*, 63: 863- 853.

Stehr, F.W., 1987. Immature insects. Kendall/Hunt Publishing Company, Dubuque, lowa; Xlll +754 pp.

Stride, G.O., 1958. On the courtship behaviour of a tropical mimetic butterfly, *Nypolimnas misippus. Int. Congr. Ent.*, 2 (10): 419-424.

Talbot, G. 1939. The Fauna of British India, including Ceylon and Burma. Butterfliles. I. Today and Tomorrow, Delhi, 600 pp.

Talbot, G., 1947. The Fauna of British India, including Ceylon and Burma. Butterflies. II. Taylor and Francis, London, 506 pp.

Tennet, W.J., 1995. Field observations of the "Hill topping" phenomenon in North West Africa and an introduction to "Ravining " (Lepidoptera; Rhopalocera). *Ent. Rec.*, 106; 57-67.

Thompson, J.N. & Pellmyer, O. 1991. Evoluton of oviposition behaviour and host preference in Lepidoptera. *Ann. Rev. Ent.*,36: 65-89.

Thompson J.N. Wehling, W. & Podolsky, R. 1990, Evolutionary genetics of host use in swallowtail butterflies. *Nature*, 344: 158-160.

Tuskes, P.M. & Brower, L.P. 1978. Overwintering ecology of the Monarch butterfly. *Danaus pexippus* L. in California. *Econ. Entomol.*, 3:141-153.

Tyler, H.A. Brown, K.S. Jr. & Witson, K.H., 1994. Swallowtail butterflies of the Americas. A study in biological dynamics, ecological diversity, biosystermatics and conservation. Scientific Pub., Gainesville, 376 pp.

Urich, F.C. & Emmel, T.C. 1990. Life histories of Neotropical buterflies from Trinidad. 2. *Antirhaea philoctetes* (Lepidoptera: Nymphallidae: Morphinae). *Trop. Lepid.*, 1 (1) : 27- 32f.

Urich, F.C. & Emmel, T.C. 1991a. Life histores of Neotropical butterfliles from Trinidad. 3. *Morphopeleides insularis* (Lepidoptera: Nymphalidae: Morphinae). *Trop. Lepid.* 2 (2): 137-139.

Urich, F.C. & Emmlel, T.C. 1991b. Life histories of Neotropical butterfliles from Trinidad. 4. *Dynastor macrasiris* (Lepidoptera: Nymphalidae: Brassolinae). *Trop. Lepid.*, 141-144.

INTRODUCTION

The Indian Thar Desert especially Barmer, Jaisalmer and Jodhpur districts of Rajasthan provide extreme xeric conditions to the insect fauna with 100-450 mm mean annual rainfall during July to September. But, during the past few years when the Indira Gandhi Nahar (Rajasthan canal) came into existence, in addition to the extant Gang canal (1928) and Bhakra-Sirhind feeder canal (1953), irrigating large areas of Sri Ganganagar, Bikaner and Jaisalmer districts in the Thar Desert of Rajasthan, it also invariably affected a great deal the biodiversity of various faunal forms, both large and small belonging to vertebrates and invertebrates, of the hot sandy desert. As there are lots of marshy water bodies built up intermittently along the embankments of the canal due mainly to seepage, these inadvertently provided ample opportunities for a variety of insects to profusely breed and propagate, such as the mosquitoes and other blood suckng insects.

As a consequence of these developmental and other causative changes supporting prodigious density and preponderance of vector anopheline mosquitoes, there occurred an abrupt increase in malaria cases both of *Plasmodium falciparum* and *P. vivax*, particularly the former, in this region (Tyagi, 1996). At times there were frequent malaria epidemics. Presently about a dozen species of *Anopheles*, 3 species of *Aedes* and 7 species of *Culex* mosquitoes are listed from the Thar Desert (Tyagi, 1996; Kumar & Kumar, 1996). However, as a large area in the Thar Desert remain unexplored there is every possibility that this statistics may further be improved as more species will almost certainly be inventorized and brought on record in the changed scenario of mesic conditions.

Among other blood-sucking insects, the tabanids and muscoid flies are important insect pests as they are not only causing nuisance, loss in milk production and body weight in animals by their painful bites but also act as vectors of several viral, bacterial, protozoan and filarial diseases (Krinsky, 1976; Foil *et al.*, 1994). Among these diseases, Trypanosomiasis (popularly known as 'surra disease' or, as in Rajasthan, 'tibursa disease') is quite prevalent among camels in the state (Pathak *et al.*, 1993). This disease is common during rainy seasons and affects livestock, camels, equines, .buffaloes and wild animals like tigers, dogs and hyenas. The tabanid flies belong to the family Tabanidae and are commonly known as horse flies, gad flies or clegs, whereas muscoid flies of the family, Muscidae, are commonly known as stable flies. These flies transmit surra disease mechanically. Surra disease is caused by a flagellate protozoan parasite, *Trypanosoma evansi* (Steel). This disease is clinically characterized by remittent fever leading to emaciation, anaemia, oedema, debility and uncoordinated walking and cause weight loss in a very short period among affected animals. Though these flies are very important, we know very little about them from the Thar Desert as only *Tabanus striatus* Fab. has so far been recorded from the the region.

Ironically, there is no report of blood sucking muscoid flies from the Thar Desert (Kumar & Kumar, 1996). Studies on these haematophagous flies in the Thar Desert have therefore become more important and require immediate attention particularly in view of shift and expansion in the agriculture practices as well as increase of dairy-related activities with hybrid zebu cattle.

In the light of above knowledge a survey for mosquitoes and other haematophagous flies was conducted in certain selected areas inhabited by armed forces in three districts of Rajasthan, namely, Jodhpur, Jaisalmer and Barmer, during 2001-2002. Among mosquitoes, *Anopheles jeyporiensis* and *Aedes albopictus* are reported here for the first time from the Thar Desert, raising total number of *Anopheles* species known from the Thar Desert to 13. However, we could not collect the following species of the mosquitoes during the present study which were earlier reported from the Thar Desert - *An. barbirostris* van der Wulp, *An. nigerrimus* Giles, *An. annularis* van der Wulp, *An. culicifacies* Giles, *An. d'thali* Patton, *An. splendidus* Koidzumi, and *Aedes vittatus* Bigot, *Ae. walbus* Theobald, and *Culex pseudovishnui* Colless, *Cx. bitaeniorhynchus* Giles, *Cx. sitiens* Wiedemann, *Cx. tritaeniorhynchus* Giles, *Cx. halifax* Theobald and *Cx. malayi* (Leicester). Similarly among haematophagous flies, 3 species of tabanid flies, i.e., *Atylotus agrestis, At. dorsilinea* and *Tabanus hisarensis,* are new records from the region and thus total number of the tabanids flies is raised to four. *Stomoxys calcitrans* and *Musca crassirostris* are also new records among muscoid flies which have raised their number to four.

As it is evident mosquito and fly fauna of the Thar Desert are undoubtedly far more richer hithero claimed to be known and, therefore, it is advocated to undertake long-term and systematic collecting to throw adequate light on the biodiversity of these blood sucking pests and vectors in the region. This exercise will also be helpful in understanding a lot about the ecology of both the pests/vectors and the diseases they transmit so that suitable control measures can be designed against them.

MATERIAL & METHODS

Extensive and intensive surveys for insect vectors and pests were made by a team of scientists from the DRDE, Gwalior and SHO, Army HQrs, Jodhpur in the allotted army units during March, September and October, 2002. Insects were collected with the help of collecting tubes (mouth-operated aspirators), canopy traps (Fig.1) and light traps. Canopy traps were used to collect horse flies and other blood sucking flies in the military farms. Larval habitats were also visited by the team, and larvae and eggs were collected there. All the material collected was brought to the laboratory for identification and further studies. These insects were identified with the help of morphological markers.

This fly was collected once from a cattle farm in Jodhpur during September. It is common in the northern and central India and has been often incriminated for surra disease. *Atylotus agrestis* is reported here for the first time from this region. *Atylotus agrestis* is recently reported from the Biological Park, Bhubaneswar (Orissa) where 13 tigers including 12 white tigers had died due to surra disease (Vijay Veer, Parashar, & Shri Prakash, 2002).

Distribution: India: common and very abundantly found across the country as well as in Sri Lanka, Pakistan, China. Palaearctic and Afrotropical regions.

2. *Atylotus dorsilinea* (Wiedemann). It is a small fly measuring 9.0-11.5 mm in length. Frons is also with two distinct callosities but basal callosity is larger than median callosity. Eye in relaxed specimens with three blue-black stripes on a dark green back ground. Sublateral pale stripes of mesonotum broader and crossing the entire scutellum. Abdominal tergite I with a median pale spot beneath scutellum. Sublateral abdominal stripes are present on tergites I-V.

Atylotus dorsilinea is collected from a cattle farm in Jodhpur during September. It is the first record of this fly from this region. This species is also capable to transmit surra disease. Life history of this fly (as *macer* Bigot) was studied at Pusa (Bihar) by Isaac (1926).

Distribution: India: most common in the southern and western parts of the country. It is also known from Pakistan and Sri Lanka.

3. *Tabanus hisarensis* Vijay Veer, Parashar & Rao. It is a small fly measuring about 8.0 mm in length. Fly can be easily identified by having broad frons (2.34 times longer than broad) with transverse and narrowly separated from eye margin basal callosity. Wings are with greyish brown tint on the cross veins. Abdomen is patterned with a pair of clear cut oblique, oval sublateral white spots.

This species is collected from Ramgarh area of Jaisalmer district. Earlier it was collected in large number from its type locality, Hisar (Haryana). This species seems to be well adapted to the arid conditions of the desert and may occur throughout the Thar Desert. This species appears to be an important fly from the point of view of transmission of the surra disease. Thus, it deserves further studies on biology and transmission potential.

Distribution: India: Hisar (Haryana), Gwalior (Madhya Pradesh), Jaisalmer (Rajasthan; *New locality record*).

4. *Tabanus striatus* Fabricius (Fig. 3) *T. striatus* is a medium-sized fly, 10-14 mm long. Frons 5.0- 5.5 times longer than broad and nearly parallel. Eyes without bands. Abdomen trivittate (three striped) with dorsal median pale stripe which is absent on tergum II and lateral stripes on tergites I-III.

T. striatus was not collected during the present survey though it was reported from Jodhpur (Kumar & Kumar, 1996). It has been incriminated in the transmission of surra disease (Basu, Menon & Sen Gupta, 1952).

Fig. 1. Canopy trap (after Vijay Veer *et al.*, 2002)

Fig. 2. *Atylotus agrestis*, adult female (after Vijay Veer *et al.*, 2002)

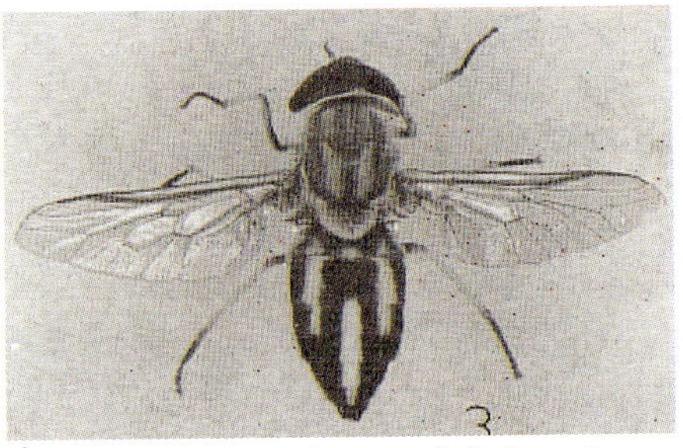

Fig. 3. *Tabanus striatus*, adult female (after Vijay Veer *et al.*, 1999)

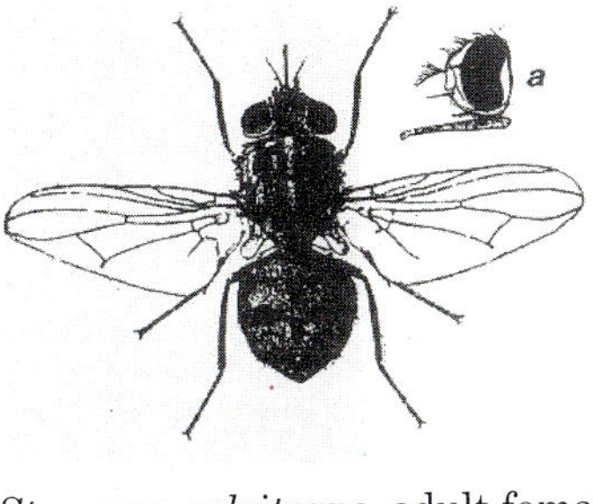

Fig. 4. *Stomoxys calcitrans,* adult female, (after Vijay Veer *et al.*, 2002)
4a- Antennal arista with rays on upper side

Distribution: India: widespread and common in the country.

5. *Stomoxys calcitrans* (Linn.) (Fig. 4). It is smaller than tabanid flies, 4-7 mm in length and looks like the common house fly, *Musca domestica,* to the untrained eyes. It is grey in colour with four black stripes on thorax and black-checkered broad markings on abdomen. Antennal arista with rays on upper side only as shown in Fig. 4a. Proboscis is projecting sword-like beneath the head.

It was collected from a cattle farm in Jodhpur during September. It is the most widely distributed and almost cosmopolitan pest of livestock. It causes debility in the host animals by extensive bloodfeeding and annoyance. It is an important mechanical vector of surra disease in the absence of tabanid flies and also of equine infectious anemia (EIA) virus.

Distribution: India: common throughout the country.

6. *Musca crassirostris* Stein. Adult *M. crassirostris* is 4.5 – 6.5 mm long with ash-grey body colour and tout proboscis with prestomal teeth which are used for scratching the sore skin so that blood flows more freely. It was collected in sufficient numbers from a cattle farm in Jodhpur during September. *Musca crassirostris* is reported here for the first time from this region.

Distribution: India: common throughout the country.

ACKNOWLEDGEMENT

Authors are highly grateful to Shri K. Sekhar, Director, Defence Research and Development Establishment, Gwalior and the concerned Army authorities for their constant support and encouragement during the present study.

REFERENCES

Basu, B.C., Menon, P.B. & Sen Gupta, C.M., 1952. Regional distribution of *Tabanus* flies in India and its relationship to the incidence of Surra. *Indian J. Vet. Sci. Anim.. Husb.* 22: 273-292.

Foil, L.D. & Hogsette, J.A., 1994. Biology and control of tabanids, stable flies and horn flies. *Rev. Sci. Tech. Off. Int. Epiz.* 13 (4): 1125-1158.

Isaac, P.V., 1926. The bionomics and life-histories of some of the common Tabanidae of Pusa. *Mem Dept. Agric. India (Entomol. Ser.* 9(2); 21-28.

Krinsky, W.L., 1976. Animal disease agents transmitted by horse flies and deer flies (Diptera: Tabanidae). *J. Medical Entomol.* 13(3): 225-275.

Pathak, K.M. Arora, J.K. & Kapoor, M., 1993. Camel trypanosomosis in Rajasthan. *Indian Vet. Parasitology* 49: 319-323.

Kumar, S. & Kumar, S., 1996. Diptera fauna of the Thar Desert: 241-252. In : Faunal diversity in the Thar Desert: Gaps in research (Eds. Ghosh, A.K. Baqri, Q.H. & Prakash, I.). Scientific Publishers (India), Jodhpur.

Vijay Veer, Parashar, B.D & Rao, K.M., 1999. Notes on Tabanidae (Diptera) that are surra vectors or pestiferous with description of a new species of *Tabanus* from India. *Oriental Insects* 33: 247-266.

Vijay Veer, B.D. Parashar & Shri Prakash, 2002. Tabanid and muscoid haematophagous flies, vectors of trypanosomiasis or surra disease in wild animals and livestock in Nandankanan Biological Park, Bhubaneswar (Orissa, India). *Current Science* 82 (5): 500-503.

Tyagi, B.K., 1996. Chronological and spatial distribution of Anopheline vectors of malaria in the Thar desert: 227-239. In: Faunal diversity in the Thar Desert: Gaps in research (Eds. Ghosh, A.K. Baqri, Q.H. & Prakash, I.). Scientific Publishers (India), Jodhpur.

Tyagi, B.K. & Chaudhary, R.C., 1996. Geographical diversification of human Plasmodia and malaria syndrome in the Thar Desert: 25-36. In: Faunal diversity in the Thar Desert: Gaps in research (Eds.: Ghosh, A.K. Baqri, Q.H. & Prakash, I.). Scientific Publishers (India), Jodhpur.

Tyagi, B.K. & Yadav, S.P., 2001. Bionomics of malaria vectors in two physiologically different areas of the epidemic-prone Thar Desert, north-western Rajasthan (India). *J. Arid Environments* 47: 161-172.

Tyagi, B.K., 2002. *Malaria in the Thar Desert – Facts, Figures and Future.* Agrobios (India), Jodhpur, 165 pp.

Changing Faunal Ecology in Thar Desert, pp. 143-148, 2005.
Eds.: B.K. Tyagi and Q.H. Baqri

8

DUNG BEETLES (COLEOPTERA: SCARABAEIDAE; COPRINAE) OF THE THAR DESERT OF RAJASTHAN

Ram Sewak

Desert Regional Station, Zoological Survey of India, Pali Road, Jhalamund, Jodhpur-342 005, India

Dung beetles are economically important because they play a very important role in pasture ecosystem, where they annually break down tons of animal dung and much of it incorporates into the soil to increase the soil fertility and also clean the soil surface to make the environment healthy. They also destroy the habitat of many pest flies of domestic animals which lay eggs in the dung. Some of them are also the intermediate hosts for several parasites of many domestic animals. This warrants the correct identification of these interesting creatures which make an important group of fauna in the Thar Desert of Rajasthan. An identification key leading to the tribes of the subfamily is, therefore, followed by the systematic account of all 57 species. The genera recorded with their number of species in descending order includes 7 species of the genus *Scarabaeus* Linnaeus, 4 species of *Gymnopleurus* Illiger, 3 species of *Heliocorpis* Burmeister, 4 species of *Catharsius* Hope, 7 species of *Copris* Geoffroy, 2 species of *Phalops* Erichson, 5 species of *Caccobius* Thomson, 14 species of *Onthophagus* Latreille, 3 species of *Oniticellus* Serville, 9 species of *Onitis* Fabricius and 1 species of *Chironitis*

Lansberge recorded from the Thar Desert. Representatives of the genus *Scarabaeus* Linnaeus are abundantly present in Barmer, Bikaner and Jaisalmer districts. The database will be useful as a baseline work for the taxonomists and biodiversity workers in other ecosystems in the country.

INTRODUCTION

Coleoptera, which includes beetles and weevils, is the largest order of the Class Insecta. They show exceptionally diverse adaptation to wide range of environmental conditions and habitats, and are economically important as they are both destructive and beneficiary, in nature. Members of the family Scarabaeidae (subfamily: Coprinae) commonly known as "Dung beetles" or "Dung rollers" are very useful because they are scavengers and they annually breakdown tons of animal excrement (dung) and remove them from the soil surface to make the environment clean. Much of it incorporates into the soil to increase the soil fertility, while, on the other hand, they destroy the habitat of many pest flies of domestic animals which lay eggs in the dung. Some of the species are also the intermediate hosts for the parasites of domestic animals.

The present paper offers an up-to-date inventory of the extant species of dung beetles from Barmer, Bikaner, Churu, Jaisalmer, Jalore, Jhunjhunu, Jodhpur, Hanumangarh, Nagaur, Pali, Sikar, Sirohi and Sriganganagar districts of Rajasthan's Thar Desert. A concise classificatory system is provided, including characters of subfamilies and tribes. Altogether 57 species, organized under eleven genera, have been catalogued so far.

SYSTEMATIC DATABASE OF DUNG BEETLES

Class: Insecta

Order: Coleoptera

Suborder: Polyphaga

Superfamily: Scarabaeoidea

Family: Scarabaeidae

Subfamily: Coprinae

Key to the tribes of Subfamily Coprinae

1. Middle coxae not widely separated; middle tibia with one terminal spur .... Scarabaeini

1a. Middle coxae widely separated, middle tibia with two terminal spurs 2

2. Posterior legs extremely long, the tarsi filiform .... Sisyphini

2a. Basal joint of the hind tarsus much longer than the second 3

3. Posterior legs not extremely long, tarsi more or less flat and tapering .... Coprini

3a. Basal joint of the hind tarsus not much longer than the second .... Panelini

Tribe I: Scarabaeini

It consists of two genera *Scarabaeus* and *Gymnopleurus*. Eleven species have been recorded from Thar Desert.

GENUS 1: *SCARABAEUS* LINNAEUS

1. *Scarabaeus sacer* Linnaeus
2. *Scarabaeus gangatinus* Castelnau
3. *Scarabaeus brahminus* Castelnau
4. *Scarabaeus cristatus* Fabricius
5. *Scarabaeus andrewesi* Felsche
6. *Scarabaeus erichsoni* Harold
7. *Scarabaeus devotus* Redtenbucher

GENUS 2: *GYMNOPLEURUS* ILLIGER

8. *Gymnopleurus cyaneus* Fabricius
9. *Gymnopleurus miliaris* Fabricius
10. *Gymnopleurus koenigi* Fabricius
11. *Gymnopleurus bicallosus* Felsche

Tribe II: Coprini

Tribe Coprini consists of the most commonly found dung beetles which are known with an ample of morphological variation in size and form. A total of 46 species identified under eleven genera, were collected from Thar Desert

GENUS 3: *HELIOCORPIS* BURMEISTER

12. *Heliocorpis gigas* Linnaeus
13. *Heliocorpis tyrannus* Thomson
14. *Heliocorpis dominus* Bates

GENUS 4: *CATHARSIUS* HOPE

15. Catharsius platypus Sharp
16. *Catharsius molossus* Linnaeus
17. *Catharsius pithecius* Fabricius
18. *Catharsius inermis* Castelnau

GENUS 5: *COPRIS* GEOFFROY

19. *Copris indicus* Gillet

20. *Copris repertus* Walker
21. *Copris delicates* Arrow
22. *Copris corpulentus* Gillet
23. *Copris numa* Lansberge
24. *Copris cribratus* Gillet
25. *Copris furciceps* Felsche

GENUS 6: *PHALOPS* ERICHSON

26. *Phalops divisus* Wiedmann
27. *Phalops olivaceus* Lansberge

GENUS 7: *CACCOBIUS* THOMSON

28. *Caccobius torticornis* Arrow
29. *Caccobius unicornis* Fabricius
30. *Caccobius himalayanus* Jek.
31. *Caccobius pantherinus* Arrow
32. *Caccobius denticollis* Harold

GENUS 8: *ONTHOPHAGUS* LATREILLE

33. *Onthophagus fuscopunctatus* Fabricius
34. *Onthophagus catta* Fabricius
35. *Onthophagus bonasus* Fabricius
36. *Onthophagus seniculus* Fabricius
37. *Onthophagus ephippioderus* Arrow
38. *Onthophagus kuluensis* Bates
39. *Onthophagus armatus* Blanchard
40. *Onthophagus semicinctus* D'orbigny
41. *Onthophagus hamaticeps* Arrow
42. *Onthophagus tragus* Fabricius
43. *Onthophagus ensifer* Boucomont
44. *Onthophagus frugivorus* Arrow

GENUS 9: *ONITICELLUS* SERVILLE

45. *Oniticellus pallipes* Fabricius
46. *Oniticellus spinipes* Roth
47. *Oniticellus cinctus* Fabricius

GENUS 10: *ONITIS* FABRICIUS

48. *Onitis siva* Gillet
49. *Onitis lama* Lansberge
50. *Onitis falcatus* Wulfen
51. *Onitis philemon* Fabricius
52. *Onitis singhalensis* Lansberge
53. *Onitis subopacus* Arrow
54. *Onitis virens* Lansberge
55. *Onitis castaneus* Redtenbucher
56. *Onitis brahma* Lansberge

GENUS 11: *CHIRONITIS* REDTENBUCHER

57. *Chironitis indicus* Lansberge

SUMMARY

Altogether fifty seven species belonging to eleven genera have been recorded from the Thar Desert of Rajasthan. These were collected from the dung of cow, buffalo, camel, horse, donkey, goat, sheep, deer and human feacal matter. The representatives of the genus *Scarabaeus* Linnaeus and *Gymnoplerus* Iliger, or ball rollers and the representatives of the genus *Heliocopris* Burmeister, *Catharsius* Hope, *Copris* Geoffroy, *Phalops* Erichson, *Caccobius* Thomson, *Onthophagus* Latreille, *Oniticellus* Servile, *Onitis* Fabricius and *Chironits* Lansberge were found under the dung burrows. The species of the genus *Gymnoplerus* Illger are working in pairs while the species of the genus *Scarabaeus* Linnaeus are mostly working singly to pushing the ball. Males are generaly distinguishable from females by presence of horns on the head and prothorax in the male while cephalic and prothoracic protuberances characterize the female. The species of the genus *Scarabaeus* Linnaeus are abundantly present in Barmer, Bikaner and Jaisalmer districts of Thar.

ACKNOWLEDGEMENT

The author is grateful to Dr. J.R.B. Alfred, Director, Zoological Survey of India, Kolkata for encouragement to undertake this study. Thanks are also due to Dr. Q.H Baqri, Ex. Additional Director, Desert Regional Station, Zoological Survey of India, Jodhpur for his valuable suggestions and to Dr. N.S. Rathore, Joint Director, Desert Regional Station, Zoological Survey of India, Jodhpur for providing necessary facilities to carryout the work.

REFERENCES

Arrow, G. J., 1931. Fauna of British India including Ceylon and Burma (Coleoptera: Lamellicornia: Coprinae). Taylor and Francis, London.

Balthasar, V., 1963. *Monographie der Scarabaeidae und Aphodiidae der Palaearktischen und Orientalischen Region* (Coleoptera: Lamellicornia), Verlag der Tscheshoslowakischen Akademic der Wissenschften, Prague; 1: 1-39; 2 and 1-627.

Biswas, S., 1978. Studies on the Scrab beetles (Coleoptera: Scarabaeidae) of North-East India: A new species and notes on other Indian species of subgenus Strandius, genus *Onthophagus*. *J. Bombay nat. Hist. Soc.* 75(3): 911-913.

Biswas, S., 1978. Studies on the Scrab beetles (Coleoptera: Scarabaeidae) of North-East India. Part II: Three new species and two new record from India, *J. Bombay nat. Hist. Soc.* 76: 339-344.

Biswas, S. and Chatterjee, S.K., 1985. Insecta: Coleoptera: Scarabaeidae: Coprinae. *Rec. Zool. Surv., India,* 82 (1-4): 147-177.

Gorden, R.D. and Oppenheimer, J.R., 1975. Taxonomy and Ecology of two species of Indian *Onthophagus* (Coleoptera: Scarabaeidae). *Oriental Insects* 9(4): 495-501.

Sewak, R., 1985. On a collection of Dung beetles (Coleoptera: Scarabaeidae: Coprinae) from Gujarat, India. *Oikassay* 2(2): 33-35.

Sewak, R., 1986. On a collection of Dung beetles (Coleoptera: Scarabaeidae: Coprinae) from Rajasthan, India. *Oikassay* 3(1): 11-15.

Sewak, R., 1991. Dung beetles (Coleoptera: Scarabaeidae: Coprinae) from five districts of western Uttar Pradesh. *Oikassay* 8(1&2): 25-27.

Sewak, R., 2004. Dung beetles (Coleoptera: Scarabaeidae: Coprinae) of India with special reference to Arunanchl Pradesh, Uttar Pradesh, Gujarat and Rajasthan. (In: *Advancements in Insect Biodiversity.* Ed. Rajiv K. Gupta), Agrobios, Jodhpur, 249-288 pp.

Sewak, R. (In press). Dung beetles (Insecta: Coleoptera: Scarabaeidae: Coprinae). Rec. Zool. Surv. India, State Fauna Series 8: Fauna of Gujarat, Part II.

Changing Faunal Ecology in Thar Desert, pp. 149-171, 2005.
Eds.: B.K. Tyagi and Q.H. Baqri

9

CHANGING ECOLOGY AND FAUNAL DIVERSITY OF AMPHIBIANS AND REPTILES IN THE THAR DESERT OF RAJASTHAN, INDIA

R.C. Sharma and Shalini Gaur

Desert Regional Station, Zoological Survey of India, Jhalamand Chauraha, New Pali Road, Jodhpur -342 005 (Rajasthan), India

Present studies are based on the observations recorded at flat, gravel, sandy and rocky habitats in Jaisalmer, Nagaur, Pali, Jodhpur Jalor, Sikar, Jhunjhunu, Churu, Bikaner, Barmer, Hanumangarh and Sri Ganganagar districts in the Thar Desert, Rajasthan. Impact has been assessed on independently evolved herpetological systems of this vast desert area in the light of recent ecological conditions in the surroundings of the Indira Gandhi Nahar Pariyojana (IGNP), the sewan grass areas and the Ghagger bed region. Attempts have been made to understand the pattern of change on community structures and dispersal behaviours of various extant species of amphibians and reptiles. The present field studies have significantly improved our knowledge on the comparative ecological interaction of relatively assembled species of the herpetofauna of Rajasthan's Thar Desert which comprises one of the most extensive and interesting natural ecological niches.

INTRODUCTION

The Rajasthan and Gujarat part of desert is the portion of the Great Indian Desert, which is also called as the Thar Desert that spreads in north-western India, and Sind and Punjab in Pakistan. This desert is the easternmost portion of the Great Palaearetic Desert extending from North Africa via Palestine, Arabia and Iran (Map 1). In India this desert mainly includes Barmer, Jaisalmer, Bikaner and Sri Ganganagar districts of Rajasthan and has got considerable impact on Gujarat, Punjab and Haryana (Map 2). Here the rainfall is scanty, generally below 30 cm. annually, and the land is open which is partly sandy, gravel and rocky. Vegetation is xerophytic, mainly comprising the *Acacia, Prosopis, Zizyphus* and *Capparis* species. The annual temperature varies between – 2^{0}C and 49^{0}C . Rajasthan desert is quite rich in animal life but at present we will keep the scope limited up to amphibians and reptiles only, which play an important role in regulating ecological balance as well as conservation of nature. The faunal elements of the Thar Desert are mainly of the palaearctico-oriental origin and exhibit a remarkable diversity in their saltatorial habitat. These include fossorial desert species capable of burrowing in soft sand and certain cursorial species which are capable of running effectively on soft soil, rocks and gravel terrestrial areas of desert. The desert amphibians and reptiles have developed suitable anatomical modifications in relation to the thermo-regulation, water-balance, sense of hearing, photoperiodism, sight, predation, mimicry and defence *etc.* Numerous lizards and snakes constitute an extremely conspicuous element of the vertebrate fauna of Rajasthan desert and they are poikilothermous creatures, obtaining their body heat absolutely from the external environment and are incapable of producing their own heat internally by means of oxidation metabolism.

Many researchers who contributed towards the knowledge of herpetofauna of theThar are Blanford (1879), who had given elaborate account of reptiles and amphibians of the Indian region, Dave (1960), Krishna (1975), Krishna & Dave (1960), Sunder Singh (1960), who had described many aspects of ecology of the Thar Desert of Rajasthan; Smith (1931, 1935 & 1943) have given the detailed account of Reptiles of Indian region including Thar. Minton (1966) given the elaborate account of Reptiles and Amphibians of Pakistan and Thar; Mansukhani and Murthy written the fauna of Rajasthan Desert. Prakash (1971, 1972 & 1973) described the ecology of various reptiles of Thar desert Sharma (1974-2003) given detailed account of Reptiles in his faunal work and Hand book. Sharma and Vazirani (1977) thrown some light on food and feeding Habits of Rajasthan reptiles including Thar and Biswas and Sanyal (1977) written the Rajasthan fauna of reptiles including the Thar Desert.

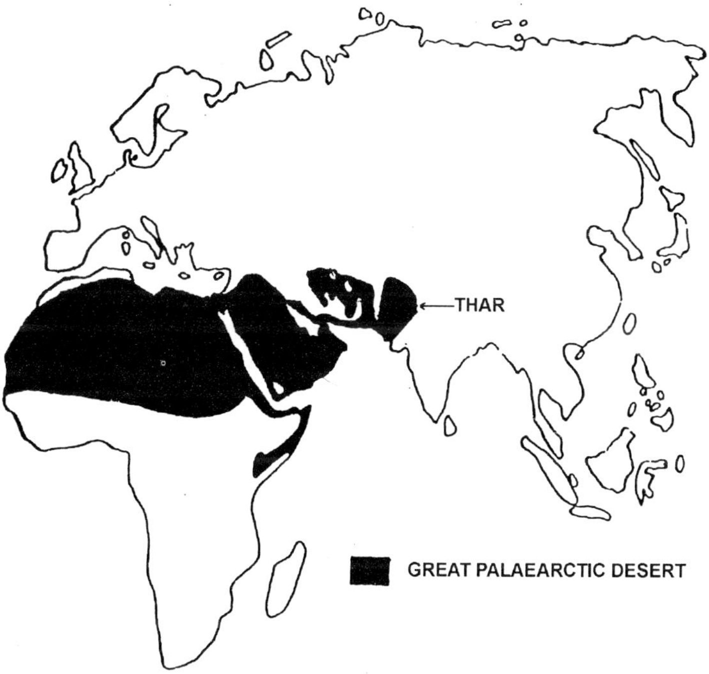

Map. 1.

The changed ecological conditions in the command area of IGNP, in the vicinity of Ghaggar bed and sewan grass areas has brought significant changes in the herpetofauna of Rajasthan Thar desert. This change has effected the relative diversity of the saltatorial species by causing hinderences in their foraging and reproductive abilities. Impact of these changes has been expressed in Table 1. The ecological communities of such desert species are not designed by natural selection as this phenomenon is restricted only to the individual organisms. This is evident that the natural selection operated by changes brought in by definite reproductive ability of the individual species. Abruptly changed ecology in the desert environment has brought about remarkable antagonistic interactions amongst different species and in their populations, producing in the process a direct bearing on inter-specific relationship by way of competition, predation and parasitism. In long term such changes in climate are bound to impact nutrient cycles, disturbance frequency and dispersal modes in such a vast and significant geographic desert area. Consequently, the changed environment has considerably effected microhabitat uses and foraging methods. Seasonal and daily changes in temperature and moisture contents have drastically influenced the temporal activities of amphibians and reptiles as this has also effected their space utilization, dispersal, survival and reproductive cycles. Main factors which have affected the herpetofauna in the changed Thar Desert environment are:

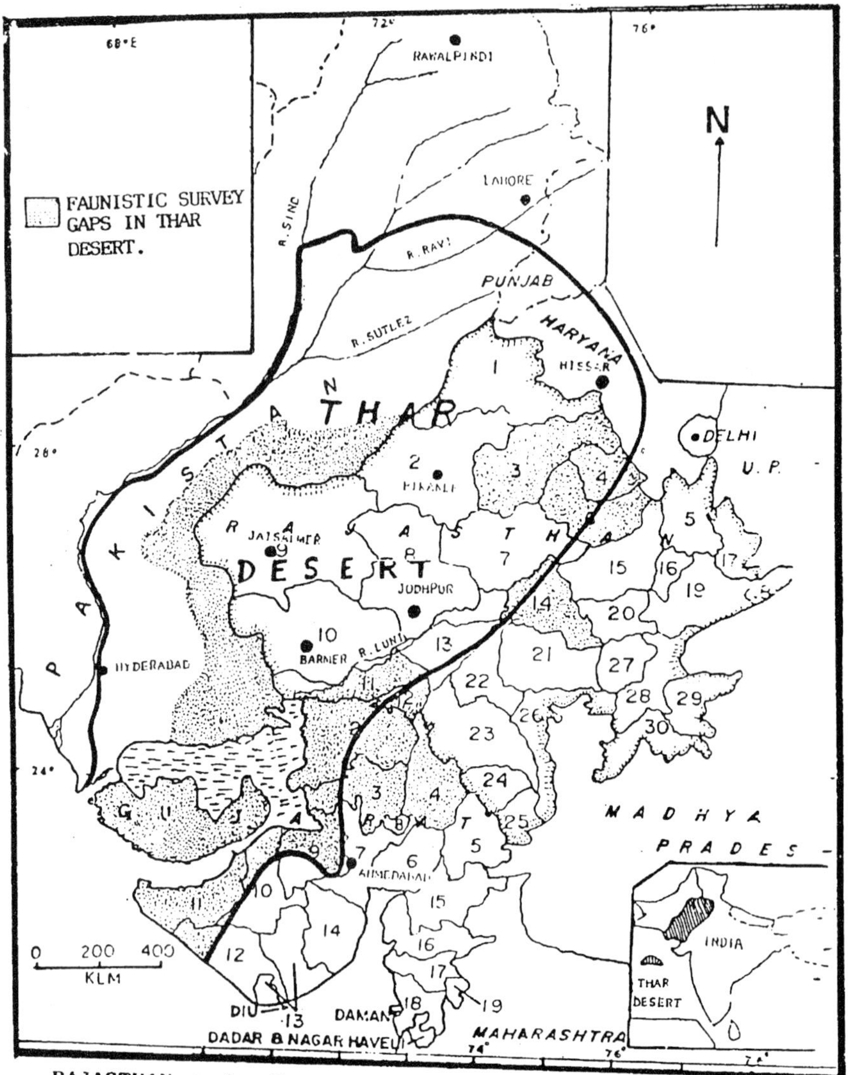

RAJASTHAN : 1. Ganganagar, 2. Bikaner, 3. Churu, 4. Jhunjhunu, 5. Alwar, 6. Sikar, 7. Nagaur, 8. Jodhpur, 9. Jaisalmer, 10. Barmer, 11. Jalore, 12. Sirohi, 13. Pali, 14. Ajmer, 15. Jaipur, 16. Dossa, 17. Bharatpur, 18. Dholpur, 19. Sawai Madhopur, 20. Tonk, 21. Bhilwara, 22. Rajsamand, 23. Udaipur, 24. Dungarpur, 25. Banswara, 26. Chittorgarh, 27. Bundi, 28. Kota, 29. Bara, 30. Jhalawar.

GUJARAT : 1. Kachchh, 2. Banaskantha, 3. Mehesana, 4. Sabarkantha, 5. Panchmahal, 6. Kheda, 7. Ahmedabad, 8. Gandhinagar, 9. Surendranagar, 10. Rajkot, 11. Jamnagar, 12. Junagarh, 13. Amreli, 14. Bhavnagar, 15. Vadodara, 16. Bharuch, 17. Surat, 18. Valsad, 19. Dangs.

Map. 2.

1. ***Afforestation in vicinity of IGNP and other areas***: Along both sides of the complete course of 650 km. Indra Gandhi canal, large scale plantation of trees, herbs and shrubs have stabilized the shifting of sanddunes, and thus played an effective role in the dispersal of fossorial-saltatorial species of lizards and snakes. These resident species were not used to changed moist habitats and as such ultimately either perished in large numbers or shifted to other adjoining desert areas. As the speed of migration of reptiles is slow, the rate of their survival is quite low and the sudden change has affected their ecological balance on account of the change in habitat and food. Now dense vegetation in the canal zone include: *Acacia nilotica, Delbergia sissoo, Prosopis cineraria, Eucalyptus camaldulensis, Zizyphus mauritana and Tecomella undulata.*

2. ***Formation of wetlands by the seepage in the Indira Gandhi canal:*** This has brought noteworthy changes in the faunal components of amphibians, testudines, water loving lizards and fresh water snakes.

3. ***Changed crop pattern:*** On account of this change numerous species of lizards and snakes, which are prevailing now, were not reported earlier.

EFFECT OF CHANGED ECOLOGY ON HERPETOFAUNA OF RAJASTHAN DESERT

Methodology and observations

1. Field Survey

The normal period of activity of amphibians and reptiles in Thar extends from March to October-November. As such numerous extensive field surveys, conducted in Thar districts of Rajasthan revealed interesting findings indicating the impact of the change on the original endemic species of Thar and on the species of amphibians and reptiles of the adjoining oriental realm. The observations were recorded in half to many sq. Kms., where change in environment has come in force. The observations were recorded on the basis of the species richness in various micro and macro habitats. The findings are based on the observations on the species which are active above ground; which live in burrows, tunnels, caves, rock crevices, under stones, hollow of trees and burrow in soft soil; arboreal species; wetland species and marshland species. Observation on the nocturnal species were recorded by sighting them with the help of Petromax light, flood lights and search lights with the help of Jeep battery. Observations were taken on the time of activity, associated fauna and flora, rainfall, temperature, wind-velocity and soil structure etc. Habitats like sand dunes, gravel land, rocky areas, clay areas, muddy soil plots, pools of water and canals etc. were observed.

2. Collection

(i) For terrestrial and burrowing species of amphibians, lizards, snakes and testudines conventional methods like by hand picking and by digging the burrows in ground with the help of spades and digging iron rods were used.

(ii) For aquatic species of amphibians, turtles and snakes suitable hand nets and cast nets were used.

(iii) For difficult unapproachable habitats, air gun was used for larger species of frogs, lizards and snakes.

(iv) For nocturnal terrestrial species of amphibians, lizards and snakes torch and jeep flood lights were used. Abrupt flood of light causes temporary blindness in most of the amphibian and reptile species, they become motionless for a moment and can be caught easily with the help of long forceps.

I. Observations on amphibians: (Plates 1-2)

The general amphibian fauna of Thar Desert represents the transitional element and is composed of the admixture of species from Palaearctic and Oriental regions.

1. Population : (Species wise)

(i) Euphlyctis cyanophlyctis:

The maximum population of the frogs of this species was recorded in Bikaner, Jaisalmer, Jodhpur, Nagaur, Barmer, Pali and Sirohi districts. Population was less in Sri Ganganagar, Churu and Jhunjhunu districts.

This species is fairly common in running water as well as in stagnant pools of stationary water. They feed here on aquatic invertebrates, water beetles and aquatic crustaceans.

(ii) Heplobatrachus tigerinus

The maximum population of this species was recorded in Sikar, Sirohi and Nagaur districts. The minimum population of this species was found in Sri Ganganagar, Bikaner, Jodhpur, Pali, Jalore and Barmer districts.

(iii) Limnonectes limnocharis

Few examples were collectcted from Pali and Sirohi districts but population seems to be quite thin.

(iv) Euphlyctis hexadactyla

It is a rare species in the Thar desert and has been recorded only from Jodhpur and Nagaur districts.

(v) Tomopterna breviceps

A rare species in the Thar and so far has been collected from Pali, Jodhpur and Nagaur districts.

(vi) Microhyla ornata

This tiny microhylid has so far been recorded from Pali, Jodhpur and Sirohi districts only and the population of this species seems to be quite thin in the Rajasthan Thar.

(vii) Bufo melanostictus

Only one example of this toad species has so far been collected from Jodhpur district.

(viii) Bufo stomaticus

This species of toad is widely distributed in Sri Ganganagar, Bikaner, Jaisalmer, Jodhpur and Naguar districts.

2. Impact of Changing faunal ecology

The effect of human activity such as the IGNP has tremendously influenced amphibian populations and their distribution in the Thar Desert. The intervention by man in this natural desert environment has affected the resident amphibian fauna in following ways :

(i) Drastically

Changes in sewan grass areas in Jaisalmer, Ghagger bed in Rangmahal area of Suratgarh and introduction of Indra Gandhi Nahar Project has spoiled natural balance and food chain of amphibians, especially the destruction of natural habitats of Bufonids (*Bufo stomaticus* mainly) and Microhylids (*Microhyla ornata*), in the Thar and as such brought a devastating effect in their living conditions.

(ii) Favourably

As IGNP area has provided different outlets in surroundings which provides a very good shelter by creating new and favourable habitats for most of the above mentioned amphibian species. Due to scarcity of water in these areas people use to store water in small tanks for different purposes which also become a shelter for different amphibians.

3. Observations on Testudines

Chelonians which comprises the fresh water species like *Hardella thurgi*, *Lissemys punctata punctata* and *Trionyx gangeticus* manifest a moderate diversity in the changed ecological scenario of the Thar Desert (Plate 2); the first named species *Hardella thurgi* being endemic to Thar desert. The changed environment will certainly effect favourably by increasing the range of these water loving species of turtles up to IGNP command area. A single species of land tortoise, *Geochelone elegans* has been recorded on the eastern fringe of the Thar and at a work-site near Bar (Pali district). This diurnal tortoise was found to be active during the morning and evening hours. The changed crop pattern and grassland extensions in stabilized sand-dune areas have improved the population of this land tortoise. This land tortoise feeds on *Bajra* leaves and seasonal vegetables like cabbage, cauliflower, beans and corn.

4. Observation on Lizards and Snakes : (Plates 3-8)

Lizards and snakes form an extremely distinct element of the vertebrate fauna of the Thar desert. Nature has not provided these creatures with the process of oxidative metabolism. As such squamatons can not produce their own heat internally and obtain their body heat absolutely from the external environment. They are low energy creatures and are extremely sensitive to the environment and habitat in which they live. Certain species of snakes and lizards can adjust nicely under the harsh and unstable hostile conditions of the desert. By means of this thermal device most of the desert lizards and snakes are adapted to conserve water and energy by retarding their movements and body activity during the heat of day. This process is also useful and helps a lot during the food shortage, drought and heat waves. Lizards are most suitably adapted to the desert environment as they are insectivorous (except *Uromastix hardwicki*, Plate 3 & 4) and generally take their food items (insects) intact and do not masticate at all. Indian Thar desert exhibits moderate diversity in lizards.

Family Gekkonidae (Plate 2) is represented by the species like *Stenodactylus orientalis, Cyrtodactylus scaber, Hemidactylus brooki, H. leschenaulti and H. flaviviridis*. Family Eublepharidae (Plate 3) is represented by Fat-tailed lizard *Eublepharis macularius* which is widespread in certain pockets of the desert but no where it is common. This species is an arch enemy of scorpions, *Buthus tamulus* and is immune to the assaults of scorpion sting. Family Agamidae (Plate 3) in the Thar includes *Agama agilis, Calotes versicolor, Phrynocephalus laungwalensis and Uromastix hardwicki*. Family Scincidae (Plate 4) is well represented by the species like *Mabuya carinata, M. macularia, Riopa punctata, Eumeces taeniolatus and Ophiomorus raithami*. Family Lacertidae (Plate 5) is represented by thick population of *Acanthodactylus cantoris cantrois* and *Eremias guttulata watsonana* as well as modrate populations of *Ophisops microlepis and O. jerdoni*. Family Varanidae (Plates 5) is represented by two species of monitar lizards, namely, *Varanus griseus* and *V. bengalensis*. Out of the above mentioned lizards, two species namely *Stenodactylus orientalis* (Plate 2) and *Phrynocephalus laungwalensis* are absolutely endemic to the Thar desert.

Plate 1

Bufo melanostictus (Schneider)

Bufo stomaticus (Lutken)

Hplobatrachus tigerinus (Daudin)

Tomopterna breviceps (Schneider)

Limnonectes limnocharis (Gravenhorst)

Euphlyctis cyanophlyctis (Schneider)

Plate 2

Hardella thurgi (Gray)

Lissemys punctata punctata (Lacepede)

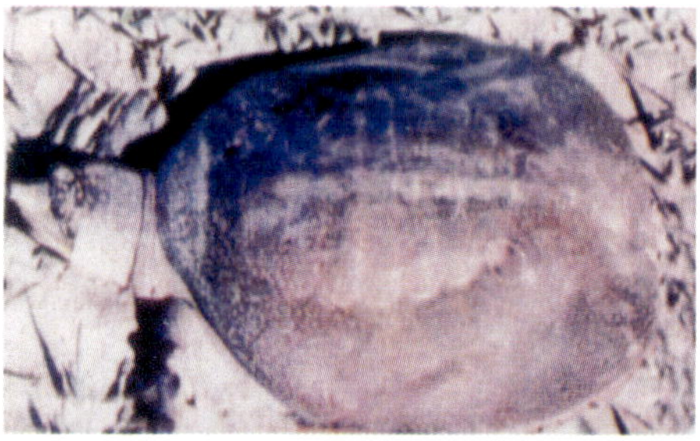

Trionyx gangeticus Cuvier

Geochelone elegans (Schoepff)

← *Stenodactylus orientalis* Blanford

Hemidactylus brooki Gray →

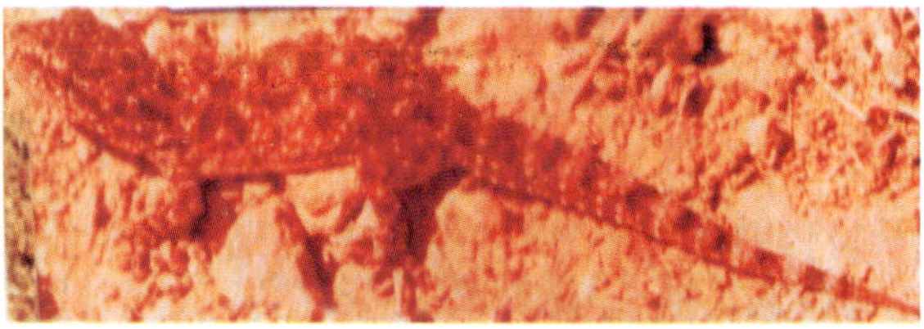

Hemidactylus leschenaulti Dum. & Bibr.

Crytodactylus scaber (Heyden)

Plate 3

Eublepharis macularius Blyth

Agama agilis Olivier

Calotes versicolor (Daudin)

Uromastix hardwicki Gray

Plate 4

Mabuya carinata (Schneider)

Riopa punctata (Linnaeus)

Ophiomorus raithmai Anderson & Leviton

Plate 5

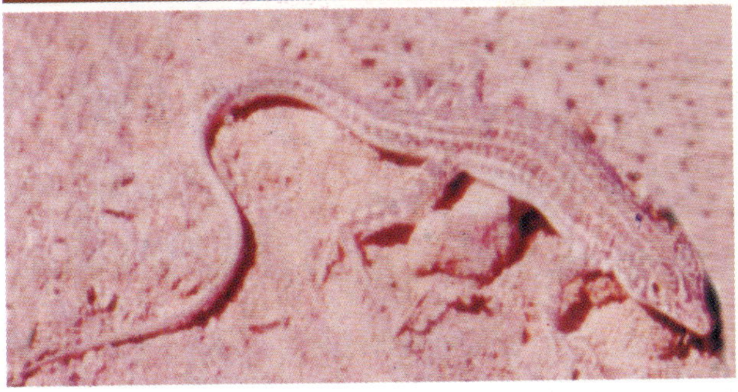

Acanthodactylus cantoris cantoris Gunther

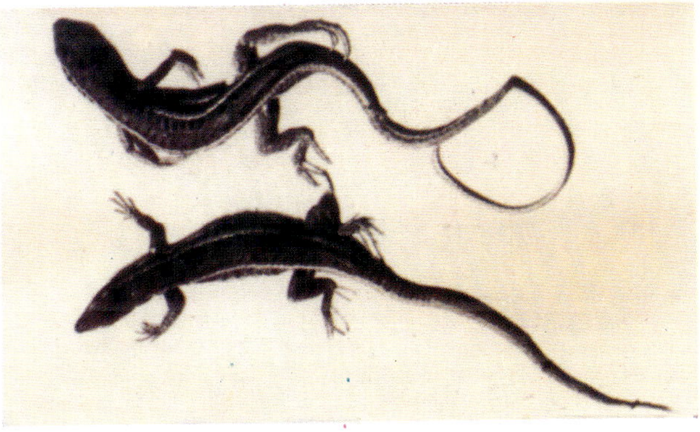

Ophisops jerdoni Blyth

Varanus griseus koniecznyi Mertens

Varanus bengalensis (Linnaeus)

Plate 6

Ramphotyphlops braminus (Daudin)

← *Eryx johni* (Russell)

Eryx conicus (Schneider)

Plate 7

Argyrogena ventromaculatus (Gray & Hardwicke)

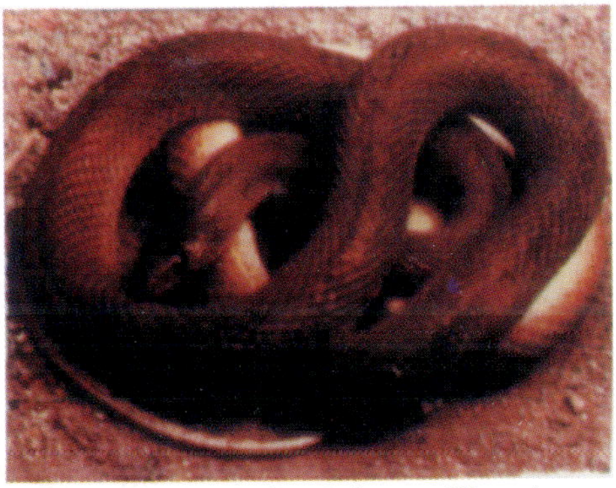

Argyrogena fasciolatus (Shaw)

Amphiesma stolata (Linnaeus)

Sphalerosophis diadema diadema (Schlegel)

Sphalerosophis articeps (Fischer)

Plate 8

Ptyas mucosus (Linnaeus)

Boiga trigonata (Schneider)

Psammophis schokari (Forskal)

Naja naja naja (Linnaeus)

Naja naja oxiana Eichwald

Echis carinatus (Schneider)

Bungarus caeruleus (Schneider)

Many snake species which serve as an inevitably indispensable component of the Thar faunal elements include *Eryx johni, E. conicus* (Plate 6), *Lytorhynchus paradoxus, Elaphe helena, Sphalerosophis diadema, Psammophis schokari* (Plate 8), *Argyrogena ventromaculatus* (Plate 7), *Boiga trigonata* (Plate 8), *Bungarus caeruleus* (Plate 8), *Naja naja naja* (Plate 8), *N. n. oxiana* (Plate 8) and *Echis carinatus* (Plate 8). The colubrid snake *Lytorhynchus paradoxus* is the only snake species which is endemic to the Thar. The snake fauna of Thar is predominantly well defined and has wide range of desert species many of which express a striking feature of having their original home in north central or Peninsular India.

RESULTS AND DISCUSSION

Amphibians

Numerous factors are responsible for formation of the faunal elements of amphibians in sandy, gravel and rocky habitats of the Thar Desert, depending on the geographical accessibility. If we trace the phylogeny, enough evidences are available to support the view that genus *Bufo* had its original home in the Nearctic region (North America) during Cenozoic period (approx. 67 million years ago). This genus spread gradually and effectively in Eurasia (Palaerctic region) and ultimately settled properly in Indo-malayan region. Early fossil records indicated the presence of *Bufo* in Indo-Malayan region and Himalayan uplands. Gradually numerous toad species of family Bufonidae like *Bufo parietalis, B. himalayanus, B. viridis, B. stomaticus, B. fergusoni, B. olivaceous, B. surdus, B. raddei and B. latastei* were distributed in complete Palaerctic region and subsequently intruded in Gangetic plains and Himalayan foot-hills through Irano-Turanian subregions of Palearctic region. Ultimately three of these species, namely, *Bufo viridis, B. melanostictus* and *B. stomaticus* reached Gujarat and the Thar Desert of India and Pakistan.

The genus *Microhyla* of family Microhylidae is tropical in origin but it gradually reached the plains of Punjab, Gujarat and north-western desert part of Indian Sub-continent, now known as Thar Desert. The original home of the family Ranidae (Frogs) is tropical portion of Ethiopean region (Africa) from where the radiation of numerous species of frogs started in Coenozoic period and they settled in various parts of the world. Besides, numerous species of frogs had migrated to various other parts of the Indian sub-continent. Species like *Heplobatrachus tigerinus, Euphlyctis cyanophylctis, Limnonectus limnocharis* and *Tomopterna breviceps* migrated to the sub-tropical areas of Indian subcontinent and also settled in course of time in certain suitable pockets of the Thar Desert.

Recent changes caused by Indra Gandhi Nahar Project seemed to have come about drastic changes in the distributional pattern of amphibians in the Thar Desert region. Recently, amphibians in the desert have been shown to possess an affinity to well- watered areas and are nicely settling down there. Though the Thar's amphibian fauna is poor in diversity is nevertheless quite rich in populations of selected species like *Bufo stomaticus* and *Euphlyctis cyanophlyctis*. Many parts of the Thar were originally poor grasslands, which in course of time have become excellent habitats under the impact of canal system for sheltering and nourishing all the four species of frogs of family Ranidae. Extensive canalization during last many decades has carried the Himalayan water to the far flung areas of Jaisalmer, Bikaner, Barmer, Nagaur, Hanumangarh, Sri Ganganagar and Pali districts of the Thar desert, which were earlier barren and monotonous. This has given a tremendous help to the resident amphibian species. In addition to the already existing species of amphibians (both frogs and toads) there has become an strong possibility of settlement of even the less known species like *Euphlyctis hexadactyla* and *Microhyla ornata*. At present, both these species are available mainly along the extreme eastern boundary of the Indian Thar.

Reptiles

During recent years significant changes are noticed in the reptile fauna of the Thar desert on account of the changed ecological conditions by the formation of Indra Gandhi Nahar Project. This has significantly affected the relative diversity of most of the saltatorial-fossorial species of testudines, lizards and snakes. Adaptation, which plays a significant role on the ecological communities of reptiles and the abrupt changes, has brought gradual changes in resident species of reptiles, namely, (i) Testudines: *Hardella thurgi, Geoclemys hamiltoni Lissemys punctata, Chitra indica, Trionyx gangeticus and Geochelone elegans,* (ii) Lizards: *Cyrtodactylus scaber Hemidactylus brooki, Eublepharis macularius, Agama agilis, Uromastix hardwicki, Mabuya carinata, Eumeces taeniolatus, Acanthodactylus cantoris, Varanus griseus* and *V. bengalensis* (iii) Snakes: *Sphalerosophis diadema, Argyrogena ventromaculatus, Boiga trigonata, Psammophis schokari, Bungarus caeruleus, Naja naja* and *Echis carinatus*.

Impact of the Changing Ecology on the Amphibians and Reptiles of the Thar Desert of Rajasthan (India)

S.No.	CLASS, ORDER, FAMILY, SPECIES	Distribution	Time of Activity	Habitat	Habits	Factors Responsible for Ecological change	Previous Status (Before Changes)	Present Status (After change)	Management Required (Suggested)
CLASS : AMPHIBIA ORDER: Anura FAMILY : Bufonidae :									
1	*Bufo melanostictus* (Schneider)	Jodhpur & Sirohi Districts	Nocturnal	Plain moist areas around water bodies and wet lands (Lives in moist, shady places, under roots, bricks and pebbles)	Insectivorcus, Eggs are arranged into strings, tadpoles form deep pigmentation	It is a water loving species IGNP will certainly increase its population and extension of range in desert part	Rare	Population increase and habitat gain	1. Certain areas must be protected for its breeding in Urban Land. 2. Mechanization of agriculture should be restricted to prevent the destruction of their abodes. 3. There should be a restriction on the use of pesticides and fumigants in their habitats and feeding grounds (agriculture fields) 4. There should be a restricted use for scientific biological experiment

2.	*Bufo stomaticus* (Lutken)	Widely distributed in Bikaner, Nagaur, Jaisalmer & Sri Ganganagar districts.	Nocturnal	Grassy areas, Lives under stones; (in holes and crevices)	Lives on insects and crustaceans	It is a water loving species IGNP will certainly increase its population and extension of range in desert part	Abundant	- do -	- do -
FAM : RANIDAE :									
3.	*Euphlyctis hexadactyla* (Lesson)	Jodhpur & Nagaur Districts	Diurnal	Muddy bottom of larger tanks and ponds	Secretive, agile and insectivorous	- do -	Rare in Thar	- do -	- do -
4.	*Euphlyctis cyanophlyctis* (Schneider)	Barmer, Jaisalmer, Jodhpur, Nagaur, Pali, Sikar, Jhunjhunu and Sri Ganganagar Districts.	Diurnal	Running and stagnant water bodies with muddy or rocky bottom with plenty of aquatic vegetation.	Agile, fast swimmer, Lives on aquatic invertebrates (beetles, odonate larvae and small crustaceans)	- do -	Abundant	Remarkable increase in population and range	- do -
5.	*Heplobatrachus tigerinus* (Daudin)	Sri Ganganagar, Hanumangarh, Nagaur, Jodhpur, Jalore, Pali, Sirohi and Sikar Districts	Diurnal	Muddy bottom of ponds; crevices in rocky water bodies	Fast on land and is a most active swimmer, prefers mainly small to moderate sized water bodies with muddy bottom.	- do -	Moderate populations were available	- do -	- do -
6.	*Limnonectus limnocharis* (Gravenhorst)	Nagaur, Pali, Jodhpur, Sirohi districts	Diurnal	Marshy wet lands and shallow pools of water	Mainly insectivorous	- do -	Moderate population in Nagaur & Jodhpur districts; rare elsewhere in Thar	Population has increased considerably	- do -

7.	*Tomopterna breviceps* (Schneider)	Jodhpur, Pali and Nagaur districts	Diurnal	Subterranean muddy tunnels under the stones near pools of water	Insectivorcus burrowing toad like sluggish frog.	It is a water loving species IGNP will certainly increase its population and extension of range in desert part	Rare	Not known so far. but population will certainly increase.	- do -
8.	*Microhyla ornata* (Dumeril & Bibron)	Pali, Jodhpur and Sirohi district	Diurnal	Small pools cf water	Insectivorous (mainly ar:ts and small inse:ts)	- do -	Rare	- do -	- do -
CL : REPTILIA OR : TESTUDINES FAM : EMYDIDAE									
9.	*Hardella thurgi* (Gray)	Jaisalmer, Barmer and Sri Ganganagar districts	Diurnal	Stagnant ponds, canals and backwaters (Avoids fast flowing wate:)	Entirely herbivorous, docile but active	In normal adjo-ining areas of IGNP this turtle will be benefited but seepage and sim formation will have drastic effect	Normal populations were available	Adverse effect on account of natural habitat loss	Population must be conserved in suitable areas without the effect of seepage and sim.
FAM : TESTUDINIDAE									
10.	*Geochelone elegans* (Schoepff)	Sirohi and Jodhpur districts	Diurnal	Rocky grassy foot hills of Aravallis adjoining to Thar	Entirely herbivorous; mates from June-October, 3-6 eggs are laid.	Area of distribution is quite far from IGNP There will not be any effect	Indeterminate	Suspected to be threatened by habitat loss.	Monitoring must be done at trade levels, its breeding must be promoted in fenced areas.

FAM : TRIONYCHIDAE									
11.	*Lissemys punctata punctata* (Lacepede)	Jodhpur and Pali Districts	Diurnal	Shallow, muddy ditches, lakes and marshes with plenty of aquatic vegetation	Entirely carnivorous feeds on frogs, fishes, shrimps and snails, 12 eggs are laid in a single clutch.	It is a water loving species so canal system will certainly have beneficial role.	Suspected to be threatened by man and habitat loss.	Population can increase and there will be a effect on its range.	After gathering exact data on its population ecology reserve areas must be demarcated for its protection.
12.	*Trionyx gangeticus* Cuvier	Jodhpur and Pali Districts	Diurnal	Large water bodies of Thar	Highly aquatic, basks on sandy banks or rests in shallow water; omnivorus, scavenger, feeds on aquatic vegetation, fish, molluscs, Frogs & crustacean, Breeds from September to October	It is a water loving species so canal system will certainly have beneficial role.	Endangered on account of its excessive killing for flesh and eggs for food; Habitat destruction	Not known	- Do -
OR : SQUAMATA		SUB OR : SAURIA (LIZARDS)				FAM : GEKKONIDAE			
13.	*Stenodactylus orientalis* Blanford	Jaisalmer, Jodhpur and Barmer Districts	Nocturnal	Inhabits barran Sand-dune areas	Most docile, saltatorial, Fossorial; Swift, Skillful burrowing, insectivorous breeds from March to May	It is strictly a desert species. IGNP has brought an adverse effect in areas where seepage and sim has occurred	Was in abundance	Became vulnerable on account of habitat loss in Bikaner district and Suratgarh	Flooding on accounts of seepage must be checked so that sim (Flooding and marsh formation can be controlled) Suitable management plans must be followed with the help of suitable agencies like AFRI.

14.	*Hemidactylus brooki* Gray	Jodhpur, Jaisalmer, Nagaur, Pali, Sirohi Districts	Nocturnal	House lizard in many areas; main habitat comprises rocky areas in desert forest.	Insectivorous; breeds in summer, two spherical eggs are laid.	Effect will be adverse in command areas of IGNP	Abundant	There is a decline in its population in adjoining areas of canal system.	- Do -
15.	*Hemidactylus leschenaulti* Dum. & Bibr.	Barmer, Nagaur, Jaisalmer & Sirohi Districts	Diurnal	Large trees of banyan, tamarind and other such trees	Insectivorous, arboreal ; breeding takes place from March to May	Effect will be adverse in command areas of IGNP	Abundant	There is a decline in its population in adjoining areas of canal system.	Flooding on accounts of seepage must be checked so that sim (Flooding and marsh formation can be controlled) Suitable management plans must be followed with the help of suitable agencies like AFRI.
16.	*Hemidactylus flaviviridis* Ruppell	Barmer, Bikaner, Churu, Jaisalmer, Jodhpur, Nagaur & Pali Districts	Nocturnal	Inhabits the buildings	Strongly Insectivorous, most agile pugnacious, marked climber; breeding takes place from March to August	No effect will take place	- Do -	- Do -	It is well protected by its dwelling habits.

17.	*Hemidactylus triedrus* (Daudin)	Jodhpur District	Nocturnal	Semi arid, rocky sandy areas where it lives in rodent and lizard burrows, crevices in trees, rocks, termite mounds etc.	Insectivorous docile, breeds from May to July	Effect will be adverse in command areas of IGNP	Common	-Do-	It is already in a safer range of distribution
18.	*Cyrtodactylus scaber* (Heyden)	Bikaner, Jodhpur & Jaisalmer Districts	Noctu-rnal	Desert areas, dry grassland, rocky hill sides, inhabit old house.	Active, insecti-vorous and make an attempt to bite.	The main habitat is desert. IGNP has brought an adverse effect in many areas of its range where seepage and sim has occurred.	Was in abundance	Vulnerable in natural desert areas on account of habitat loss and change in food chain.	Suitable mana-gement plans must be enforced with the help of AFRI to check sim and seepage.
FAM : EUBLEPHARIDAE									
19.	*Eublepharis macularius* Blyth	Jaisalmer, Barmer, Bikaner, Nagaur and Pali districts	Noctu-rnal	Inhabits rocky desert areas but avoid sand dunes	Most pugnacious and bites readily, insectivorous mainly but also devours scorpions and small geekos.	- Do -	was common in the areas of its range.	Undeterminate	- Do -
FAM : AGAMIDAE									
20.	*Calotes versicolor* (Daudin)	Barmer, Nagaur, Bikaner, Pali, Jaisalmer, Sirohi, Jodhpur district	Diurnal	Live on shrubs and trees near desert areas	Strongly arboreal and excellent climbers, insectivorous, breeds from May to September	No effect will take place	Most Common	No Effect	It is well protected by its strong arboreal habits.

21.	*Agama agilis* Olivier	Barmer, Bikaner, Jaisalmer, Jodhpur, Nagaur and Sri Gangangar Districts	Diurnal	Arid desert rocky, sandy areas with some xerophytic vegetation	Most agile, climbs on small bushes for basking Insectivorous, breeds between May to August	IGNP has brought drastic effect by destroying its habitat and food chain.	Most Common	Vulnerable now in throughout its range on account of habitat loss.	Suitable manag-ement plnas must be followed with the help of AFRI to checked sim and destruction of sewan grass pastures.
22.	*Uromastix hardwicki* Gray	Pali, Jodhpur, Bikaner, Jaisalmer, Jalore and Sri Ganganagar Districts	Diurnal	Lives in hard gravel or rocky soil patches surrounded by grass and xerophytic vegetation.	Agile, strong burrower, most docile, Perfectly vegetarian; breeding from March to July	IGNP has brought drastic effect by destroying its habitat and food chain.	Most Common	Vulnerable now in throughout its range on account of habitat loss.	Suitable management plnas must be followed with the help of AFRI to check sim and destruction of sewan grass pastures.
23.	*Phrynocephalus laungwalensis* Sharma	Jaisalmer district	Diurnal	Sandy barren desert parts of Jaisalmer district	Insectivorous, saltatorial fossorial	-Do-	-Do-	-Do-	-Do-
FAM : SCINCIDAE									
24.	*Mabuya carinata* (Schneider)	Barmer, Pali, Jodhpur districts	Diurnal	Inhabits terrestrial grassy, vegetated areas	Insectivorous Oviparous (Lays 22 or 23 eggs in a clutch)	In Pali district it is unaffected, adverse effect on population noticed in IGNP areas.	Very Common	Has become vulnerable in Jodhpur and Barmer Districts	Flooding on account of seepage must be checked, sim formation in canal areas must be controlled.
25.	*Mabuya macularia* (Blyth)	Barmer, Pali, Jodhpur districts	Diurnal	-Do-	Insectivorous breeds in April to May (Laid 3-4 eggs)	-Do-	-Do-	-Do-	-Do-

26.	*Ophiomorus raithmai* Anderson & Leviton	Barmer, Jodhpur and Nagaur districts	Nocturnal	Sand dwelling, saltatorial	Insectivorous, oviparous	Adversely effected in IGNP comm and areas	-Do-	-Do-	-Do-
27.	*Eumeces taeniolatus* (Blyth)	Barmer, Jaisalmer districts	Nocturnal	Sandy borrows and under stones in rocky habitats	Most secretive, agile and insectivorous	Adverse effect on its population in IGNP areas	No where common	Becoming rare on account of habitat loss	Flooding in IGNP areas must be checked.
FAM : LACERTIDAE									
28.	*Acanthodactylus cantoris cantoris* Gunther	Jodhpur, Nagaur, Barmer, Jaisalmer and Sirohi Districts	Diurnal	Sandy inland areas in dense xerophytic bushes	Burrowing, insectivorous, oviparous (Laid 3-5 eggs)	Adverse effect on its population in IGNP areas	Very Common	Population has decreased	Flooding in IGNP areas must be checked.
29.	*Ophisops microlepis* Blandford	Jodhpur and Pali districts	Diurnal	Soft soil areas in dense clusters of xerophytic vegetation	Burrowing, insectivorous, 7 eggs are laid, breeds upto September.	- Do -	- Do -	- Do -	- Do -
30.	*Ophisops jerdoni* Blyth	Jaisalmer, Jodhpur and Sirohi districts	Diurnal	- Do -	Burrowing; insectivorous	- Do -	- Do -	- Do -	- Do -
FAM : VARANIDAE									
31.	*Varanus griseus Koniecznyi* Mertens	Bikaner and Jaisalmer districts	Mainly Diurnal	Undulating sandy grounds with sparse vegetation	Burrowing, food comprises Invertebrates and small vertebrates, breeds from May to September (15-20 eggs are laid); if needed lizard attains great speed	-Do-	Common	Population has declined (Vulnerable)	Sim formation in IGNP command areas must be controlled, Destruction of sewan grass pastures must be stopped.

32.	*Varanus bengalensis* (Linnaeus)	Jodhpur, Pali, Nagaur, Bikaner, Sirohi and Jaisalmer districts.	Diurnal	Lives in burrows, nullahs, clifs (in the rock crevices) under stones and boulders; in dense vegetation; around marshes, ponds, canals hollow of trees.	Burrowing, movements are fast ; feeds mainly on beetles and other insects, small mammals, snakes and lizards, breeds from June to September (14-29 eggs are laid)	Adverse effect on its population in IGNP areas	Vulnerable	Endangered	Sim formation in IGNP command areas must be controlled, Destruction of sewan grass pastures must be stopped.
OR : SQUAMATA			SUB OR : SERPENTES (SNAKES)			FAM : TYPHLOPIDAE			
33.	*Ramphotyphlops braminus* (Daudin)	Jodhpur and Bikaner Districts	Diurnal	Subterranean burrows in soft soil, decaying leaf litter and wood.	Burrowing, food comprises worms and soft bodied arthropods, oviparous, breeds from May to June (2-8 eggs are laid)	Adversely effected in IGNP command areas	Very Common	Projects have effected and population has declined in IGNP areas	-Do-
FAM : LEPTOTYPHLOPIDAE									
34.	*Leptotyphlops macrorhynchus* Jan	Jodhpur District	Diurnal	Live in inter connected tunn-els in termite infested areas	Fast burrowing, feed on termites, soil arthropods and their eggs.	-Do-	-Do-	-Do-	-Do-
FAM : BOIDAE									
35.	*Eryx johni* (Russell)	Jodhpur, Pali and Bikaner Districts	Noctur-nal	Found in flat arid areas with loose clay soil with sparse grass	Strongly burro-wing, sluggish, constricter, feeds on small mammals, birds and arthropods, viviparous (5 to 14 young ones are produced)	Adversely effected in IGNP area and sewan grass lands	Very Common	Population has declined	Sim formation in IGNP command areas must be controlled, Destruction of sewan grass pastures must be stopped.

36.	*Eryx conicus* (Schneider)	Jodhpur and Pali Districts	Nocturnal	Inhabits loose sand or loose silty soil.	Strong burrower, pugnacious, bites savagely; feeds on small mammals, birds, reptiles and amphibians, Viviparous (6-11 young ones are produced breeds from May to July	-Do-	-Do-	-Do-	-Do-
FAM : COLUBRIDAE									
37.	*Argyrogena ventromaculatus* (Gray & Hardwicke)	Jodhpur, Bikaner and Jaisalmer Districts	Nocturnal	Inhabits rocky sandy and clay areas of desert where sparse grass and scrub vegetation is available.	It is a strong climber and partly arboreal; its main food comprises the small lizards, birds and small mammal, breeds from April to September, oviparous (6 to 9 eggs are laid in a clutch)	-Do-	-Do-	-Do-	-Do-
38	*Argyrogena fasciolatus* (Shaw)	Barmer and Jaisalmer Districts	Nocturnal	Inhabits arid areas with loose soil, slight grass and other xerophytic vegetation, also in agriculture fields.	Secretive, pugnacious, feeds on small rodent and lizards.	Adversely effected in IGNP area and sewan grass lands	Very Common	Population has declined	Sim formation in IGNP command areas must be controlled, Destruction of sewan grass pastures must be stopped.

39.	*Amphiesma stolata* (Linnaeus)	Jodhpur, Barmer, Jaisalmer, Bikaner, Nagaur, Sri Ganganagar and Sirohi districts	Diurnal	Moist grassy terrestrial areas of the desert (under stones, wooden logs	Gentle snake, never bites, feeds on frogs, toads, lizards & small mammals, breeds in May (10 eggs are laid)	Adversely effected IGNP areas and sewan grass lands.	Fairly common	Decline in population	Sim formation in IGNP must be controlled destructions of seven grass pastures must be stopped.
40.	*Sphalerosophis diadema diadema* (Schlegel)	Jodhpur and Nagaur districts	Diurnal	Inhabits rocky arid areas with plenty of crevices in rocks and xerophytic bushes in the surroundings	Docile snake, feeds on rodents and other small mammals, lizards frogs small birds and their eggs.	Not effected	-Do-	Fairly common, Population is not affected	Skin trade should not be allowed.
41.	*Sphalerosophis atriceps* (Fischer)	Nagaur district	Noctu-rnal	Inhabits dense scrubby areas in Thar, lives in burrows, rock crevices.	Most active; bites on provocation, feeds on mice, rats and birds, ovipa-rous (68-78 eggs are laid in May)	Not effected	-Do-	-Do-	-Do-
42.	*Sphalerosophis arenarius* (Boulenger)	Bikaner and Jodhpur districts	Diurnal	-Do-	Food is same like *S.atriceps*	Not effected	-Do-	-Do-	-Do-
43.	*Ptyas mucosus* (Linnaeus)	Jodhpur, Pali, Jaisalmer, Barmer, Bikaner, Sri Ganganagar and Sirohi districts	Diurnal	Inhabits plains agriculture fields, grass lands, trees and hollow in them, under dead wooden logs and under stones	Most active; feeds on all small vertebrates, oviparous breeds from April to September	Already effected in IGNP areas and Sewan grass lands	Fairly common	Decline in Population	Sim formation in IGNP, must be controlled; destruction of sewan grass pastures should be stopped and skin trade should be stopped.

44.	*Lycodon aulicus* (Linnaeus)	Jodhpur district	Nocturnal	Lives in human habitations	Pugnacious ; good climber; feeds on mice, lizards frogs, oviparous (Laid 3-11 eggs in a clutch	Not effected	-Do-	Still fairly common	Not needed
45.	*Boiga trigonata* (Schneider)	Jodhpur district	Noctu-rnal	Lives in small to medium bushes and trees	Arboreal, terres-trial, feeds on lizards, small mammals, birds and their eggs and large insects, ovip-arous (6-12 eggs are laid) breeds from April-August.	Not effected	-Do-	-Do-	-Do-
46.	*Psammophis schokari* (Forskal)	Jodhpur and Bikaner districts	Diurnal	Sandy desert with sparse vegetation	Food comprises lizards and birds	Not effected	-Do-	-Do-	-Do-
47.	*Xenochrophis piscator* (Schneider)	Sirohi, Pali and Jodhpur districts	Diurnal	Marshy areas, water bodies, canals with plenty of aquatic vegetation	Vicious snake, most agile, feeds mainly on fishes and frogs, ovipa-rous, breeds from November to May (8 to 90 eggs are laid in a clutch)	Not effected as it is on aquatic species	-Do-	-Do-	-Do-
FAM : ELAPIDAE									
48.	*Naja naja naja* (Linnaeus)	Jodhpur, Jaisalmer, Nagaur, Pali and Sriganganagar districts	Diurnal	Inhabits agriculture fields, grassy plots, rodent holes, roots of trees and dens of animals	Most intelligent and pugnacious species; feeds on all vertebrates which the snake can over power oviparous (9-45 eggs are laid in May to June)	Not effected as the snake has got quick shifting tendency	Fairly common	Decline in population on account of habitat destruction	-Do-

49.	*Naja naja oxiana* (Eichwald)	Jodhpur district	Diurnal	-Do-	-Do-	-Do-	-Do-	-Do-	-Do-
50.	*Bungarus caeruleus* (Schneider)	Jodhpur, Bikaner and Jaisalmer districts	Noctu-rnal	Confined to grass land, alluvial, rocky and marshy areas in Thar.	It is a docile snake, feeds on small mammals, amphibians and snakes, oviparous (6-10 eggs are laid)	-Do-	Not so common	Becoming rare on account of habitat loss	-Do-
FAM : VIPERIDAE									
51.	*Echis carinatus* (Schneider)	Jodhpur, Pali, Jaisalmer, Bikaner, Nagaur, and Sirohi districts	Noctu-rnal	Available in a variety of habitats like sandy, alluvial rocky, arid environment with grass, herbs, shrubs, xerop-hytic vegetation	Most pugnacious snake, feeds on mice, squirrels lizards, amphibians insect, and arthropods, viviparous, 9-14 young ones are produced	Drastically effected in IGNP area	Most Common	Decline in population on account of habitat loss	Sim formation in IGNP must be controlled, destruction of Sewan grass pastures must be stopped
52.	*Vipera russelli* (Shaw)	Jodhpur district	Noctu-rnal	Inhabits the plains and foothills	Sluggish but ill tempered. feed on rodents and frogs; viviparous (20-25 young ones are produced	out of range	—	—	—

ACKNOWLEDGEMENTS

We are grateful to Dr. Q.H. Baqri, Ex-Additional Director and Dr. N.S. Rathore, Jt. Director & Officer Incharge, Desert Regional Station, ZSI, Jodhpur for providing necessary facilities in connection with this work.

REFERENCES

Biswas, S. and Sanyal, D.P., 1977. Fauna of Rajasthan, Reptilia, *Rec. Zool, Surv. India* 73: 247-269.

Blandford, W.T., 1879, Notes on a collection of reptiles made by Major O.B. St. John. T.E. at Ajmer in Rajputana. *J. Asiat.Soc. Bengal, Calcutta* 48(2): 119-127.

Dave, K.C., 1960. Contribution to the systematic, distribution and ecology of the reptiles of the desert of Rajasthan, with special reference to the ecology of certain lizards. *Proc. 47th Indian Sci. Congr. Calcutta, Pt.* 3 (Abstract): 482-483.

Mansukhani, M.R. and Murthy, T.S.N., 1970. Fauna of Rajasthan, India, Part 6 Amphibia, *Rec Zool, Surv.* Indian 62 (1 & 2): 51-60.

Minton, S.A., 1966. A contribution to the herpetology of West Pakistan *Bull. Am. Mus. nat Hist.,* 134(2) : 27-184, text-figs. 1-12 plates 9-36, tables 1-7, maps 1-5.

Prakash, I., 1973. The ecology of Vertebrates of the Indian desert. In *ecology and Biogeography of India,* 369-420 pp. Hagua (Dr. W. Junk).

Sharma, R.C., 1974. A report on the snakes of Rajasthan. Indian *Journal of Zoology, Bhopal* 2: 19-28.

Sharma, R.C., 1996. Herpetology of the Thar Desert. In : Faunal Diversity in Thar Desert : Gaps in Research. Ed. by A.K. Ghosh, Q.H. Baqri and I. Prakash, 297-306 Scientific Publisher, Jodhpur.

Sharma, R.C., 1998 Fauna of India. Reptilia (Testudines and Crocodilians) Vol. 1 XVI+1-196 pp. (Published-Director, ZSI, Kolkata).

Sharma R.C., 2002. The Fauna of India and the adjacent countries- Reptilia (Sauria) Vol. II : 1-430 pp. (Published Director, Z.S.I., Kolkata).

Sharma, R.C., 2003. Handbook- Indian Snakes : 1-292 (Published Director, Z.S.I. Kolkata)

Sharma, R.C. and Vazirani, T.G., 1977. Food and feeding habits of some reptiles of Rajasthan *Rec. Zool Survy. India* 73 : 77-93.

Smith, M.A., 1943. The Fauna of British India, including Ceylon and Burma. Reptilia and Amphibia Vol. III, Serpentes xii + 583 pp figs. map.

Changing Faunal Ecology in Thar Desert, pp. 173-185, 2005.
Eds.: B.K. Tyagi and Q.H. Baqri

10

EVOLUTION OF FISH HATCHERIES IN INDIA

Devendra Mohan

Department of Zoology, Jai Narain Vyas University,
Jodhpur - 342 001 (Rajasthan), India

The important economic group of fishes for fresh water aquaculture includes: Indian major carps and exotic carps. Most of these fishes do not breed in confinement though they can grow and attain sexual maturity there. For fish culture, our country was mainly dependant on natural seed collection of culturable carps till the development and standardization of hypophysation technique. A major breakthrough in fish seed production is observed with the development of various designs of hatcheries and infrastructural facilities for breeding different types of fishes. With the gradual development of induced breeding as a technology for commercial seed production, incorporation of a hatchery in a fish seed farm is also developed. The availability of synthetic hormones like ovaprim and ovatide have replaced the traditional use of pituitary extract for fish spawning and revolutionized the fish hatchery operations. With the development of these inducing agents and the knowledge of fish spawning in captivity, the fish hatcheries have also been evolved from simple mud hatchery to modern vertical jar and circular Chinese hatchery.

INTRODUCTION

Unavailability of pure seed of culturable carps is one of the limiting factors for aquaculture development in India. The fish seed industry in India still strives to a great extent on collection of eggs, spawn and fry from rivers. Apart from spawn collection from rivers, methods of bund-breeding and induced breeding have also been developed. The prevalent system with double walled cloth hapa (Alikunhi *et al,* 1964; Chaudhuri and Tripathi, 1976) have limitation in large scale commercial application. Although, the development of modern hatching technologies like Zoug jar hatchery (Marcel, 1971), Aluminium and plastic bin hatchery (Shirgur, 1973) vertical flow jars of transparent polythene (Shirgur, 1972) and Galvanized and PVC hatcheries (Dwivedi, 1976) have indeed served the purpose to some extent. But, in these methods the environmental parameters like temperature, pH and oxygen play an important role for hatching of eggs. Dwivedi and Reddy (1986) developed CIFE-D81 hatchery to breed the fish in controlled environmental system and ensured high efficiency of both breeding and hatching.

EVOLUTION OF HATCHERIES

A hatchery is the most vital component of a modern fish farm, although the traditional "Mud hatchery pit" prevailed in India from centuries, yet their actual evolution starts after the development of hypophysation technique by Houssay in 1931. Consequently, while on one hand, a large number of chemicals like HGC, LRH-A, Progesterone, Corticosteroids, Estrogen and Androgens, prostaglandin and synthetic chemicals like Ovaprim and ovatide were used, on the other hand, different designs of hatchery units were developed to work out most suitable condition for spawning of fish in captivity. These are discussed below in some detail:

1. Earthen Vessel hatchery

Baked clay vessels are used which, though cheap and easily replaceable, and also somewhat porous and hence cool due to surface evaporation, have a disadvantage of being opaque. The volume of hatching jars may vary from 1 to 200 lit with the egg holding capacity up to one lakh.

2. Fixed Hapa hatchery (Fig. 1)

Such hapas consist of an outer hapa made of thin muslin cloth (2 x 1 x 1m) and inner hapa of round mesh mosquito cloth (1.75 x 0.75 x 0.5 m). These are tied to bamboo poles fixed in a series in a pond with clear oxygenated water free from predators. Each unit of hapa can hold about 75,000 to 1,00,000 eggs. The fertilized eggs are spread uniformly on inner hapa. When the hatching occurs, hatchlings escape to outer hapa leaving behind the egg shells in inner hapa. This hatchery is still in use all over the country due to its low cost and minimum management, but its efficiency is low.

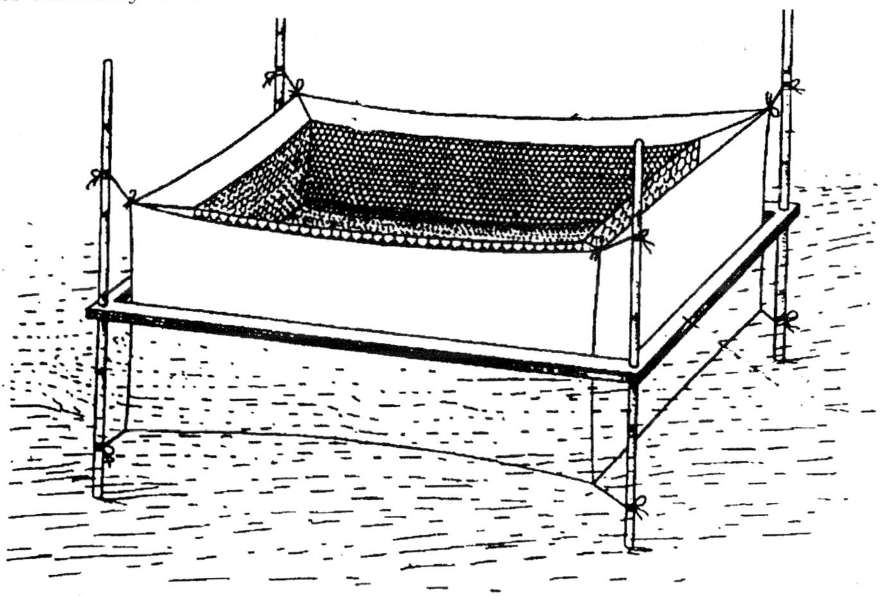

Fig. 1. Floating hapa

3. Floating hapa hatchery (Fig. 1)

This is only a slight modification of ordinary hatching hapa. In place of bamboo poles, the hapa is fitted with a light rectangular wooden frame which keeps the hapa in floating condition. The hatching technique in floating hapa is the same as in ordinary hatching hapa.

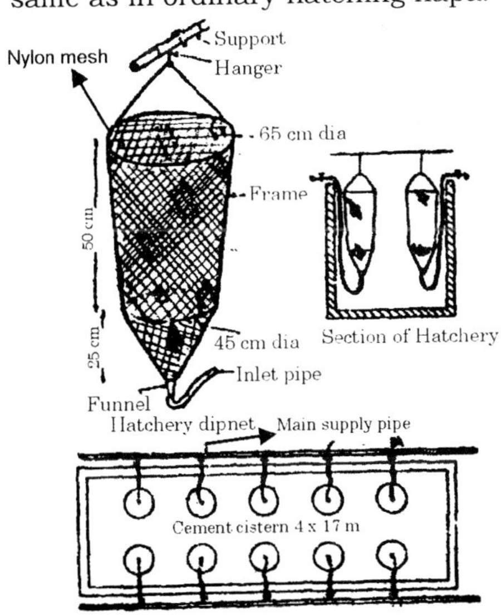

Fig. 2. Hanging dip net Hatchery

4. Hanging Dipnet hatchery (Fig. 2)

Sometimes, soft material like nylon or canvas is used to prepare funnel-shaped immersed type of incubators. These are installed or hanged in a rectangular water tank. The funnel incubators have water inlet from the bottom to keep the eggs in rotation for their development.

5. Shirgul linear bin hatchery

It contains a rectangular aluminium tub in which three egg vessels can be kept at a time of hatching. The outer aluminium tub has a size 54 x 18 x 22 inches. which has three chambers. The plunger lid is used to keep the eggs inside the vessel constantly circulating with mild water current and at the same time it prevents the eggs from escaping off the egg vessel.

6. Zoug Jar hatchery (Fig. 3)

Zoug jars are large inverted bottles open at the top, terminating at the bottom in a narrow neck. They are set up vertically, the neck terminates downwards, linked to a pressurized water pipe and separated from it by a tap-topped by a nozzle. The nozzle supports the jar and retains its equilibrium. The normal height of the jar is around 60 to 70 cm and the diameter is from 15 to 20 cm. The average flow is around 4 litres per minute. The water flows over the top of each jar and runs down.

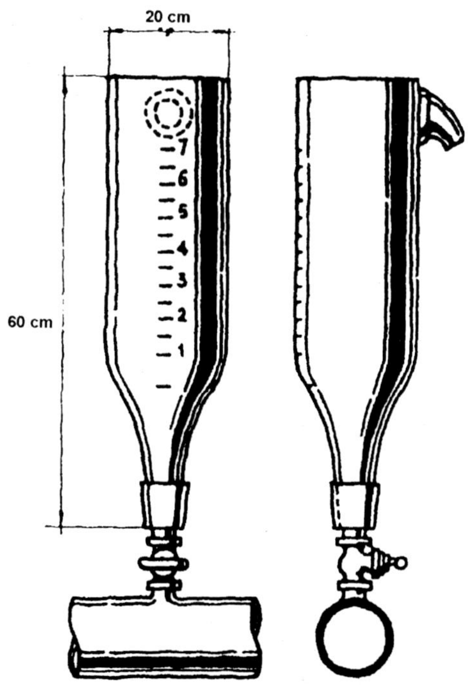

Fig. 3. Zoug Jars

7. Glass Jar/Polythene Jar hatchery (Fig. 4)

This hatchery was designed by Central Inland Fisheries Research Institute on the lines of Zoug jars to reduce the mortality rate and to increase hatching percentage. The hatchery is a self-contained unit consisting of overhead tank breeding tank, hatchery jar and the spawnery. The overhead tank is connected to hatchery jars by galvanized pipe of diameter ranging from 2.5 cm to 7.5 cm. Clean and plankton-free water is used for breeding purpose.

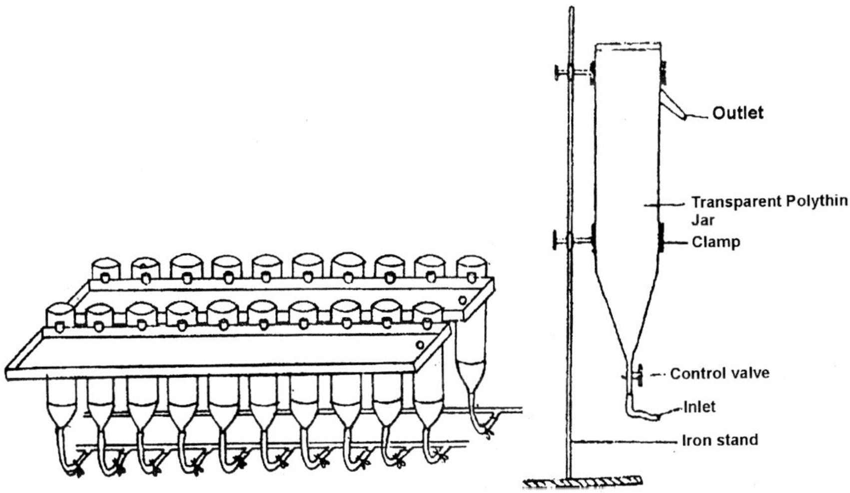

Fig. 4. Glass jar/plastic jar Hatchery

The hatching jars are cylindrical towards the top and funnel shaped towards bottom. The cylindrical part has inner diameter of 13 cm while the total length is 61.5 cm. The capacity of each jar is 6.35 lit. to accommodate 0.5 lakh developing eggs of Indian and Chinese carp at a time. The jars are vertically placed through circular holes made on wooden table. The height of table is conveniently adjusted. The size of the table can vary with the number of jars to be accommodated. The top of each jar is fitted with a galvanized ring having a beak which serves an outlet for water. The overflowing water from each jar flows into a long open galvanized iron conduit fitted lengthwise in between the two rows of jars and then flows down to spawnery. Each jar is connected at bottom to a bibcock (tap) with a rubber tube which, in turn, is fitted with respective tap on the inlet pipe.

The spawnery is a nylon hapa fitted on a frame and placed inside a cement cistern with a provision of overhead shower to keep the water cool. For hatching hardened eggs are liberated in each jar and normally hatching take place in 12-15 hours at 26-27 ^{0}C temperature. A flow of water 600-800 ml / min is maintained for Indian major carp and 800-1000 ml / min for silver and grass carps. After completion of hatching the rate of water flow can be slightly increased for speedy escape of hatchlings (Rath, 1993).

8. CIFE D-81 hatchery (Fig. 5)

It is a portable rigid type hatchery and its components are made up of low density polyethylene (LDPE). The major constituents of the unit are (i) Breeding-cum-spawnery, (ii) Hatching unit, and (iii) Air compressor with distribution system.

Fig. 5. CIFE D-81 Carp Hatchery

The breeding unit comprises cooling tower with HDPE (High density polyethylene) overhead tank (5000 lit. capacity) and two large portable plastic pools of size 0.9-2.5 m diam. and 0.6-1.2 m high or 2000 – 3000 lit. capacity. The injected fishes are kept in the pool for breeding and after breeding the same pools are used as spawnery.

The hatching unit consists of 6 to 12 conical bottomed hatchery jar with 40 lit. capacity each, supported on iron tripod. Each jar is funnel shaped and measures 48-62 cm in height with 32-44 cm diameter at top. The bottom of bucket is connected with bibcock to overhead tank. The egg container which is made of 2 mm calico cotton netting fitted on aluminium frame is hanged on bucket so as to keep a space of 4.5 cm at sides and 7.5 cm at bottom. The rate of water flow is conveniently maintained at 6-8 lit. /min. and 8 – 10 lit./min. for Indian major carp and Chinese carp respectively. The spawn are collected in spawnery with outflow of the water through tube fitting with outlet of bucket. For proper oxygenation of water constant aeration is made inside the bucket and spawnery by ½ or 1 HP air compressor through distribution pipes and air stones. The egg capacity per jar is one lakh and its range is 3 jar system to 36 jar system. It is useful for laboratories, educational institutes and fish farms. Some of the abiotic factors remain under control resulting in the increased efficiency and low spawn mortality.

9. CIFE D-85 hatchery (Fig. 6)

This hatchery unit is similar to CIFE D-81 model except containing a larger hatchery jar (incubation chamber). The ratio of hatchery jar and spawn receiver was 4:1 in D-81 model, while in D-85 it was 1:1. The egg capacity per jar is 8 to 18 lakh. It's range can be from one jar to twelve jars. It is useful for educational institutions and fish farms.

Fig. 6. CIFE D-85 Carp Hatchery

10. CIFE D-86 hatchery (Fig. 7)

To increase hatching at reservoir site CIFE has developed this hatchery in 1986. It has following components: (i) Plastic raft pool 8 x 3 ft (egg holding capacity 19 lakh), (ii) Air lift system incorporated in pool, (iii) Sprayer aeration/circulation system, (iv) floating egg sac 19 nos. for 8ft diameter, (v) Aeriation system with air stone, (vi) ½ HP pump, (vii) ½ HP air compressor, and (viii) 8 filters to clean water.

Fig. 7. CIFE D-86 Carp Hatchery

11. Modified fish hatchery for Western Rajasthan (Fig. 8)

CIFE D-81 model of hatchery was modified by attaching a filteration and sedimentation unit, recirculation of water in breeding tank and use of hanging dipnets for hatchery (Mohan, 2000). The three chambered filteration and sedimentation unit receive total effluents from breeding and hatching units. The first chamber is provided with sand and gravel beds through which the filtered water enters into the second chamber where water rises slowly and is supplied with compressed air. The supernatant from this chamber is lead by porcelain drain to the bottom of the third chamber from where it rises slowly and the supernatant of this chamber flows by gravity to the brooders' pond to compensate the evaporation losses.

The water of the breeding pool is recirculated with ½ HP motor after injecting inducing hormone to the brooders, thus a continuous supply of water is regulated and the hanging dipnets are used for the hatching process in place of hatchery buckets to reduce the water requirement.

The component of hatchery unit are: (i) overhead and under groundwater storage tanks, (ii) breeding unit, (iii) hatching and nursery unit, and (iv) filteration and sedimentation unit, and (v) brooder stocking tanks. Hatching of about 4 lakh eggs can be conducted at a time in this unit. The minimum quantity of water is used for a breeding operation. So, the unit is suitable in the desert areas where quality water is limited.

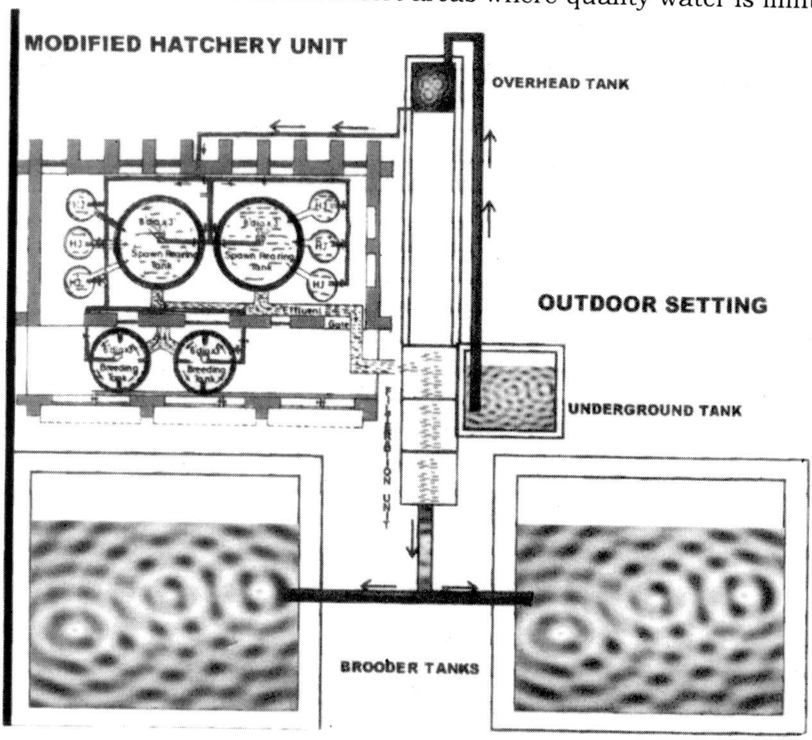

Fig. 8. Modified D-81 Hatchery for Western Rajasthan

12. "Sen Model": portable mini circular hatchery (Fig. 9)

It is very similar to Chinese hatchery with a portable hatching unit and is made of cylindrical G.I. sheet drum 0.8m diameter and 0.67 m height mounted on iron frame. At the center of drum through circular hole a PVC pipe of 5 cm diameter is fitted. The PVC pipe is 60 cm high and serves as a spill over of waste water. The pipe has 3 detachable sections to adjust height of water according to bulk of eggs. On the interior surface, at the bottom of the drum another PVC pipe of same diameter is fitted with it horizontally with an elbow joint to drain out waste water. Inside the drum a concentric collar ring of GI sheet is fitted at the bottom which is 27.5 cm in diameter. and 15 cm in height. A cylindrical grill (65 cm high) of iron mesh is fitted with the collar and this forms the inner chamber which is covered with 1/40th in. nylon mesh cloth. Under the outer chamber a concentric pipe line of 1.25 cm diameter runs at the radius of 27.5 cm from the center of the drum. Eight copper pipes each of 12.5 cm length and 8 mm diameter are welded with this pipe and passed inside the drum, which are bent unidirectionally at an angle of 60^0 towards the base 5 cm above bottom. The circular pipe below, with a T-joint and an adjustment valve, is connected with the overhead water supply tank. Six to ten lakh eggs can be accommodated at a time in the outer chamber with an average of 75% hatching rate. An L-shaped pipe fitted at one corner of the bottom of the outer chamber is meant for collecting spawn. The mouth of this pipe is kept closed when the hatchery unit is in operation. The hatchery requires only 7-9 lit. of water flow/min. This hatchery unit which is quite popular in East and West Godavari districts of Andhra Pradesh is highly suitable for small scale farmer due to its low cost. Four ranges of models are available with an egg-holding capacity of 4-30 lakh. The components of this unit are hatchery tank, breeding tank, oxygenation pool, overhead tank, nursery rearing pool and pumps.

Fig. 9. Sen Model of mini potable circular hatchery

Fig. 10. Chinese circular Hatchery (Masonary structure)

13. Circular hatchery of Masonry Structure (Fig. 11)

The circular hatchery of masonry structure evolved in China during 1970 and, thereafter, was introduced in India and many other countries. The hatching is more effective in this system due to uniform distribution of oxygen by centrifugal flow of water. A large quantity of filtered and iron magnesium free water is generally required for effective functioning of this hatchery.

Fig. 11. Incubation tank of Chinese Hatchery

A self-contained small scale unit of hatchery comprises: (i) an overhead water storage tank, (ii) circular spawning pool, (iii) two circular incubation pools, and (iv) a spawnery.

The overhead tank is masonry structure of 30,000 to 50,000 lit. capacity constructed about 3.5m above the base of spawning pool. The spawning pool which is a clock shape structure has two parts, namely, (a) Circular (6-8m diameter to 1.2-1.5m height) breeding chamber proper, and (b) Rectangular (3 x 2.5 x 1.6 m) egg collection chamber connected to each other by 10 cm diameter underground GI pipe. This pipe, which has connection with a centrally placed orifice of the breeding pool serves dual purpose as the main outlet of water of breeding pool and the passage for automatic transfer of eggs by slope gradient from breeding chamber to egg collection chamber. The orifice is covered by 25 mm meshed wire to prevent access of the breeders and to make free passage of eggs. The water from the egg collection chamber flows out by another 10 cm diameter pipe provided with screw valve arrangement system to control water level of both breeding pool and egg collection chamber at a time. The eggs are collected in egg collection net having narrow and broad mouths. The narrow mouth is tied at the mouth of pipe in the egg collection chamber. The broad mouth is just tied with rope and left dipping under water with sinkers. At the time of egg collection the broader mouth is untied and eggs are taken out by mug and bucket and released in incubation pool. The bottom slope of the circular tank is 1:30 towards the orifice. The inlet of breeding pool, a 10cm diam. G.I. pipe emerging out the overhead tank, branches into 3 L shaped 5cm diameter pipes with screw valves for controlled clockwise flow of water inside the spawning pool. These pipes are placed 15 cm above the bottom to maintain better circulation of water. Another branch of the main inlet which connects the circular spray pipe (4cm diameter) spreading along the inner surface of the pool near top, is fitted with number of sprayers at regular intervals at an angle of 40^0 towards the center for showering and keeping the water cool.

The incubation or hatching pool is a circular cemented tank measuring 3-4.5m diameter and 1.2-1.5 m deep. The bottom slope from the peripheral wall to central orifice is 1:25. The orifice is a 10cm diameter GI turn down the outlet pipe. The hatching pool is conveniently constructed 30 cm below the spawning pool so as to facilitate automatic transfer of eggs from the breeding pool to hatching pool. The hatching pool consists of two chambers and both the chambers are partitioned by a monofilament cotton marking cloth GARFAB (GHP 40-40/20) fitted across on the vertically and horizontally arranged cross iron rods on 28 cm thick x 20 cm high circular wall of the inner chamber. The hatching pool is directly connected to overhead tank by 7.5 or 10.0 cm diameter G.I. pipe which after bring branched into two 5 cm pipes joins the two underground circular 5 cm pipes in the outer chamber. About 21 – 24 number of duck mouths (each 2 cm diameter) pipes fitted in two rows above the ground at 90^0 angle in regular intervals of clockwise direction are present to maintain a centrifugal force of water at desired speed (45-60 lit/min) for sieving out disintegrated egg shells after hatching through the partition cloth into the inner chamber and thereafter through the outlet pipe. A spawn collection pipe of 10 cm diameter connects the spawnery with the outer chamber of the hatching pool. The water inlets pipes and the spawn transfer pipes are provided with screw valves to control the speed of inflow water and facilitate discharge as and when required.

The spawnery, a rectangular cemented tank (3m x 2m x 1m), is constructed close to hatching pool. A spawn collection pipe of 10 cm diameter fitted with screw valve emerging out of the outer chamber of hatching pool opens into the spawnery for automatic draining of the spawn by slope gradient. A water supply pipe of 10 cm dia with screw valve arising from overhead tank directly connects the spawnery with arrangements of a number of sprayers placed longitudinally over the spawnery for cooling purpose. An outlet pipe of 10 cm diam. is fitted at the base for complete drainage.

Now a days modern hatcheries like CIFE D models and circular Chinese hatcheries are mainly used in fish farms. These units have some advantages like (i) circular breeding tanks with continuous flow system, (ii) adoptable to large operation, (iii) dead spaces are absent which make uniform distribution of dissolved oxygen hence hatching is more effective, (iv) water quality can be improved by additional aeration. Disadvantage in most of the modern hatcheries is that they require a large quantity of water and the separation of egg shells from hatching is not very effective.

BREEDING TECHNIQUES

In all types of hatcheries mature males and females of culturable carps are used for induced breeding. The technique was developed by Houssay of Argentina in 1931 and was applied in Brazil by Von Ihering *et al.* 1937. In India, the first attempt of induced spawning was made in 1937 by Khan (1938) using pituitary extract. Afterwards, a large number of substances were used for fish spawning like Human Chorionic Gonadotropin, salmon Gonadotropin, SG-G 100 (Lam 1982), Hypothalmic extract (LH-RH) (Lam *et al.* 1975). In 1975, a highly effective luteinizing releasing hormone analogue (LRH-A) was synthesized. Peter (1980) observed that LRH-A is more effective in teleost. Now, the synthetic chemicals like ovaprim containing calibrated amount of salmon gonadotropin releasing hormone and domperidone, a dopamine antagonist dissolved in an organic solvent, are commercially available. The latter is produced by Syndel laboratories of Canada. An indigenous product is also commercially available called ovatide. These two chemicals are now very frequently used to breed fishes. In modern circular hatcheries, a continuous flow of water is maintained to simulate reverine conditions and showering for rainy conditions. The compressed air is also supplied in some of the hatcheries. Chlorine-free crystal clear, plankton and sand free fresh water is used for fish spawning. After breeding, the eggs are transferred to incubation pools (Chinese hatcheries) and in hatching buckets (CIFE Models). After hatching, the eggs are retained in the spawn collection tank till stocking to the nursery pond.

REFERENCES

Alikunhi, K.H., A. Dutta, V.D. Singh, A. Singh and G.P. Dubey 1964. Observations on the breeding of carps in bundhs, near Nowgong, Madhya Pradesh during July-August, 1964. *Bull. Cent. Inst. Fish. Educ. Bombay* 1 : 22.

Chaudhuri, H. and S.D. Tripathi, 1976. Problems of warm water fish seed production. Advances in Aquaculture 127-134. Paper presented at *FAO Technical Conference on Aquaculture, Kyoto, Japan,* 26 May – 2 June, 1976.

Dwivedi, S.N., 1976. Indigenous vertical jar hatchery. *J. Indian. Fish. Assoc.* 6 (1 & 2) 140.

Dwivedi, S.N. and A.K. Reddy, 1986. Fish breeding in a controlled environment. Carp hatchery CIFE D 81. *Aquaculture* 54 (1 & 2) 27-36.

Houssay, B.A. 1931. Accion sexual de la hipofissi en los percesy replites. *Rev. Soc. Orgent. Bio.* 6: 686 – 688.

Khan, H., 1938. Ovulation in fish (effect of administration of anterior pituitary gland) *Curr. Sci.* 7 (5) 233-234.

Lam, T.J., S. Pandey and W.S. Hoar 1975. Induction of ovulation in gold fish by synthetic luteinizing hormone releasing hormone (LHRH). *Can J. Zool.* 53: 1139 – 1192.

Lam, T.J. 1982. Application of endocrinology to fish culture. *Can J. Fish Aqua. Sci.* 39: 111 – 137.

Marcel, H. 1971. Text book of fish culture – breeding and cultivation of fish. *Fishing News Book Ltd. Farmharm Survey, England.* 437 pp.

Mohan, D. 2000. Modified fish hatchery for the stressed ecosystem of Western Rajasthan. *Aquacult.* 1 (2) 111 – 119.

Peter, R.E. 1980. Serum gonadotropin levels in mature male gold fish in response to Lutinizing hormone – releasing hormone (LH – RH) and des gly 10 – D – Ala – LHRH ethylamide. *Can. J. Zool* 58: 1100 – 1104.

Rath, R.K., 1993. *Fresh water aquaculture.* Scientific Publishers (India), Jodhpur 493 pp.

Shirgur, G.A. 1972. Observations on efficiency of vertical flow jars (glass and transparent polythine) for hatching of major carp eggs. *J. Indian Fish. Assn.* 2(1- 2) 90 – 101.

Shrigur, G.A. 1973. Observation on efficiency of large capacity vertical flow bins for hatching major carp eggs. *J. Indian Fish. Assn.* 3: 80 – 85.

Von Ihering, K. 1937. A method of inducing fish to spawn. *Prog. Fish. Cult.* 24: 15 – 16.

Changing Faunal Ecology in Thar Desert, pp. 187-229, 2005.
Eds.: B.K. Tyagi and Q.H. Baqri

11

DIVERSITY, ABUNDANCE AND DOMINANCE OF AVIAN SPECIES IN THE THAR DESERT OF RAJASTHAN

C. Sivaperuman, Sumit Dookia, P.L. Kankane and Qaiser H. Baqri

Desert Regional Station, Zoological Survey of India, Pali Road, Jodhpur-342005, India

INTRODUCTION

Interestingly, a number of studies have been conducted on the avifauna of Thar Desert but these were mainly either of Kutch and Saurashtra part of Gujarat or from Sind which is now in Pakistan (Ticehurst, 1922a, 1922b, 1923a, 1923b, 1923c 1923d, 1924a & 1924b). As far as Rajasthan part of the Thar Desert is concerned, Adam (1873 & 1874) was the first to report birds of Sambhar Lake and its vicinity. However, our previous knowledge of birds of the area is derived from the following two main sources. The first is the work of Whistler (1938) who had conducted systematic surveys during the year 1933-34 in erstwhile Jodhpur State, which roughly constituted Jodhpur, Jalore, Nagaur, and Pali districts of present state of Rajasthan. The other study was conducted by Rahmani (1996 & 1997) on avian fauna in Jodhpur, Barmer, Bikaner, Churu, Ganganagar and Nagaur districts during 1993-94. Besides, there are few auxiliary reports like: Sangha & Kulshreshtra (1993); Sangha (1994); Soni (1994); Rana, Jain & Tripathi (1996); Kumar & Musavi (2001); Vohra and Goyal (1992); Changani (2002). Recently

Sangha (2002) published supplementary notes on the avifauna of the Thar Desert. He has further updated the list by Rahmani (1997), with some additional records.

However, most of the studies cited above provide the account of birds of a particular locality and/or habitat. Hence, we have made an attempt under the project to cover as much habitats as possible so that we can have a complete picture of the avifauna of the study area. Since the Thar Desert, located in the north-western part of India, is predominantly sandy hot desert which extends from and Saharan countries and middle east in the west, its avian fauna is also the admixture of Saharan, Palaearctic and Oriental elements.

According to Rahmani (1996) the bird fauna is fast changing in the Thar Desert, mainly because of the Indira Gandhi Nahar Pariyojna (IGNP). The present study also reveals that a number of forest and aquatic birds that had never been reported earlier are now seen frequently in the area, mainly because of the Indira Gandhi Pariyojna.

The present study was conducted in the Thar Desert of Rajasthan during the years 2000-2003 under the project sponsored by the Ministry of Environment and Forests, Govt. of India, with the following objectives: To conduct intensive and extensive surveys to study the avifauna including their relative abundance and dominance, to map out the distribution of threatened species of birds, and to assess impact of ecological changes on avifauna in the vicinity of Indira Gandhi Nahar Pariyojna (IGNP) area, Ghaggar bed and Sewan grass (*Lasiurus sindicus*) areas. The information collected during the field surveys are being furnished in this paper which reports 271 species belonging to 55 families under 17 Orders of Aves, of which 23 species have been recorded for the first time from the Thar desert of Rajasthan. Besides diversity, relative abundance and richness of species (order wise, family wise, habitat wise, month wise) have also been calculated and presented.

MATERIAL AND METHODOLOGY

Under this project, birds in the Thar Desert were studied throughout the year, except August, in different habitats like sandy areas, stable and shifting type of sand dunes, rocky areas, gravel, sewan grass, lakes and tanks of saline and fresh water, canal areas and agricultural fields etc. which provide excellent shelter to birds in the Thar Desert. The avian fauna includes the resident birds which live here throughout the year, and also the migratory birds, which come to pass winter and return to their cold native countries before the commencement of summer.

Species Composition

In each habitat (grass land, scrub land, wetlands, canal side plantation, etc.) of the study area, appropriate number of transects were laid depending on the continuity of the habitat type. Each line transect was walked / travelled in the day time at a constant pace when bird activity was at its highest. Bird species were estimated by line transects and total count methods (Burnham *et al.*, 1980; Hoves and Bakewell, 1989). Birds were identified using binoculars (10 X 50, 7 X 35) with the help of standard field guides and reference books (Ali and Ripley, 1983; Grimmett *et al.*, 1998).

Species Richness, Abundance and Diversity

Bird species richness (number of species), abundance (number of individuals) and diversity were calculated from the census data and field observations. Diversity measures the variation in the richness and abundance. Hence, the study on diversity is considered important in ecological studies because such study helps us to know the stability and health of an ecosystem. Diversity index combines the information on multiple species into a single number. These indices provide easily understandable measures of diversity. Richness indices like R1 and R2, Hill's number N1 and N2, diversity indices like Shannon-Weiner (H') and Simpson's (λ), Evenness measures like E1, E2 have been calculated using the computer program SPDIVERS.BAS developed by Ludwig and Reynolds (1988).

1. RICHNESS INDEX

Margalef index (1958)

$$R1 = S\text{-}1/\log(n)$$

Menhinick index (1964)

$$R2 = S/\sqrt{n}$$

2. HILL'S NUMBERS

Hill's diversity N1 calculated from Shannon-Weiner index

$$N1 = e^{H'}$$

Hill's diversity N2 calculated from Simpson's index

$$N2 = 1/\lambda$$

3. SHANNON-WEINER DIVERSITY INDEX (H')

$$H' = -\sum p_i \log p_i$$

where p_i = the proportional abundance of the ith species (n_i/N).

4. SIMPSON'S INDEX (λ)

$$D = \sum(n_i(n_i-1)/(N(N-1)$$

Where n_i = the number of individuals in the ith species, and N = the total number of individuals.

RESULTS

The information provided here is based on 154 surveys conducted in almost all the habitats throughout the year except in August, by different members of the research team of the project from November, 2000 to May, 2003 in 12 districts (Barmer, Bikaner, Churu, Ganganagar, Hanumangarh, Jaisalmer, Jalore, Jodhpur, Nagaur, Pali, Sirohi and Sikar) constituting Rajasthan part of the Thar desert. The details about these localities and their geographic distribution are being furnished in Map 1. **A total of 271 Species of birds have been identified** in the study area, besides their district wise distribution pattern was also recorded (Table 3). Out of these species, 23 have been recorded for the first time from the Thar desrt of Rajasthan.

Distribution

Based on data collected from 148 localities surveyed under the project, the distribution records of all the 271 species have been analysed. This information has been used to map out the distribution of some of the avian species, specially distribution maps of the species listed under Schedule-I of the Wildlife (Protection) Act, 1972 (Plates 1-6). These maps will not only be helpful in the study of distribution of birds in the area but also in formulating the management plan for the endangered species.

The Indira Gandhi Nahar Pariyojana (IGNP), which brings Himalayan water through Harike barrage to the thirsty Rajasthan part of the Thar Desert, is now the life line to Rajasthan Thar inhabitants. Though the IGNP has not become fully operational, large-scale changes have already taken place at landscape level due to plantation of tall tree species in canal area, besides creation of wetlands (some of them are perennial) owing to escape reservoirs designed along the canal and also due to seepage of water. Moreover, availability of water through network of canals has changed the farming practices, *i.e.,* from traditional crops like millet, guar, etc to commercial crops. In addition, plantation work carried out by forest department especially in the hilly areas of the Thar has resulted in good vegetation, which offers a very conducive environment to number of winter visitors in this part of the Thar. The resultant effect is the introduction of new avian species in the area.

The most interesting finding is the recording of **23 avian species for the first time from the study area.** These 23 species are spread over eleven families and five orders.

In order to prepare baseline data we have drawn a comprehensive checklist using most of the above cited references. The systematic accounts and other details (number of birds, locality etc.) of these newly recorded species of birds are given below. Common and scientific names were after Manakadan and Pittie (2001).

NEW RECORDS LISTED ACCORDING TO SYSTEMATIC POSITION

Order **Anseriformes**

Family Anatidae

Marbaled duck ***Marmaronetta angustirostris***

Ten individuals were seen at 1140 RD, Escape Reservoir, Jaisalmer, on 23rd February 2001

Order **Ciconiiformes**

Family **Accipitridae**

Osprey ***Pandion haliaetus***

One individual was seen on Digga lake, Mohangarh, Jaisalmer on 24th February 2001

Order **Pelecaniformes**

Family **Pelecanidae**

Dalmatian Pelican ***Pelecanus crispus***

Winter visitor to the area sighted more than 200 Pelicans, on 15th January 2002, in the eastern part of the desert at Sardarsamand in Pali district.

Order **Charadriiformes**

Family **Scolopacidae**

Bar-tailed Godwit ***Limosa lapponica***

Recorded twice in two wetlands, first time nearly 300 birds on 13th February 2001 in the eastern part of the desert at Sardarsamand in Pali district and second time on 7th September 2002 in the northern part at Badopal in Hanumangarh district.

Buff-breasted Sandpiper ***Tryngites subruficollis***

Sighted once in the northern part of the study area in Jamsar Canal area on 30th March 2001 in Ganganagar district.

Family **Burhhinidae**

Great-Stone Plover ***Esacus recurvirostris***

Sighted 11 birds on 13th February 2001 in the eastern part of the desert at Sardarsamand in Pali district.

Order **Columbiformes**

Family **Columbidae**

Oriental Turtle Dove ***Streptopelia orientalis***

Recorded once in the heart of the desert in a scrub forest near Osian in Jodhpur district on 26th Sept 2000.

Spotted Dove ***Streptopelia chinensis***

Recorded twice, four birds at Chang forest reserve, Pali district on 14th Feb. 2002 and 6 birds at Mungthala in Sirohi district on 12th Feb. 2003.

Order **Piciformes**

Family **Capitonidae**

Crimson-throated Barbet ***Megalaima rubricapilla***

Recorded twice in the north-eastern part of the desert once in deciduous forest in Harsh Parbat near Sikar in Sikar district and second time near Indrapura in Jhunjhunun district during first week of November 2001.

Order **Passeriformes**

Family **Aludidae**

European Calendra Lark ***Melanocorypha calandra***

Sighted 17 birds in the western part of the desert in a sand dune at Sam/Sudasari area in Jaisalmer district during last week of February 2001.

Family **Hirundinidae**

Northern House Martin ***Delichon urbica***

Recorded twice, in IGNP area, in the western part of the desert; seven birds were sighted at Digha escape reservoir in Jaisalmer district during last week of February 2001 and second time eight birds sighted in Bikaner district at 750 RD escape reservoir during first week of December 2001.

Family **Pycnonotidae**

Red Whiskered Bulbul ***Pycnonotus jocosus***

Recorded four times in western part; first sighting was near Chandan in Jaisalmer district where more than 20 birds were perching on a tree on 28th January 2002; second time, on 29th January 2002, a pair was sighted near a bush at Sam in Jaisalmer; third time a pair perching on a tree at Digha escape reservoir on 30th January; and fourth time a pair perching on a tree at 507 RD escape reservoir near Sansardeshar in Bikaner district on 10th February 2002.

Family **Laniidae**

Southern Grey Shrike ***Lanius meridionalis***

Recorded four times from almost four corners of the desert; first time sighted near Indrapura in Jhunjhunun district during first week of November 2000; second time from Rangmahal, in Ganganagar district on 31st March 2001; third time one bird in the sand dunes near Sam in Jaisalmer district on 29th January 2002; and fourth time four birds on 14th March 2002 in sandy plains near Panchla Sidh in Nagaur district.

Family **Muscicapidae**

Grey-headed canary flycatcher ***Culicicapa ceylonensis***

Five individuals were seen at 582 RD, Chhattargarh, Bikaner on 3rd December 2000

White-capped water Redstart ***Haimarrornis leucocephalus***

At 507 RD head Bikaner on 4th December 2000

Eurosian Reed-Warbler ***Acrocephalus scirpaceus***

Sighted eight birds at 1140 RD escape reservoir during last week of February 2001 in Jaisalmer district.

Rusty-tailed Flycatcher ***Muscicapa ruficauda***

Sighted twice in the northern part of the desert. First time in a deciduous forest at Harsh Parbat in Sikar district and second time in the agriculture field near Indrapura in Jhunjhunun district in first week of November 2000.

Smoky Warbler ***Phylloscopus fuligiventer***

Sighted a pair on 13th February 2001 at Sardarsamand in Pali district.

Franklin's Prinia ***Prinia hodgsonii***

Sighted two birds in the rocky hills of Harsh Parbat in Sikar district on 12th April 2001.

White-throated Fantail-Flycatcher ***Rhipidura albicolis***

Sighted once in the deciduous forest of Harsh Parbat during the first week of November 2000 in Sikar district.

White-headed Babbler ***Turdoides affinis***

Recorded twice from the northern part of the desert. First time 22 birds were sighted in a grassland near Suratgarh in Ganganagar district on 6th September 2002 and second time 12 birds in Kola forest farm in Hanumangarh district on 9th September 2002 .

Family **Sturnidae**

Jungle Myna ***Acridotheres fuscus***

Recorded thrice from the three corners of the desert. First time ten birds were sighted on 20th June in Rajasthan Agriculture University Field Station near Jawai dam in Pali district; second time eight birds were sighted in a stony area on 21st June 2002 near Merta City in Nagaur; and third time 12 birds were sighted on 11th July 2002 in a *Oran* (sacred place) named Bhadriya in Jaisalmer district.

Family **Zosteropidae**

Oriental white-eye ***Zosterops palpebrosus***

Recorded twice; one bird at Gadra road on 19th May 2003, in Barmer district and eight birds at Chatargarh in Dec. 2000 in Bikaner district.

Systematic list of Species showing Extension of Range

Order Anseriformes

Family Anatidae

Ferruginous Pochard *Aythya nyroca*

Family Charadridae

White-tailed Lapwing *Vanellus leucurus*

Family Laridae

Black-headed Gull *Larus ridibundus*

River tern *Sterna aurantia*

Order Galliformes

Family Phasianidae

Black Francolin *Francolinus francolinus*

Order Coraciiformes

Family Alcedinidae

Pied kingfisher *Ceryle rudis*

Common kingfisher *Alcedo atthis*

Order Piciformes

Family Picidae

Black-rumped Flameback *Dinopium benghalense*

Order Passeriformes

Family Muscicapidae

White-browed Fantail *Rhipidura aureola*
Oriental Magpie Robin *Copsychus saularis*

Avian Diversity

During the present study, a total of 272 species of birds belonging to 17 orders and 55 families have been recorded from the Thar Desert of Rajasthan. Out of these, 223 species were resident and 49 migrant species.

District Wise Distribution

The information on district wise distribution of the bird community has been analysed and furnished in the Table 3. As per analysis, only three species of birds were recorded in all the districts of Thar in Rajasthan, *viz.* like Red-wattled Lapwing, Eurasian Collared-Dove and Red-vented Bulbul.

Species Richness and Abundance

Order wise richness and abundance

Of all the species belonging to 17 orders recorded, highest species richness and abundance were observed of order Passeriformes followed by Charadriiformes (Fig. 1)

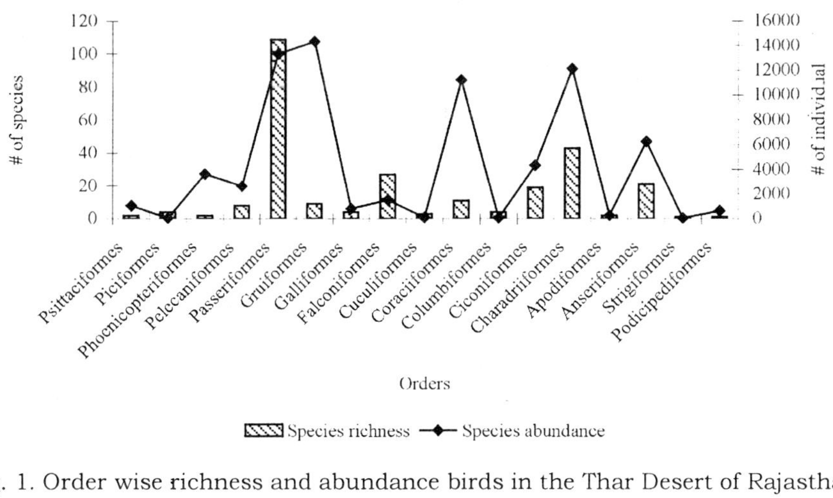

Fig. 1. Order wise richness and abundance birds in the Thar Desert of Rajasthan

Family wise richness and abundance

Out of fifty-five families of birds recorded from the Thar Desert, species richness was highest in the family Muscicapidae and abundance in Rallidae (Fig. 2).

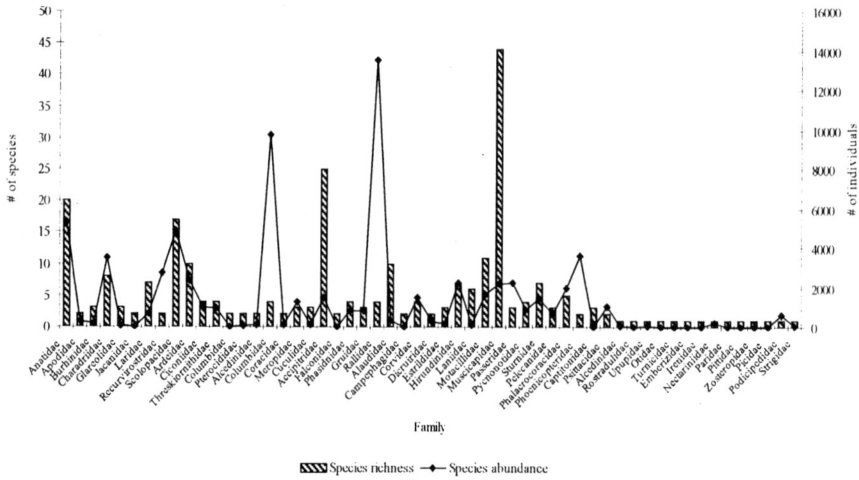

Fig. 2. Family wise species richness and abundance of birds in the Thar Desert of Rajasthan

Month wise richness and abundance

Species richness was highest in the month of January followed by February during the study period (four years). Bird species abundance was observed maximum in the month of February and minimum in July (Fig. 3).

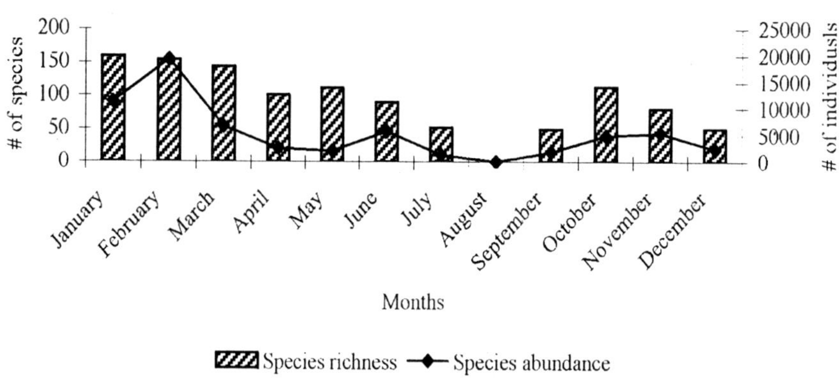

Fig. 3. Species richness and abundance of birds in different months in the Thar Desert of Rajasthan

Habitat wise richness and abundance

Species richness and abundance was highest in freshwater annual and freshwater perennial habitats (Fig. 4).

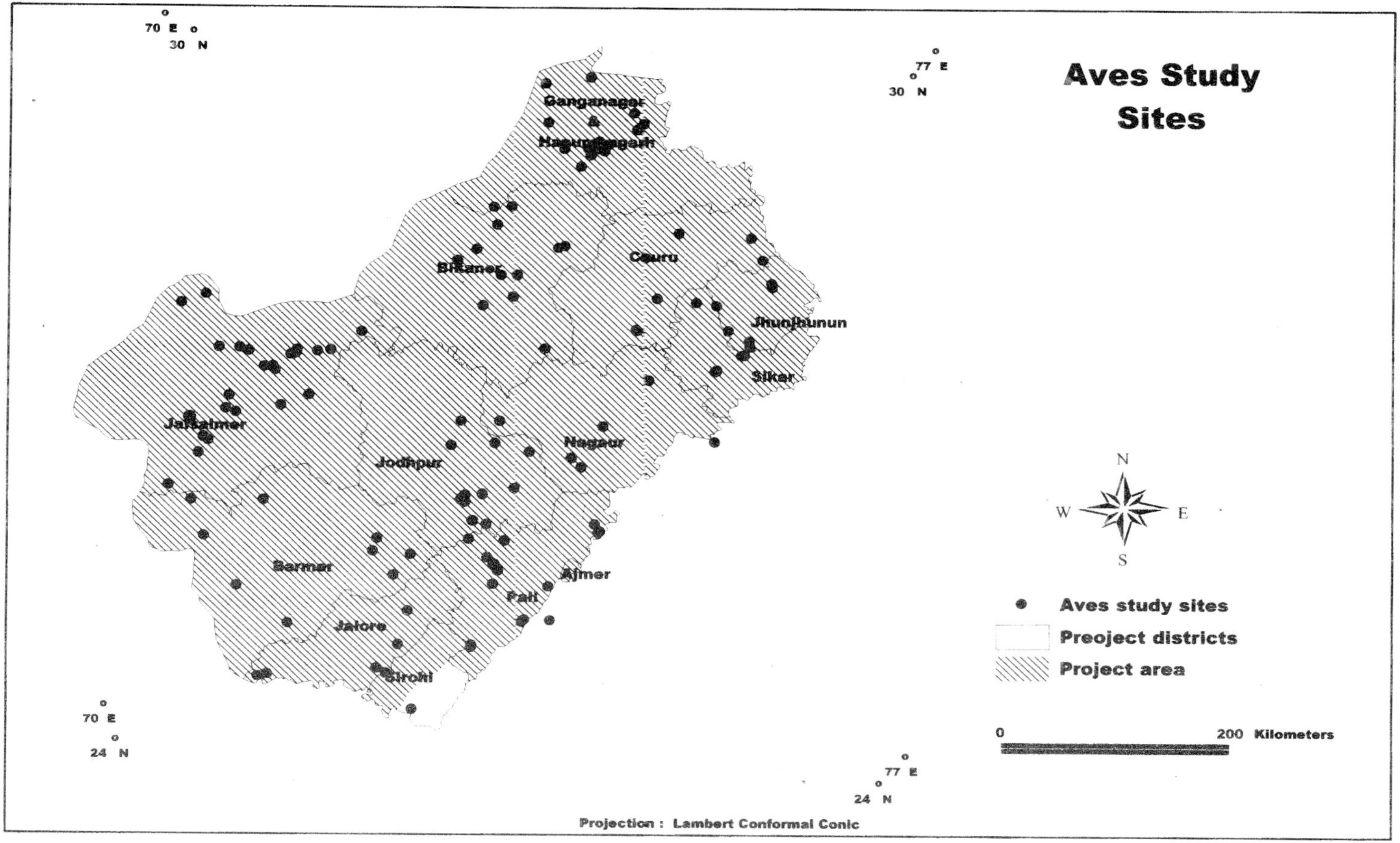
Aves Study Sites
70° E 30° N
77° E 30° N
70° E 24° N
77° E 24° N
Ganganagar & Hanumangarh
Bikaner
Cauru
Jhunjhunun
Sikar
Jaisalmer
Jodhpur
Nagaur
Barmer
Jalore
Sirohi
Pali
Ajmer
N
W
E
S
Aves study sites
Preoject districts
Project area
0
200 Kilometers
Projection : Lambert Conformal Conic

Map 1

Sightings of Birds Listed Under Schedules of WI (P) Act From Thar Desert

Bonelli's Eagle
Hieraaetus fasciatus

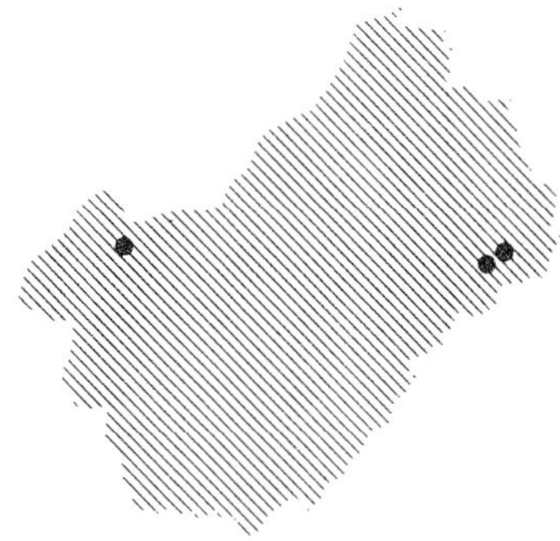

Eastern Imperial Eagle
Aquila heliaca

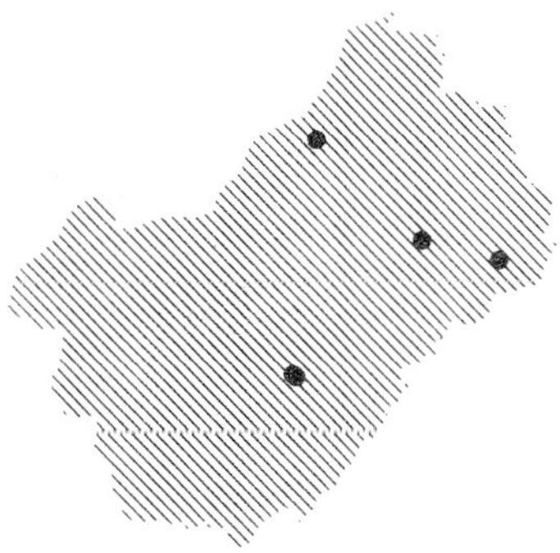

Greater Spotted Eagle
Aquila clanga

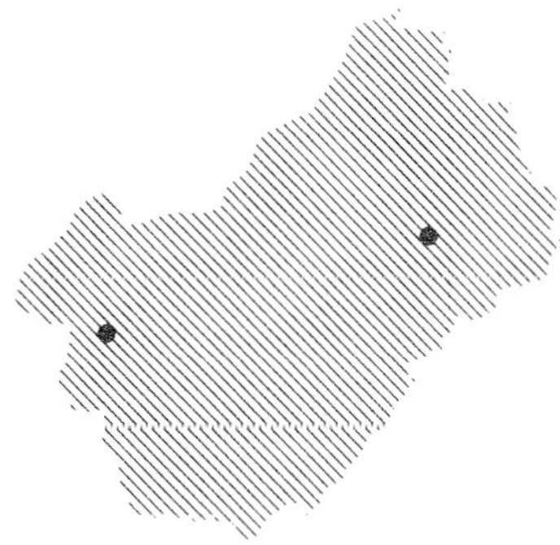

Lesser Spotted Eagle
Aquila pomarina

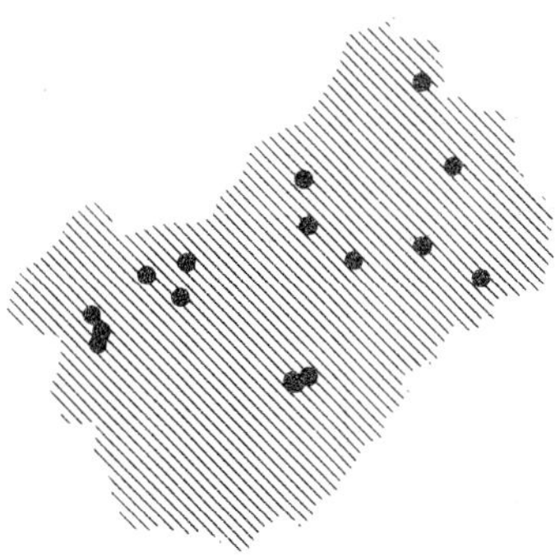

Tawny Eagle
Aquila rapax

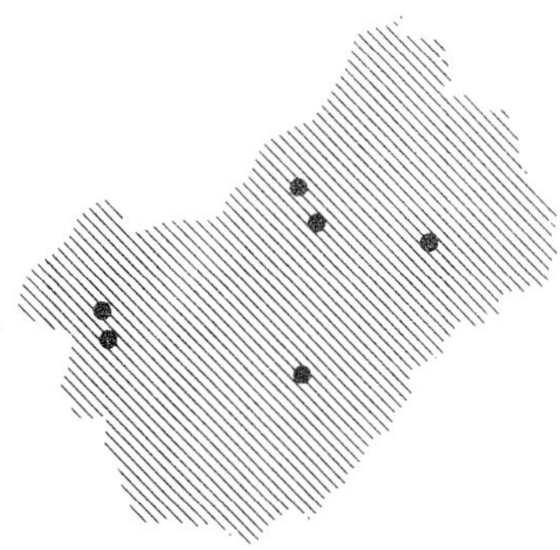

Steppe Eagle
Aquila nipalensis

Plate 1

Sightings of Birds Listed Under Schedules of WI (P) Act From Thar Desert

Short-toed Snake Eagle
Circaetus gallica

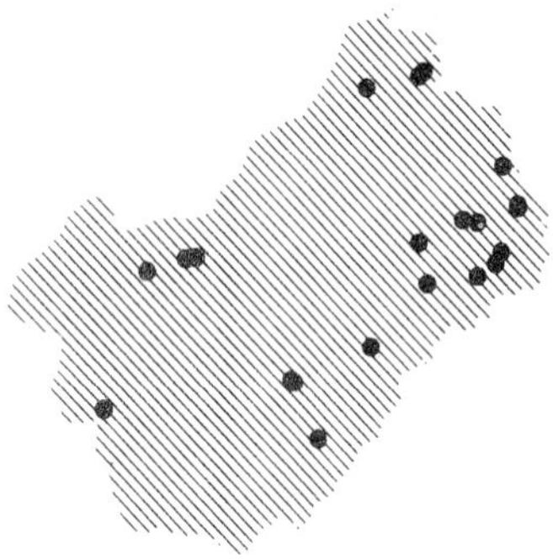

Black Kite
Milvus migrans

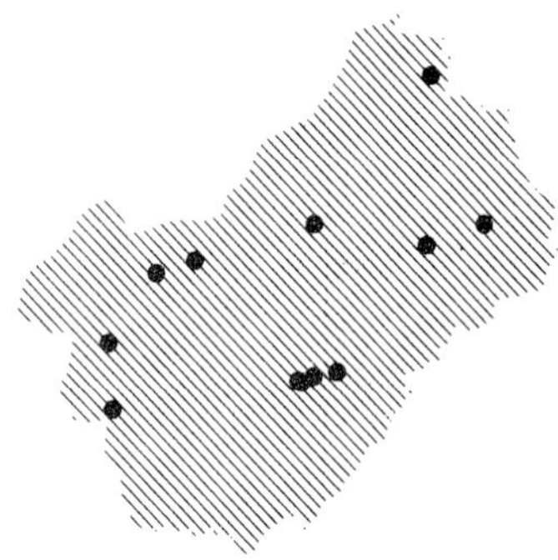

Black-shouldered Kite
Elanus caeruleus

Eurasian Sparrowhawk
Accipiter nisus

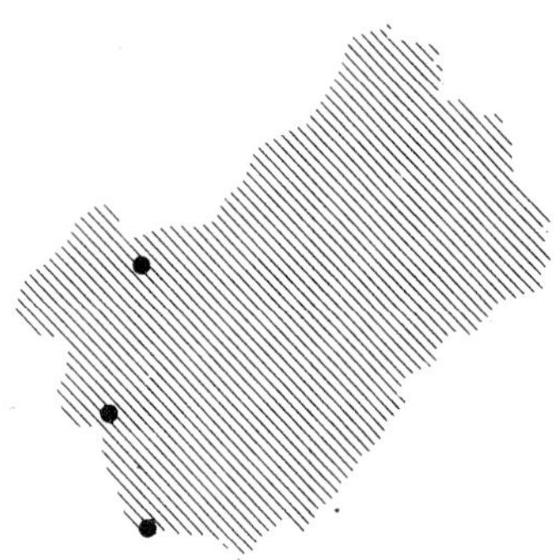

Osprey
Pandion haliaetus

Plate 2

Sightings of Birds Listed Under Schedules of WI (P) Act From Thar Desert

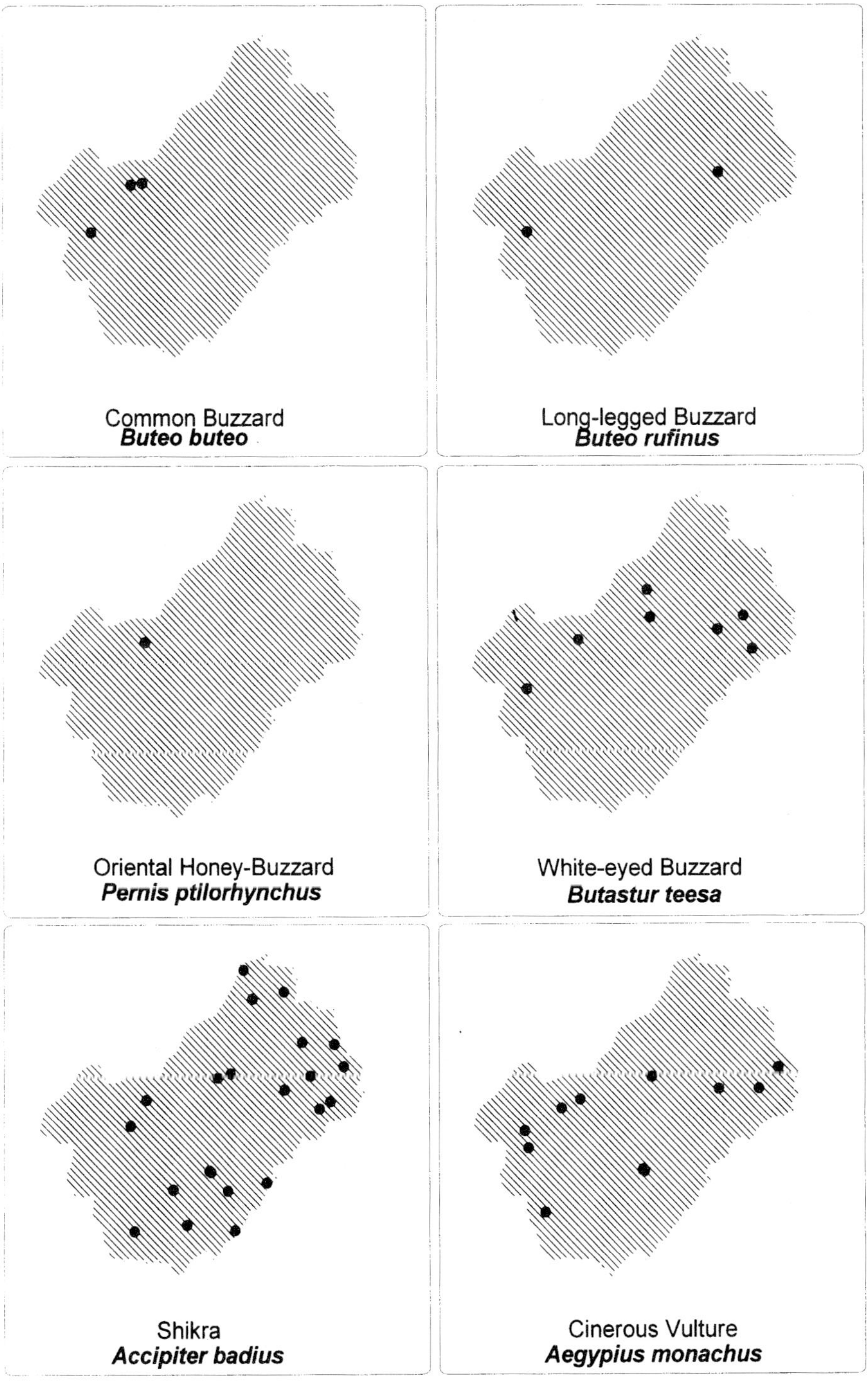

Plate 3

Sightings of Birds Listed Under Schedules of WI (P) Act From Thar Desert

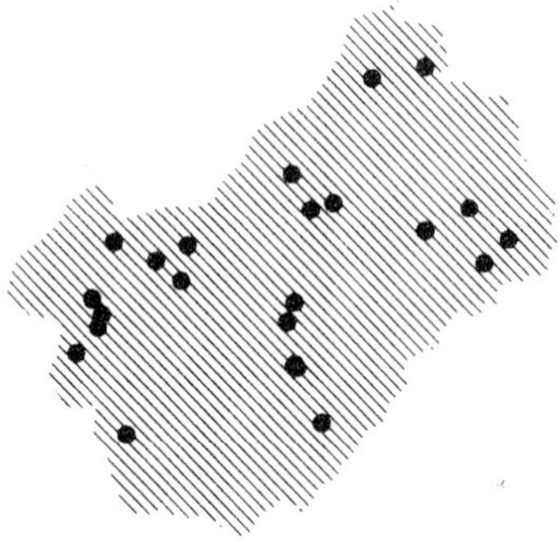

Egyptian Vulture
Neophron percnopterus

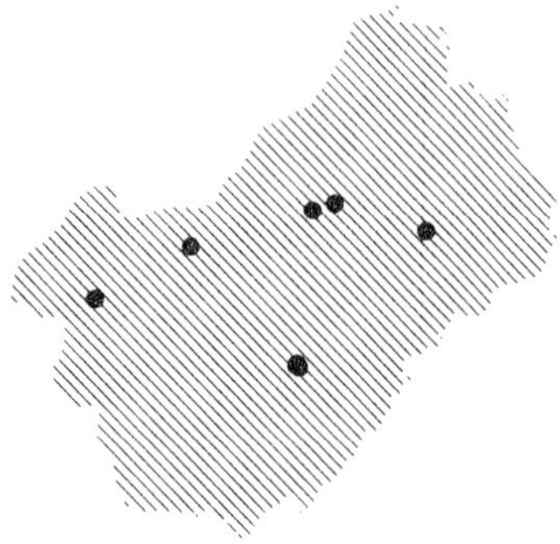

Eurasian Griffon
Gyps fulvus

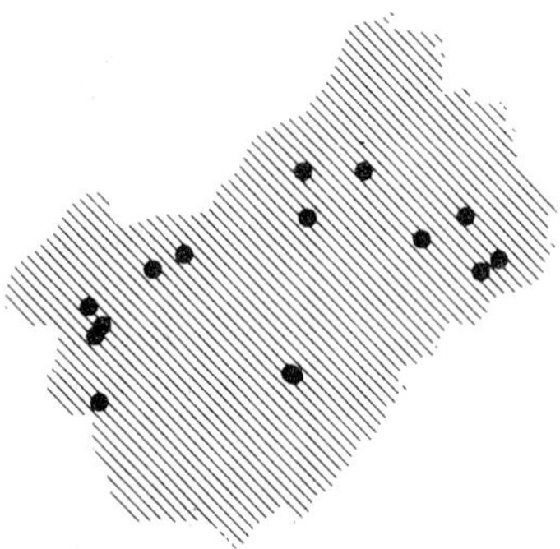

Indian White-backed Vulture
Gyps bengalensis

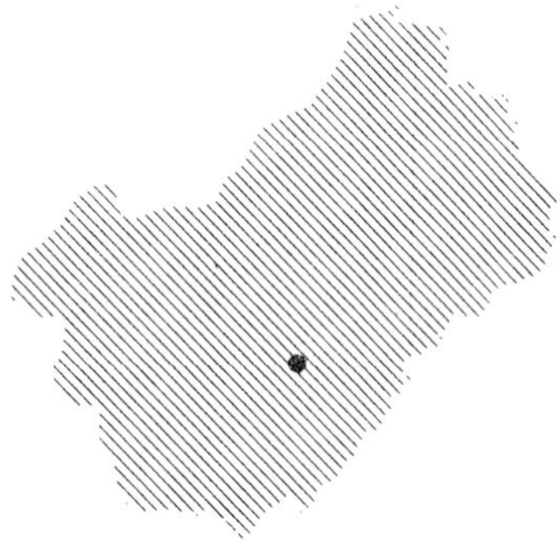

Himalayan Griffon
Gyps himalayensis

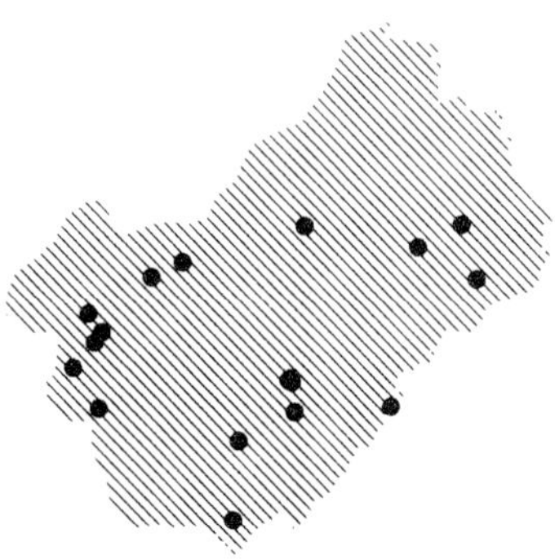

Long-billed Vulture
Gyps indicus

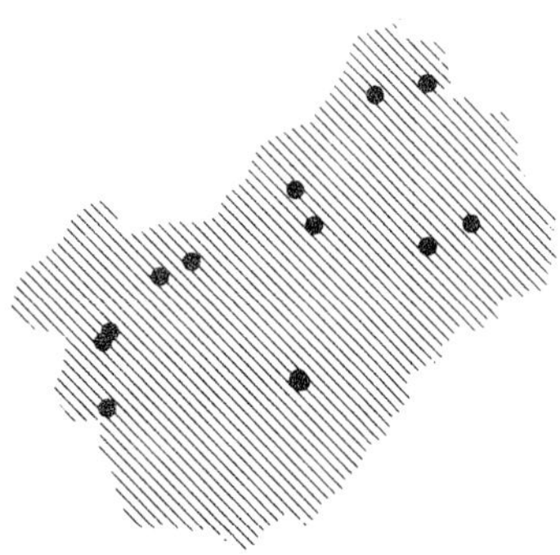

• Red-headed Vulture
Sarcogyps calvus

Plate 4

Sightings of Birds Listed Under Schedules of Wl (P) Act From Thar Desert

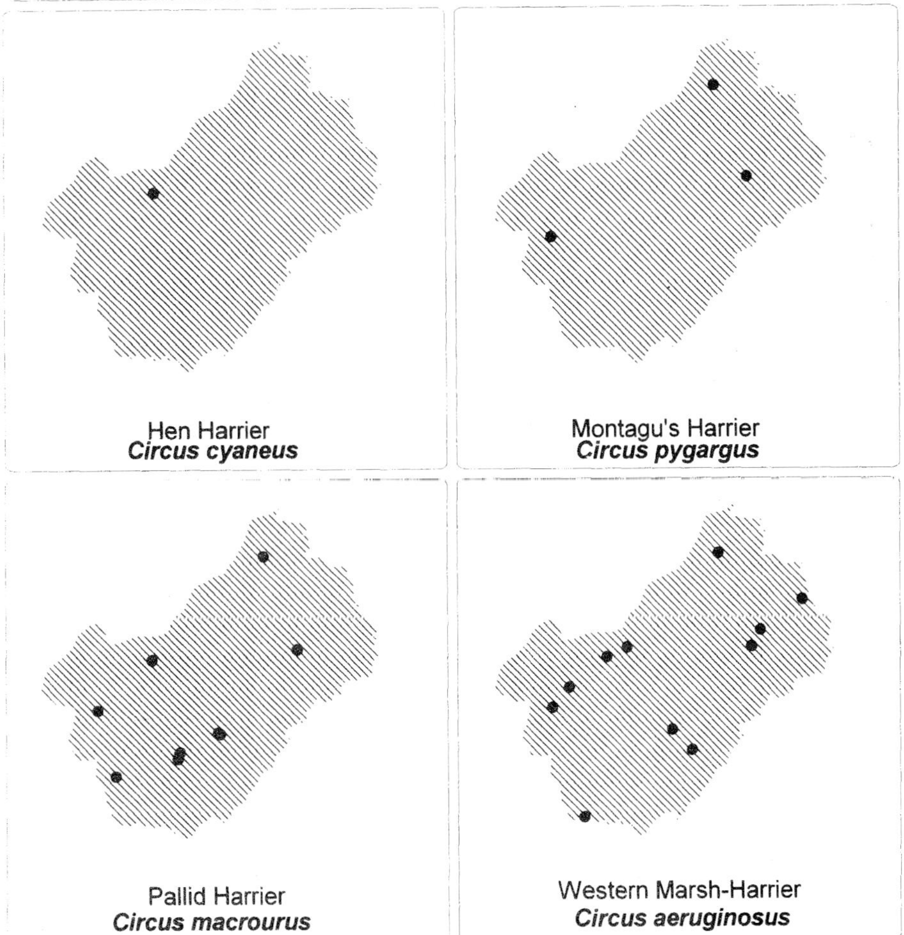

Hen Harrier
Circus cyaneus

Montagu's Harrier
Circus pygargus

Pallid Harrier
Circus macrourus

Western Marsh-Harrier
Circus aeruginosus

Plate 5

Sightings of Birds Listed Under Schedules of WI (P) Act From Thar Desert

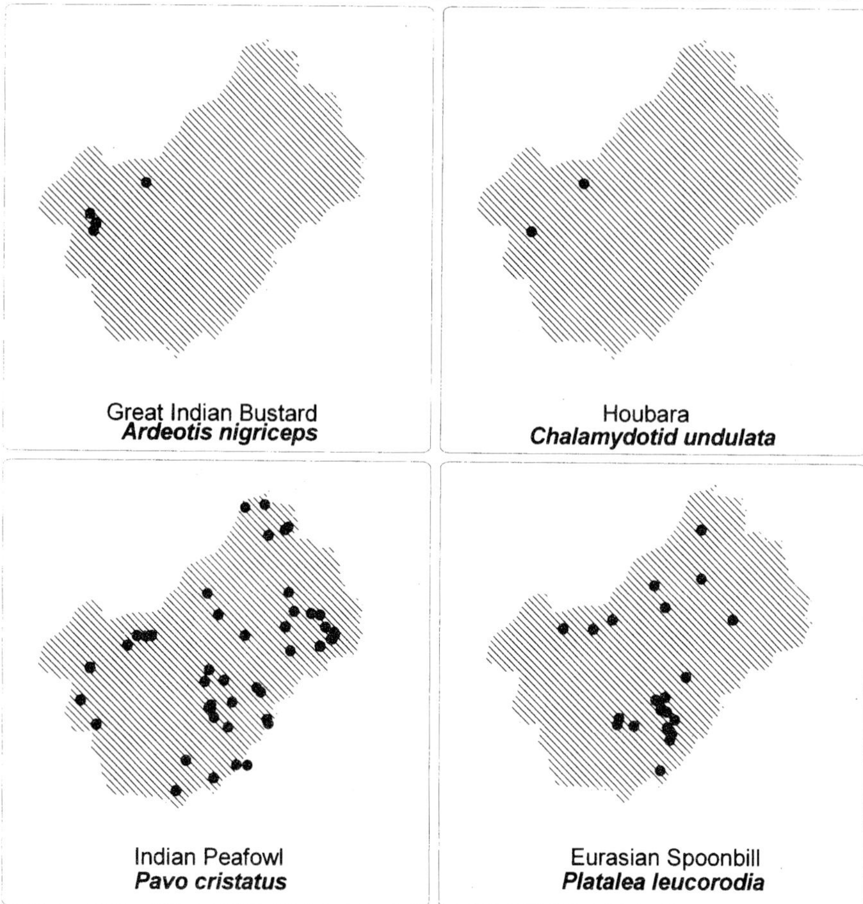

Plate 6

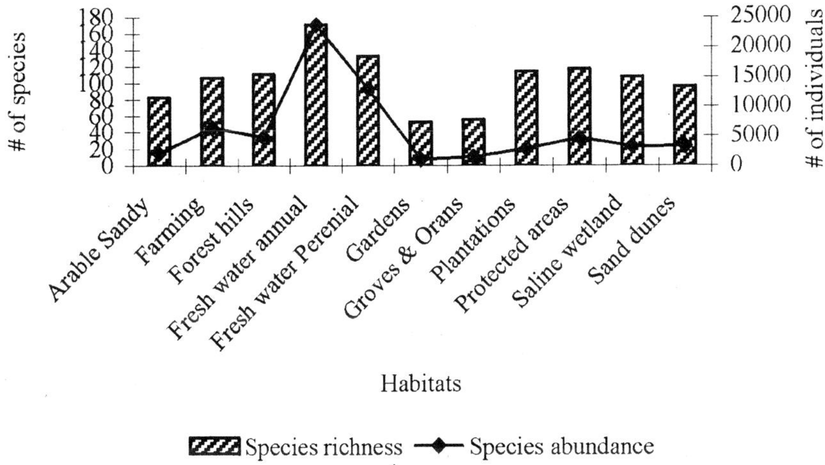

Fig. 4. Species richness and abundance in different habitats in the Thar Desert of Rajasthan

Dominance of Bird Species

The analysis of the observations reveals that Common coot was highest in dominance followed by Blue Rock Pigeon and Eurasian Collard Dove. Relative abundance of birds recorded from the Thar Desert is given in Table 1. Species richness and abundance was highest in freshwater annual and freshwater perennial habitats.

Species Diversity Indices in the Thar Desert

Overall diversity index

Most widely used diversity indices like Shannon-Weiner index of diversity, Simpson's index of diversity and Hill's numbers were estimated for the birds of Thar Desert. The analysis of the results of Richness index (R1 and R2) showing the richness of birds of the Thar Desert, Shannon index showing high values of diversity, similarly Hill's numbers also showing high values in the Thar Desert are being furnished below:

Richness Indices		Diversity Indices		Hill's numbers		Evenness Indices	
R1	R2	Simpson's	Shannon	N1	N2	E1	E2
24.22	1.01	0.05	3.94	51.36	19.24	0.70	0.19

Order wise diversity index

The study concludes that the order Passeriformes were highest in abundance followed by Charadriiformes (Fig. 5).

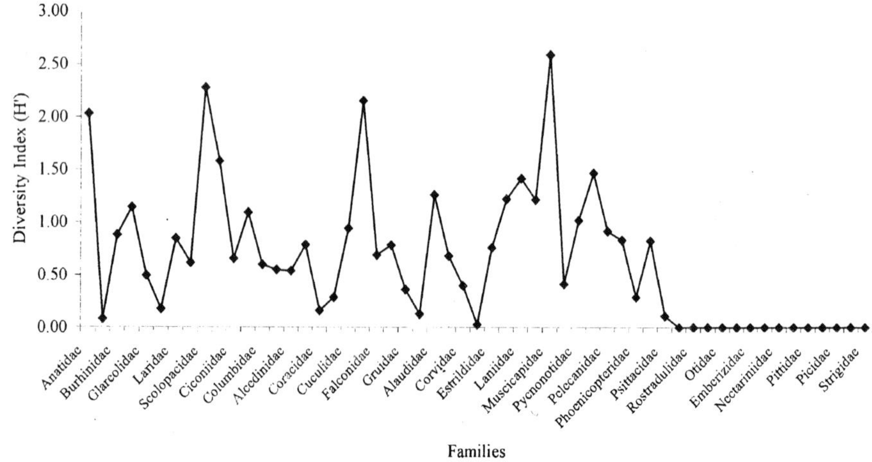

Fig. 5. Diversity index (H') of birds in different orders in the Thar Desert of Rajasthan

Family wise diversity index

The diversity index was observed highest of the family Muscicapidae, followed by Scolopacidae (Fig. 6).

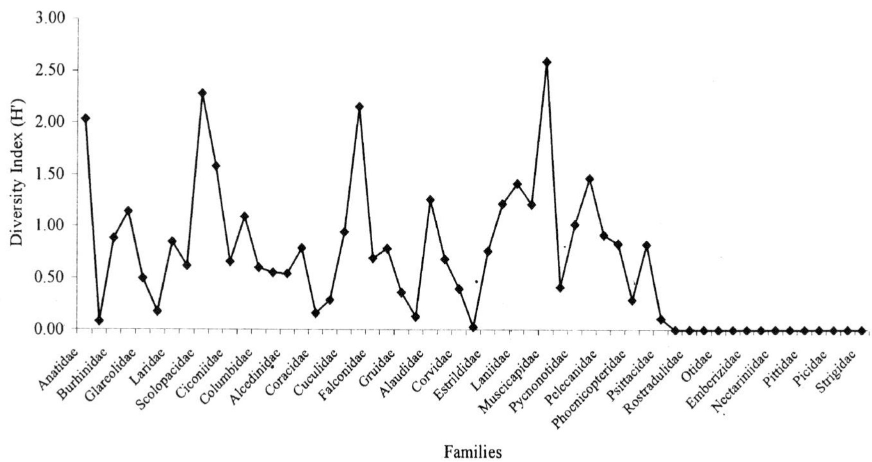

Fig. 6. Diversity index of birds in different families in the Thar Desert

Month wise diversity index

Overall diversity was highest in the month of May and October (Fig. 7). During the month of August survey was not conducted.

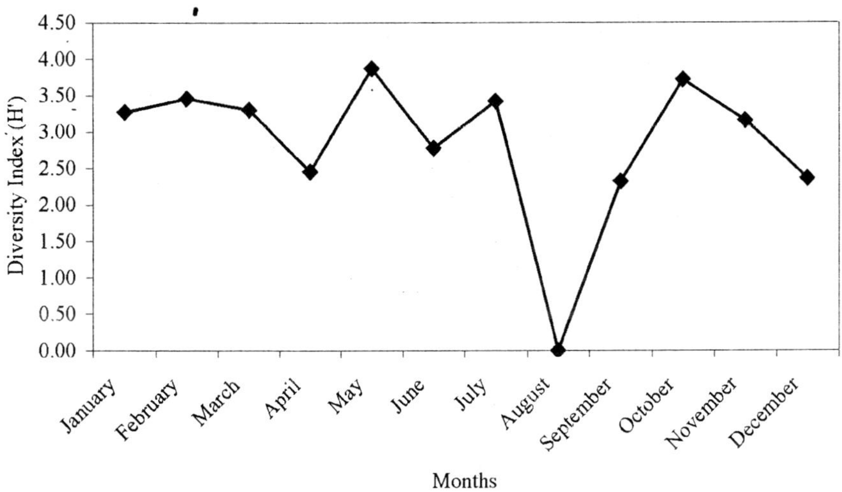

Fig. 7 Diversity index of birds in different months in the Thar Desert

Habitat wise diversity index

Among the habitats, wetland habitat has shown maximum diversity than other habitats (Fig. 8).

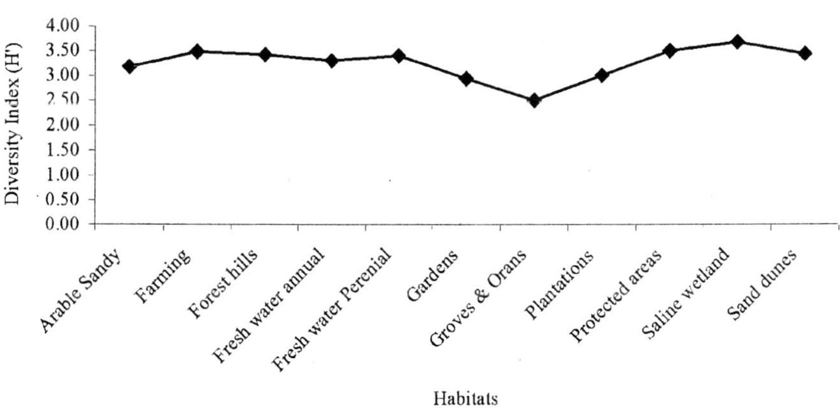

Fig. 8. Diversity index of birds in different habitats in the Thar Desert of Rajasthan

Diversity Indices of Bird Species in IGNP Area

The Richness index (R1, R2) showing the richness of birds, Shannon index showing high values of diversity, similarly Hill's numbers also showing high values in the IGNP areas.

Richness		Simpson's	Shannon	Hill'sNumber		Eveness	
R1	R2			N1	N2	E1	E2
14.04	1.78	0.11	3.21	24.86	8.98	0.67	0.21

Diversity indices of different orders of birds in the Thar Desert

The richness indices like R1 and R2 showing the richness of species belonging to different orders in the Thar Desert has been furnished in Table 4. Highest richness was in the order Passeriformes followed by Charadriiformes. Shannon index was also highest in Passeriformes and Charadriiformes. Similarly Hill's numbers like N1 and N2 also showed highest in the order Passeriformes and Charadriiformes. E1 and E2 showing evenness of birds was also found in Cuculiformes, Columbiformes and Ciconiformes.

Diversity indices of different families of birds in the Thar Desert

The information of richness indices like R1 and R2 showing the richness of the families of birds has been furnished in Table 5. The analysis of the observations reveals that highest richness was in the family Anatidae followed by Scolopacidae, while Shannon index was highest in Muscicapidae and Scolopacidae. Similarly Hill's numbers like N1 and N2 also showed highest in the family Muscicapidae. Highest evenness was observed in the family Falconidae.

Diversity indices of different months of birds in the Thar Desert

Richness indices like R1 and R2 showing the richness of the bird species during different months in the Thar Desert. Highest richness was observed in the month of January followed by February. Shannon index and Hill's numbers like N1 and N2 were also noted highest in October while highest evenness was observed during the months of July and May (Table 6).

Threatened Bird Species Recorded From The Thar Desert

During the period under study, 14 species of threatened species of birds were recorded from the Thar Desert in Rajasthan. Among these, Painted Stork was the first dominant species followed by Beach Stone Plover. Relative abundance and dominance of the threatened species are given in the Table 7.

Diversity Index of Threatened Bird Species

Highest diversity index was observed in Beach Stone Plover, followed by Spot-billed Pelican Fig. 9.

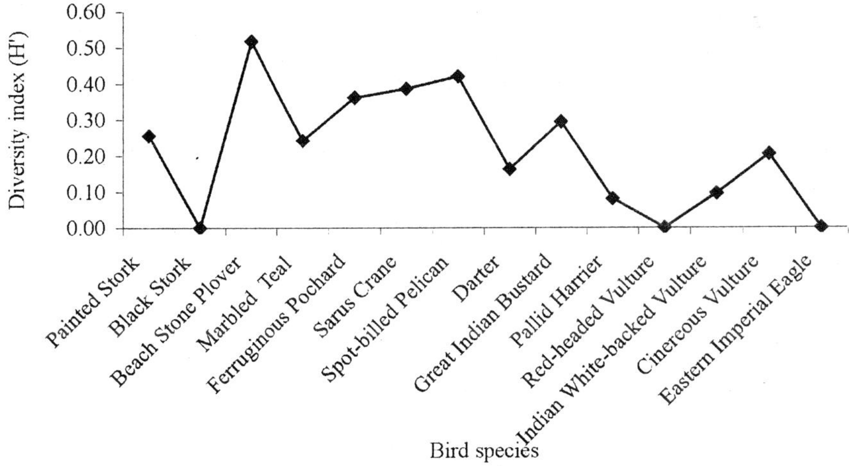

Fig. 9 Diversity index of threatened bird species

Discussion and Analysis

During the period under study, birds were observed in different niches. Some interesting observations are being noted below:

Among the *Francolinus* species, the Grey Patridge (*F. pondicerianus*) is the only species which dares to live in the hottest desert, mainly because of its low dependence on water. Peafowl (*Pavo* cristatus) and Roseringed Parakeet (*Psittacula krameri*) were recorded near the human settlements where water availability is more. The species like Indian Ring Dove (*Streptopelia decaota*) and Little Brown Dove (*Streptopelia senegalensis*) have also been recorded in large numbers near the village ponds.

Sand dunes and sandy plains are commonly found in the Barmer and Jaisalmere districts. Though these areas are characterised by strong wind and high temperature during the summer months and extreme cold in winter season, many species of birds are found in these habitats, *viz*, Crested Lark (*Galerida cristata*), Rufous-fronted Wren Warbler (*Prinia buchanai*), Desert Weatear (*Oenanthe deserti*).

The crops such as millet, wheat, sorghum, green vegetables and oil seeds are widely cultivated in plains where water is available for irrigation. The under ground water is generally used for irrigation purposes in the Thar Desert, except the IGNP command area. As a result, the soil surface generally remains moist for some time in these areas. The standing crops of these regions provide shelter to vast number of resident birds and also attract migratory birds, e.g., Rosy Pastor (*Sturnus roseus*), Pale Harrier (*Circus macrourus*). Resident birds play a major role in the agricultural economy. The insectivores like Bee-eater (*Merops orientalis* and *Merops supercicilis*) were recorded in high number in the vicinity of electric line and agricultural crops.

Though the orchards are rare but are one of the important habitats in the Thar Desert. They provide shelter for many birds during the summer months, i.e., *Psittacula krameri, Megalaima haemocephala, Dendrocitta vagabunda* and *Pycnonotus leucogenys*.

A larger number of migratory birds visit the Thar Desert every year from Himalayas and Central Asian countries. The wetlands provide good habitat for winter visitors. The depressions of escape reservoirs, resulted from the seepage of water due to mismanagement in IGNP and saline wetland (Deedwana) attract many species of migratory waders like Common Sandpiper (*Tringa hypoleucos*), Lesser Sandpiper (*Charadrius mongolus*), Eurasian Curlew (*Numenius arguata*), Curlew Sandpiper (*Charadrius ferruginea*), Whiskered Tern (*Chilidonia hybridus*), Blackheaded Gull (*Larus ridibundus*) and Great Blackheaded Gull (*Larus ichthyaetus*).

Reservoirs like Jawai Bandh and Hemaswas harbour many species of waterfowls. Because of the availability of food resources, good number of waterfowls, were recorded in these wetlands, viz. Little Cormorant (*Phalacrocorax niger*), Lesser Whistling Teal (*Dendrocygna javanica*), Darter (*Anhinga rufa*), Spot-billed Duck (*Anas poecilorhynca*) and many species of wading birds also recorded here such as Egrets (*Egretta intermedia, Egretta garzetta, Ardea alba*), Storks (*Mycteria lecucocephala, Anastomus oscitans*), Flamingo (*Phoenicopterus roseus*).

The species richness, abundance and diversity in the Thar Desert is much greater than that in the other ecosystems. While analysing the species richness and abundance in different habitats of the Thar, the wetland habitats showed the highest richness and abundance. The overall abundance of wetland species like Coot, Lesser Flamingo, Barheaded Goose, Blackwinged Stilt, Red Shank, Little Cormorant, Northern Shoveller, etc were also showing the highest dominance. Besides the wetlands, the plantations of trees in the vicinity of IGNP and gardens also showed high relative abundance of birds species.

IMPACT OF SHRINKAGE OF GRASS LANDS IN THE IGNP AREA

Grasslands of a predominantly occurring Sewan grass (*Lasiurus sindicus*) are spread in the 100-150 mm rainfall zone of Bikaner, Jaisalmer and Barmer districts. It is highly adapted to low rainfall region with highly permeable soils. This grass species is not found anywhere else except in the command area of Indira Gandhi canal in Jaisalmer and Barmer districts. Due to irrigation facilities, the vast grasslands are being transformed into crop fields in which flood irrigation is being practised. This has threatened the very survival of these grasslands because Sewan grass is adapted to low soil moisture regime though, these grasslands cadre the need of large herds of livestock, especially during summer months. Prior to IGNP, during droughts, the livestock herds used to migrate to Banaskantha (Gujarat), Malwa (Madhya Pradesh) and Ganga-Jamuna (Uttar Pradesh). Now the migration of cattles does not take place because of availability of water. In the circumstances, the livestock populations are putting additional pressure on fragile ecosystem because the Sewan grasslands are continuously shrinking while the populations of domestic animals are increasing. It is feared that with the impact of over-grazing, the sewan grassland will vanish from the Thar Desert after a few decades (Prakash, 1989). If this is allowed to happen, it will be a severe blow to the genetic diversity since *Lasiurus sindicus* is endemic to the Thar Desert.

Some other grasses, herbs and shrubs make excellent food for the livestock, and their seeds and fruits also provide nutritive food for the wild birds, mammals and other wild animals.

Effects of livestock grazing on birds

Though livestock grazing is considered to be the best economic use of both private and closed areas in the Thar Desert of Rajasthan, it significantly affects all birds associated with grassland and shrub. Birds are especially more responsive to livestock grazing, compared to other animal populations. Bird Survey data on Breeding in other countries indicate that grassland bird species are experiencing decline more than any other bird group (Millchuna *et al.*, 1998).

Therefore, we have reasons to be concerned about continuing negative effects of livestock grazing on bird fauna and thus we propose to implement the grazing policy. There is also a little doubt that domestic livestock have a controlling influence over the structure and function of the ecosystems in which they are grazing. Hence, the Government must formulate a rational policy which may have positive effect on bird populations.

There are some reasons for being careful and case specific in making our conclusions. Besides, there are variety of factors, in addition to direct effects of livestock grazing, which have been responsible for the shrinkage of the grassland ecosystem. Agricultural cultivation in former grasslands, dewatering of wetland habitats, fire suppression, introduction and spread of exotic vegetation, extirpation of native herbivores (cattle) and habitat fragmentation due to agriculture and urban sprawl have impacted the grasslands. Some, but not all, of these are directly related to livestock populations. However, all these factors have directly affected the composition of bird population in the IGNP area.

The individual bird populations will respond differently to livestock grazing, depending on their particular habitat and resource needs. Some birds respond negatively to grazing of livestock while others usually respond positively, depending on the particular plant community and the type and intensity of grazing. The principal means by which livestock grazing affects bird populations is by altering habitat structure and food availability. Grazing invariably reduces the ground cover and height of grasses at least temporarily, thus depriving many birds of the cover they need, both as a refuge from predators and as a favourable environment for roosting and nesting. Grazing related loss of grass cover also decreases the frequency and intensity of fire in grassland and shrub ecosystems. Grazing also reduces grass seed production, which significantly attracts winter bird densities.

IMPACT OF INDIRA GANDHI NAHAR PARIYOJNA (IGNP)

The major portion of the Thar Desert was coming under the princely states of Jaisalmer, Jodhpur and Bikaner (now in Rajsthan) before independence. It was the desire of every ruler to bring water to thirsty landscapes so as to provide water for drinking as well as irrigation purposes. One of the first attempts to green the desert was made by Maharaja Ganga Singh, the ruler of Bikaner to bring water from Sutlej through a canal which was later named Gang Canal. The construction of Gang canal was started in 1920 and was commissioned on 26th October 1927 while the entire work was completed in 1928. Its total length was initially 130 km which could irrigate about 1.4 lakh ha only in Bikaner state. After independence, a separate plan was chalked out by the Govt. of India to bring water to the adjacent areas to bring under cultivation so that the increased demand of water and food may be met for the growing human population. The work on this project was initiated in 1958 and the canal was named as Rajasthan canal. The excavation could commence only in 1960 after the signing of an agreement with the Govt. of Pakistan "Indus water Treaty". Initially, the area covered four districts of Rajasthan namely Ganganagar, Bikaner, Hanumangarh and Jaisalmer. Later, this canal was renamed as Indira Gandhi Nahar Project/Pariyojna (IGNP) after the (Late) Prime Minister of India, Smt. Indira Gandhi. In addition to optimising agriculture production, the project also envisages the regional development and ecological improvement by arresting the desertification. The IGNP is one of the largest and most expensive irrigation systems in the dry land in the world. Many urban and rural areas of Bikaner, Churu, Ganganagar and Jodhpur districts are getting drinking water through the IGNP canal.

The IGNP is now considered a grand endeavour to bring water from Himalayas to vast stretches of Arid Western Rajasthan. The water is diverted from the Harike Barrage in Punjab through the 204 km IGNP feeder canal and into 445 km main canal in Masitawali. The project was to be completed in three stages. Stage-I was completed in 1973 and Stage-II in 1985. The work is still under progress for the final stage. After the completion of IIIrd or final stage the main canal and feeder channels will spread over about 8000 km in length. As per planning, about 11 per cent of the Thar desert of Rajasthan will be irrigated after completion of this project.

Greater part of the main IGNP canal suffers from wind and shifting sand dunes, which block the flow of water in canals. In order to prevent the wind and shifting sand dunes, Government of Rajasthan started afforestation on both sides of the IGNP canal up to 100 m width. A large number of plant nurseries have been established along the canal at various places, e.g. Hanumangarh, Chhatargarh, Bajiv, Bhikempur, Phalodi and Mohangarh, etc. The afforestation has been completed in two stages. Stage-I was initiated in 1962, sponsored by the International Development agency (IDA) and was completed in the year 1989. In stage-one, 1,02,769 ha land has been brought under afforestation on the canal banks and roads under the sand dune stabilization and pasture development schemes. Afforestation stage-II was started in 1985 and continued up to 1996. This project was funded by World Food Program (WFP), Command Area Development (CAD) and Desert Development Program (DDP). A total of 21,101 ha area was planted up to 1990. Main species of planted trees are *Acacia nilotica, Dalbergia sissoo, Eucalyptus camaldulensis, Prosopis cineraria, Tecomella undulata* and *Zizyphus mauritiana*. Some of the aforested area provides shelter to the mammals, viz., Wild Boar, Nilgai, Jackal and Fox. Many species of residents birds using as corridor for the movement, e.g. Common Babbler, Paradise Flycatcher, Green Pigeon, etc.

Availability of canal water for irrigation has completely changed the pattern of Kharif cropping in the IGNP command area. The Groundnut *Arachis hypogea*, Cotton *Gossypium* sp., Paddy *Oryza staiva* and Surgarcane *Saccharum officinanarum* have replaced the traditional crops like Moong *Phaseolas radiatus*, Moth *Vigna acontifolia*, Guar *Cyamopsis tetragonoloba*, and Bajra *Pennisitum tyhoides*. Single cropping areas are being converted into double and triple crop areas. Sandy soil fields have been brought under regular two crops like bajra and guar in kharif while mustard, gram, wheat, barely in rabi, and vegetables. In spite of all these plus points, the IGNP has become subject of debate because the new cropping pattern has recently attracted many species of agricultural pests, (insects and nematodes) which had never been recorded earlier.

In fact, the invertebrate fauna which is consumed as food by many birds, always becomes the first victim whenever any ecosystem is disturbed, is fast changing. However, the soil moisture favours the population of earthworm, nematodes, insect and several species of bugs, snails, etc. Due to mismanagement of canal water, new water bodies have developed, which provide an ideal breeding place for the mosquitoes. If the availability of water has attracted many species of passerines birds, then on the other hand the conversion of grasslands into croplands is responsible for the reduction of the suitable habitat of native birds.

According to Rahmani (1996) one of the endemic species in the desert Stoliczka'a or White Browed Bush Chat (*Saxicola macrorhyncha*) is facing severe threat, besides Great Indian Bustards (*Ardeotis nigriceps*).

In the Canal area, the under ground water table is rising due to seepage from the canal. As a result of sub- standard construction and poor maintenance of the canal in many places inter dunal reservoirs have been formed where the vegetation cover has changed from the

Xerophytic species to Hydrophytic and Mesophytic plants. These inter dunal wetlands attract large number of waterfowls. This arid system has been supporting many species of waterfowls, waders and passerines. Overall the canal water in this area has increased the diversity of Avifauna in the Thar Desert.

A total number of 119 species of birds have been recorded from the IGNP canal areas during the period under study. Among these Common Coot was the most abundant species followed by Eurasian Collared Dove and Little Cormorant (Table 8).

Table 1. Dominance and Relative abundance of Birds Recorded from the Thar Desert of Rajasthan

Sl. No.	Common Name	Estimated Number of individual	Dominance
1	Common Coot	13186	18.20
2	Blue Rock Pigeon	6032	8.32
3	Eurasian Collared-Dove	3440	4.75
4	Lesser Flamingo	3332	4.60
5	Bar-headed Goose	2301	3.18
6	Little Ringed Plover	2010	2.77
7	House Sparrow	1988	2.74
8	Black-winged Stilt	1919	2.65
9	Common Redshank	1367	1.89
10	House Crow	1360	1.88
11	Little Cormorant	1354	1.87
12	Green Bee-eater	1146	1.58
13	Northern Shoveller	1034	1.43
14	Rose-ringed Parakeet	1030	1.42
15	Grey Wagtail	1029	1.42
16	Red-wattled Lapwing	958	1.32
17	Little Egret	932	1.29
18	Streak-throated Swallow	908	1.25
19	Pied Avocet	848	1.17
20	Red-rumped Swallow	785	1.08
21	Demoiselle Crane	727	1.00
22	Asian Open-bill Stork	704	0.97
23	Common Poachard	693	0.96
24	Black-tailed Godwit	653	0.90
25	Little Grebe	630	0.87
26	Brahminy Starling	619	0.85
27	Little Stint	618	0.85
28	Cattle Egret	582	0.80

Sl. No.	Common Name	Estimated Number of individual	Dominance
29	Common Swallow	500	0.69
30	Common Babbler	486	0.67
31	Great Cormorant	482	0.67
32	Red-vented Bulbul	477	0.66
33	River Tern	475	0.66
34	Common Teal	471	0.65
35	Wood Sandpiper	454	0.63
36	Indian Peafowl	439	0.61
37	Jungle Babbler	421	0.58
38	Steppe Eagle	405	0.56
39	Common Myna	394	0.54
40	Great White Pelican	393	0.54
41	Northern Pintail	381	0.53
42	Eurasian Spoonbill	374	0.52
43	Grey Francolin	358	0.49
44	Eurasian Griffon	325	0.45
45	Oriental White Ibis	320	0.44
46	White Wagtail	307	0.42
47	Bar-tailed Godwit	304	0.42
48	Greater Flamingo	304	0.42
49	Common Moorhen	298	0.41
50	Painted Stork	297	0.41
51	Indian Great Reed Warbler	295	0.41
52	Tufted Duck	282	0.39
53	Common Sandpiper	278	0.38
54	Medium Egret	274	0.38
55	White-eared Bulbul	267	0.37
56	Temminck's Stint	263	0.36
57	Spot-billed Duck	261	0.36
58	Egyptian Vulture	258	0.36
59	House Swift	254	0.35
60	Little Brown Dove	252	0.35
61	Black Drango	247	0.34
62	Kentish Plover	245	0.34
63	Purple Sunbird	245	0.34
64	Curlew Sandpiper	231	0.32
65	Lesser Sand Plover	220	0.30
66	Black Ibis	217	0.30
67	Spotted Redshank	216	0.30

Sl. No.	Common Name	Estimated Number of individual	Dominance
68	Baya Weaver	213	0.29
69	Eastern Skylark	203	0.28
70	Dalmatian Pelican	200	0.28
71	Bank Myna	197	0.27
72	Indian Pond-Heron	192	0.26
73	Large Egret	179	0.25
74	Gray Heron	175	0.24
75	Striated Babbler	175	0.24
76	Eurasian Wigeon	174	0.24
77	Ruff	168	0.23
78	Himalayan Griffon	150	0.21
79	Green Sandpiper	149	0.21
80	Citrine Wagtail	137	0.19
81	White Throated Munia	132	0.18
82	Brahminy Shelduck	128	0.18
83	Indian Robin	122	0.17
84	Whiskered Tern	116	0.16
85	Mallard	115	0.16
86	Gadwall	114	0.16
87	Rosy Paster	112	0.15
88	Common Greenshank	107	0.15
89	Large Grey Babbler	104	0.14
90	Indian Roller	102	0.14
91	White-browed Fantail Flycatcher	101	0.14
92	Blue-cheeked Bee-eater	97	0.13
93	Large Pied Wagtail	95	0.13
94	Beach Stone Plover	89	0.12
95	Marbled Teal	83	0.11
96	Common Crested Lark	83	0.11
97	Indian Shag	82	0.11
98	White-cheeked Bulbul	81	0.11
99	Lesser Whistling-Duck	80	0.11
100	Purple Heron	78	0.11
101	White-breasted Kingfisher	77	0.11
102	Yellow-wattled lapwing	75	0.10
103	Stone Curlew	74	0.10
104	Sarus Crane	74	0.10
105	Spot-billed Pelican	73	0.10

Sl. No.	Common Name	Estimated Number of individual	Dominance
106	Variable Wheatear	66	0.09
107	Indian Treepie	63	0.09
108	Small Pratincole	62	0.09
109	Common Starling	60	0.08
110	Jungle Crow	57	0.08
111	Cinereous Vulture	56	0.08
112	Tawny Eagle	55	0.08
113	White-headed Babbler	54	0.07
114	Great Grey Shrike	53	0.07
115	Long-billed Vulture	51	0.07
116	Red Collared-Dove	50	0.07
117	Common Hoopoe	50	0.07
118	Yellow Wagtail	49	0.07
119	Spanish Sparrow	47	0.06
120	Black-headed Gull	45	0.06
121	Asian Koel	44	0.06
122	Common Lesser White-throat	44	0.06
123	Darter	44	0.06
124	Greater Coucal	40	0.06
125	Plain Prinia	39	0.05
126	Chestnut-bellied Sandgrouse	38	0.05
127	Wire-tailed Swallow	36	0.05
128	Greater Scaup	35	0.05
129	Common Snipe	34	0.05
130	Shikra	33	0.05
131	Garganey	32	0.04
132	Lesser Pied Kingfisher	32	0.04
133	Jungle Myna	32	0.04
134	Spotted Owlet	32	0.04
135	Common Tailorbird	29	0.04
136	Ferruginous Pochard	28	0.04
137	Broadbilled Sandpiper	28	0.04
138	Black-throated Munia	28	0.04
139	Black Kite	26	0.04
140	Bay-backed Shrike	26	0.04
141	Red Whiskered Bulbul	26	0.04
142	Pied tit	25	0.03
143	Dusky Crag Martin	24	0.03

Sl. No.	Common Name	Estimated Number of individual	Dominance
144	Black Redstart	24	0.03
145	Plum-headed Parakeet	24	0.03
146	Pheasant-tailed Jacana	23	0.03
147	Small Blue Kingfisher	23	0.03
148	Red-headed Vulture	23	0.03
149	Lesser Goldenbacked Woodpecker	22	0.03
150	Indian White-backed Vulture	21	0.03
151	Purple Moorhen	21	0.03
152	Yellow-legged Buttonquail	20	0.03
153	Oriental Magpie-Robin	20	0.03
154	Spotted Munia	20	0.03
155	Ashy Prinia	20	0.03
156	Indian Chat	19	0.03
157	Graceful Prinia	19	0.03
158	Asian Pied Starling	19	0.03
159	Greater Painted Snipe	18	0.02
160	White-breasted Waterhen	18	0.02
161	Yellow-eyed Babbler	18	0.02
162	Pied Bush Chat	18	0.02
163	Western Marsh-Harrier	17	0.02
164	Common Raven	17	0.02
165	European Calendra Larke	17	0.02
166	Brown Rock Pipit	16	0.02
167	Desert Wheatear	16	0.02
168	Greenish Leaf-warbler	16	0.02
169	Great Indian Bustard	15	0.02
170	Rufous-backed Shrike	15	0.02
171	Northern House Martin	15	0.02
172	Common Quail	14	0.02
173	Black-crowned Finch Lark	14	0.02
174	Oriental White eye	14	0.02
175	Black Winged Kite	13	0.02
176	Bluethroat	13	0.02
177	Red-throated Flycatcher	13	0.02
178	Marsh Sandpiper	12	0.02
179	White-tailed lapwing	12	0.02
180	Painted Sandgrouse	12	0.02
181	Great Stone Plover	11	0.02

Sl. No.	Common Name	Estimated Number of individual	Dominance
182	Southern Grey Shrike	11	0.02
183	Spotted Dove	10	0.01
184	White-eyed Buzzard	10	0.01
185	Pallid Harrier	10	0.01
186	Eurasian Tree Pipit	10	0.01
187	Cotton Teal	9	0.01
188	Indian Courser	9	0.01
189	Pied Crested Cuckoo	9	0.01
190	Tawny Pipit	9	0.01
191	Ashy-crowned Finch Lark	9	0.01
192	Grey-headed Flycatcher	9	0.01
193	Eurasian Curlew	8	0.01
194	Yellow-legged Gull	8	0.01
195	White-necked Stork	8	0.01
196	Lagger Falcon	8	0.01
197	Common Crane	8	0.01
198	Brown Shrike	8	0.01
199	Eurasian Reed-Warbler	8	0.01
200	Coppersmith Barbet	·8	0.01
201	Short-toed Snake Eagle	7	0.01
202	Common Kestrel	7	0.01
203	Booted Warbler	7	0.01
204	Eastern Calendra Lark	7	0.01
205	Desert Warbler	7	0.01
206	Pallas's Gull	6	0.01
207	Red-headed Vulture	6	0.01
208	Rufous-fronted Prinia	6	0.01
209	Asian Paradise-Flycatcher	6	0.01
210	Osprey	5	0.01
211	Rufoustailed Finch Lark	5	0.01
212	Common Swift	4	0.01
213	Oriental Turtle Dove	4	0.01
214	Eurasian Roller	4	0.01
215	Great Spotted Eagle	4	0.01
216	Eastern Imperial Eagle	4	0.01
217	Small Minivet	4	0.01
218	Pygmy Cormorant	4	0.01
219	Glossy Ibis	3	+

Sl. No.	Common Name	Estimated Number of individual	Dominance
220	Black Francolin	3	+
221	Greater Whitethroat	3	+
222	Common Woodshrike	3	+
223	Rusty-tailed Flycatcher	3	+
224	Crimson-throated Barbet	3	+
225	Warbler	3	+
226	Red-crested Pochard	2	+
227	Comb Duck	2	+
228	Common Shelduck	2	+
229	Creamcoloured Coursor	2	+
230	Gull-billed Tern	2	+
231	Northern Lapwing	2	+
232	Greater Sand Plover	2	+
233	Little Green Heron	2	+
234	Black Bittern	2	+
235	European Bee-eater	2	+
236	Eurasian Sparro-hawk	2	+
237	Black Kite	2	+
238	Marshall's Iora	2	+
239	Water Pipit	2	+
240	Greater Short-Toed Lark	2	+
241	Red-winged Bush Lark	2	+
242	Common Chiffc Chaff	2	+
243	Smoky Warbler	2	+
244	Dusky Warbler	2	+
245	Bengal Bush Lark	2	+
246	Plain Leaf-Warbler	2	+
247	Paddyfield Pipit	2	+
248	Plain Martin	2	+
249	Northern Weather	2	+
250	Thickbilled Warbler	2	+
251	Brown-headed Gull	1	+
252	Bronze-winged Jacana	1	+
253	Buff-breasted Sandpiper	1	+
254	Black Stork	1	+
255	Black-crowned Night Heron	1	+
256	Chestnut Bittern	1	+
257	Common Buzzard	1	+

Sl. No.	Common Name	Estimated Number of individual	Dominance
258	Montagu's Harrier	1	+
259	Bonelli's eagle	1	+
260	Rufous-bellied Babbler	1	+
261	Rufous-tailed Scrub-Robin	1	+
262	Crested Bunting	1	+
263	Rufous-tailed Wheatear	1	+
264	Indian Pitta	1	+
265	Orphean Warbler	1	+
266	White-capped Redstart	1	+
267	Rufous-tailed Shrike	1	+
268	Franklin's Prinia	1	+
269	White-throated Fantail Flytcather	1	+
270	Ashy Drango	1	+
271	Brown-headed Barbet	1	+
Total		72,457	99.96

+ = Less than 0.01

Table 2. Orders of Birds and their Status Recorded from the Thar Desert of Rajasthan

Sl. No.	Order	Status		Total	%
		R	M		
1	Podicipediformes	01	--	01	0.37
2	Anseriformes	11	10	21	7.72
3	Phoenicopteriformes	02	--	02	0.74
4	Pelecaniformes	06	02	08	2.94
5	Ciconiformes	18	01	19	6.99
6	Charadriiformes	14	29	43	15.81
7	Falconiformes	23	04	27	9.93
8	Gruiformes	09	--	09	3.31
9	Galliformes	04	--	04	1.47
10	Psittaciformes	02	--	02	0.74
11	Piciformes	04	--	04	1.47
12	Cuculiformes	03	--	03	1.10
13	Coraciiformes	12	--	12	4.41
14	Columbiformes	04	--	04	1.47
15	Apodiformes	02	--	02	0.74
16	Strigiformes	01	--	01	0.37
17	Passeriformes	107	03	110	40.44
Total		223	49	272	100

Table 3. Distribution of Birds in Different District of Thar Desert of Rajasthan

Sl. No.	Common Name	Barmer	Bikaner	Churu	Hanumangarh	Jaisalmer	Jalore	Jhunjhunu	Jodhpur	Nagaur	Pali	Sikar	Sirohi	Sri Ganganagar
1.	Bar-headed Goose	-	-	-	-	-	-	-	P	-	-	-	-	-
2.	Brahminy Shelduck	-	-	-	-	-	-	-	-	-	P	-	-	-
3.	Comb Duck	-	-	-	-	-	-	-	-	-	-	-	P	-
4.	Common Pochard	-	-	-	-	P	-	-	P	-	P	-	-	-
5.	Common Shelduck	-	-	P	-	-	-	-	-	-	P	-	-	-
6.	Common Teal	-	-	P	-	P	P	-	P	-	P	-	-	-
7.	Cotton Teal	-	-	P	-	P	-	-	-	-	-	-	-	-
8.	Eurasian Wigeon	-	P	-	-	-	P	-	-	-	-	-	P	-
9.	Ferruginous Poachard	-	-	P	P	-	P	-	-	-	-	-	-	-
10.	Gadwall	-	-	P	-	-	-	-	-	-	-	-	-	-
11.	Garganey	P	-	-	-	-	-	-	-	-	-	-	-	P
12.	Lesser Pied Kingfisher	-	-	P	-	P	-	-	P	-	-	-	-	-
13.	Lesser Whistling-Duck	-	-	-	-	P	P	-	-	-	-	-	P	-
14.	Mallard	-	-	-	-	P	-	-	-	-	-	-	-	-
15.	Marbled Teal	P	-	-	-	-	-	-	-	-	-	-	-	P
16.	Northern Pintail	-	-	P	-	-	P	-	-	-	P	-	-	-
17.	Northern Shoveller	-	-	P	-	P	P	-	P	-	P	-	-	-
18.	Red-crested Pochard	-	-	-	-	-	-	-	-	P	-	-	-	P
19.	Spot-billed Duck	P	-	P	-	P	P	P	P	-	P	-	P	-
20.	Tufted Duck	-	-	P	-	P	P	-	P	-	-	-	-	-
21.	Greater Scaup	-	-	-	-	-	-	-	-	-	P	-	-	-
22.	Common Swift	P	-	-	-	-	P	-	-	-	-	-	-	P
23.	House Swift	P	-	-	P	P	P	P	P	P	-	-	-	P
24.	Bar-tailed Godwit	-	-	-	-	P	-	-	-	-	-	-	-	-
25.	Beach Stone Plover	-	-	-	-	-	-	-	P	-	-	-	-	-
26.	Black-headed Gull	-	-	-	-	-	-	-	-	-	-	-	-	P
27.	Black-tailed Godwit	-	-	-	-	P	-	-	P	-	-	-	-	-
28.	Black-winged Stilt	-	-	P	-	P	P	P	P	-	P	-	P	P
29.	Broadbilled Sandpiper	-	-	-	-	P	-	-	-	-	-	-	-	-
30.	Bronze-winged Jacana	-	-	-	-	-	-	-	P	-	-	-	-	-

31.	Brown-headed Gull	-	-	-	-	P	-	-	-	-	-	-	-	-
32.	Buff-breasted Sandpiper	-	-	-	-	-	-	P	-	P	-	-	-	-
33.	Common Greenshank	-	-	-	-	-	-	-	P	-	P	-	-	-
34.	Common Redshank	-	-	P	-	P	-	-	P	-	P	-	-	-
35.	Common Sandpiper	-	-	P	-	P	P	P	P	-	P	-	-	-
36.	Common Snipe	-	-	-	-	-	-	-	P	-	-	-	-	-
37.	Cream Coloured Coursor	-	-	-	-	-	-	-	-	-	-	-	-	P
38.	Curlew Sandpiper	-	-	-	-	P	-	-	-	-	-	-	-	-
39.	Eurasian Curlew	-	-	-	-	P	P	-	P	-	-	-	-	-
40.	Great Stone Plover	P	-	-	-	-	P	-	-	-	-	-	-	-
41.	Greater Painted Snipe	P	-	-	-	-	-	-	-	-	-	-	-	-
42.	Greater Sand Plover	-	-	-	-	-	-	P	-	-	-	-	-	-
43.	Green Sandpiper	-	-	-	-	P	-	-	-	-	P	-	-	-
44.	Gull-billed Tern	P	-	-	-	-	-	-	P	-	-	-	P	-
45.	Indian Courser	P	-	-	-	-	-	-	P	-	-	-	-	-
46.	Kentish Plover	P	-	-	-	-	-	P	-	-	-	-	-	-
47.	Lesser Sand Plover	P	-	-	-	-	-	-	P	-	-	-	-	P
48.	Little Ringed Plover	-	-	P	-	P	P	-	P	-	P	-	-	-
49.	Little Stint	-	-	P	-	P	-	-	-	-	P	-	-	-
50.	Marsh Sandpiper	-	-	-	-	-	-	-	-	-	-	-	-	P
51.	Northern Lapwing	-	-	-	-	P	-	-	-	-	-	-	-	-
52.	Pallas's Gull	-	-	-	P	-	-	-	-	-	-	P	-	-
53.	Pheasant-tailed Jacana	-	-	P	P	-	-	-	P	-	-	-	-	-
54.	Pied Avocet	P		P		P			P		P			
55.	Red-wattled Lapwing	P	P	P	P	P	P	P	P	P	P	P	P	P
56.	River Tern	-	-	-	P	P	-	-	P	-	-	-	-	-
57.	Ruff	-	-	-	-	-	-	-	-	-	P	-	-	-
58.	Small Pratincole	-	-	-	P	-	-	-	-	-	-	-	-	-
59.	Spotted Redshank	-	-	-	-	-	-	-	P	-	P	-	-	-
60.	Stone Curlew	-	-	P	-	-	P	-	-	-	-	-	-	-
61.	Temminck's Stint	-	-	-	-	P	-	-	P	-	-	-	-	-
62.	Whiskered Tern	-	-	-	-	P	-	-	P	-	-	-	-	-
63.	White-tailed lapwing	-	-	-	-	-	-	-	-	-	-	-	-	P
64.	Wood Sandpiper	-	-	P	-	P	P	P	P	-	P	-	-	-
65.	Yellow-legged Gull	-	-	-	-	P	-	-	-	-	-	-	-	P
66.	Yellow-wattled lapwing	-	-	-	-	-	-	-	-	-	-	-	-	P
67.	Asian Openbill-Stork	-	-	P	-	-	-	P	P	-	-	-	-	-
68.	Black Bittern	-	-	-	-	-	-	-	-	-	-	-	P	-
69.	Black Ibis	-	-	-	P	P	P	-	-	P	P	-	-	P

70.	Black Stork	-	-	-	-	-	-	-	-	-	-	-	-	P
71.	Black-crowned Night Heron	-	-	-	-	-	-	-	-	-	-	-	P	-
72.	Cattle Egret	P	P	P	P	P	-	P	P	-	P	P	P	P
73.	Chestnut Bittern	-	-	-	-	-	-	-	-	-	P	-	-	-
74.	Glossy Ibis	-	-	-	-	-	-	-	-	-	-	-	-	P
75.	Gray Heron	P	-	P	P	P	-	-	P	-	P	-	P	-
76.	Indian Pond-Heron	P	-	P	P	P	-	P	P	-	P	P	P	P
77.	Large Egret	-	-	-	-	P	-	-	P	-	P	-	P	-
78.	Little Egret	P	-	P	P	P	P	P	P	-	P	-	-	-
79.	Little Green Heron	-	-	-	-	-	-	-	-	-	-	-	-	-
80.	Median Egret	-	-	P	P	P	P	-	P	-	P	-	-	-
81.	Oriental White Ibis	-	-	P	P	P	-	P	P	-	-	-	P	-
82.	Painted Stork	-	-	P	P	P	-	-	P	-	P	-	-	-
83.	Purple Heron	P	-	P	P	-	P	-	P	-	P	P	-	-
84.	Eurasian Spoonbill	P	-	P	P	-	P	-	P	-	-	-	-	-
85.	White-necked Stork	-	-	-	-	-	-	-	-	-	-	-	-	P
86.	Chestnut-bellied Sandgrouse	-	-	-	-	-	-	-	-	P	-	-	-	-
87.	Oriental Turtle Dove	-	-	-	-	-	-	-	-	-	-	-	P	-
88.	Painted Sandgrouse	-	-	-	P	-	-	P	-	-	-	-	-	-
89.	Spotted Dove	-	-	P	-	-	-	-	-	P	-	-	-	-
90.	Blue Rock Pigeon	P	P	P	P	P	P	P	P	P	P	-	P	P
91.	Blue-cheeked Bee-eater	P	-	P	P	P	P	-	P	P	-	-	P	P
92.	Common Hoopoe	P	-	P	-	-	P	P	P	P	P	-	P	P
93.	Eurasian Collared-Dove	P	P	P	P	P	P	P	P	P	P	P	P	P
94.	Eurasian Roller	-	P	-	-	-	-	-	-	-	-	P	-	-
95.	European Bee-eater	-	-	-	-	-	-	-	-	P	-	-	-	-
96.	Green Bee-eater	P	P	P	P	P	P	P	P	P	P	P	-	P
97.	Indian Roller	P	-	P	P	P	-	P	P	P	P	-	P	P
98.	Little Brown Dove	P	P	P	P	P	-	-	P	P	P	P	P	P
99.	Red Collared-Dove	P	-	-	-	-	-	P	-	-	P	-	-	-
100.	Small Blue Kingfisher	-	-	P	-	P	-	-	-	-	P	-	-	-
101.	White-breasted Kingfisher	-	-	P	P	P	-	-	P	-	P	P	-	-
102.	Asian Koel	P	P	-	-	P	P	P	-	P	P	P	P	-
103.	Greater Coucal	P	-	P	-	P	-	P	P	P	P	-	-	P
104.	Pied Crested Cuckoo	P	-	-	-	-	-	-	P	-	-	P	-	P
105.	Black Kite	P	P	-	P	P	-	P	P	-	-	-	P	P
106.	Red Kite	-	-	-	-	P	-	-	-	-	-	-	-	-
107.	Black Winged Kite	P	-	-	P	-	-	-	P	-	-	-	P	P
108.	Bonelli's eagle	-	-	-	P	-	-	-	-	-	-	-	-	-

109.	Cinereous Vulture	P	-	-	-	-	P	-	P	P	-	-	-	-
110.	Common Buzzard	-	-	-	-	-	-	-	-	-	-	P	-	-
111.	Common Kestrel	-	P	-	-	-	-	-	P	-	-	-	-	-
112.	Eastern Imperial Eagle	-	-	-	-	-	-	-	-	-	-	-	-	P
113.	Egyptian Vulture	P	-	-	-	P	P	-	P	-	-	-	P	-
114.	Eurasian Griffon	-	-	-	-	P	P	-	-	P	-	-	-	-
115.	Eurasian Sparrow-hawk	-	-	-	-	-	-	-	P	-	-	-	-	-
116.	Great Spotted Eagle	-	-	-	-	P	-	-	-	-	-	-	-	-
117.	Himalayan Griffon	-	-	-	-	-	P	-	-	-	-	-	-	-
118.	Indian White-backed Vulture	P	-	-	-	-	-	-	-	-	-	-	-	P
119.	Lagger Falcon	-	-	-	-	-	P	-	-	-	-	-	P	-
120.	Long-billed Vulture	P	-	P	P	-	P	-	-	P	-	-	-	P
121.	Montagu's Harrier	P	-	-	-	-	-	-	-	-	-	-	-	-
122.	Osprey	-	-	-	-	-	-	-	P	-	-	-	-	P
123.	Pallid Harrier	-	-	P	-	-	-	-	P	-	-	-	-	-
124.	Red-headed Vulture	P	-	-	-	-	P	-	-	-	-	-	-	P
125.	Peregrine Falcon	-	P	-	-	-	-	-	-	-	-	-	-	-
126.	Shikra	-	P	-	-	P	-	P	P	P	P	-	-	-
127.	Short-toed Snake Eagle	-	-	-	P	P	-	-	-	P	-	-	-	-
128.	Steppe Eagle	-	-	-	-	-	P	-	-	-	-	-	-	-
129.	Tawny Eagle	-	-	-	P	P	-	P	P	-	-	-	P	-
130.	Western Marsh-Harrier	-	-	P	-	-	-	P	-	-	P	-	-	-
131.	White-eyed Buzzard	-	-	P	-	-	-	-	-	-	-	-	P	-
132.	Black Francolin	-	-	-	-	-	-	-	-	-	-	-	-	P
133.	Common Quail	P	-	-	-	-	-	P	P	-	-	-	-	-
134.	Grey Francolin	P	-	P	P	P	-	P	P	P	P	P	P	P
135.	Indian Peafowl	P	-	P	P	P	P	P	-	P	P	-	P	P
136.	Common Coot	-	-	P	-	P	P	-	P	-	P	-	P	-
137.	Common Crane	-	-											P
138.	Common Moorhen	-	-	P	P	-	P	-	P	-	P	P	-	-
139.	Demoiselle Crane	-	-	-	-	P	P	-	P	P	-	-	-	-
140.	Great Indian Bustard	-	-	-	-	P	-	-	-	-	-	-	-	-
141.	Purple Moorhen	-	-	P	-	-	-	-	-	-	P	-	-	-
142.	Sarus Crane	-	-	-	-	-	-	-	P	-	-	-	-	-
143.	White-breasted Waterhen	-	-	-	-	-	-	P	-	-	P	-	-	P
144.	Yellow-legged Buttonquail	-	-	-	-	-	-	-	-	P	-	-	-	P
145.	Ashy Drango	-	-	-	-	-	-	-	-	-	-	-	P	-
146.	Ashy Prinia	-	-	-	P	P	-	-	P	-	-	-	-	P
147.	Ashycrowned Finch Lark	-	-	-	-	P	-	-	-	P	P	-	-	-

148.	Asian Paradise-Flycatcher	-	-	-	-	-	-	P	P	-	-	-	-	-
149.	Asian Pied Starling	P	-	-	-	-	-	P	P	P	P	-	-	-
150.	Bank Myna	P	P	P	P	P	P	P	P	P	P	P	-	P
151.	Baya Weaver	-	-	P	-	-	-	P	P	P	-	-	-	P
152.	Bay-backed Shrike	P	-	-	P	P	-	-	-	P	P	-	-	P
153.	Bengal Bush_Lark	-	-	-	-	P	-	-	-	-	-	-	-	-
154.	Brahminy Starling	P	-	-	P	P	-	-	P	P	P	-	P	P
155.	Black Drango	P	P	P	P	P	-	P	P	P	P	P	P	P
156.	Black Redstart	-	-	P	-	P	-	-	-	-	P	P	-	-
157.	Black-crowned Finch Lark	-	-	-	-	-	-	-	P	-	-	-	-	P
158.	Black-throated Munia	-	-	-	-	-	-	-	-	-	-	-	-	P
159.	Bluethroat	-	-	P	P	-	-	-	P	-	-	P	-	-
160.	Booted Warbler	-	-	-	-	-	P	-	-	-	-	-	-	-
161.	Brown Rock Pipit	-	-	-	-	-	P	-	-	-	-	-	-	-
162.	Brown Shrike	-	-	-	-	-	-	-	-	-	-	-	-	P
163.	Citrine Wagtail	-	-	P	-	-	-	-	-	-	-	-	-	-
164.	Common Babbler	P	P	P	P	P	P	-	P	P	P	P	P	P
165.	Common Chiffc Chaff	-	-	-	-	-	-	-	-	-	P	-	-	-
166.	Common Crested Lark	P	-	-	-	P	P	P	P	-	P	-	P	P
167.	Common Lesser White-throat	-	-	P	-	P	-	-	P	-	-	P	-	-
168.	Common Myna	P	P	P	P	P	P	-	P	P	P	-	P	P
169.	Common Raven	P	-	P	-	P	-	-	P	-	-	-	-	-
170.	Common Starling	-	-	P	-	P	-	-	P	-	-	-	-	-
171.	Common Swallow	P	-	P	-	P	-	P	P	P	P	-	P	P
172.	Common Tailorbird	-	-	P	-	P	-	-	P	P	-	-	-	-
173.	Common Woodshrike	-	-	-	-	-	-	-	-	P	P	-	-	-
174.	Crested Bunting	-	-	P	-	-	-	-	P	-	-	-	-	P
175.	Desert Warbler	P	-	-	-	-	-	-	P	P	P	-	-	-
176.	Desert Wheatear	-	-	-	P	-	-	-	-	-	P	-	-	P
177.	Dusky Crag Martin	-	-	P	-	-	-	-	-	-	-	-	-	-
178.	Dusky Warbler	-	-	P	-	-	-	-	-	-	-	P	-	-
179.	Eastern Calendra Lark	P	-	-	-	-	P	-	-	-	-	-	-	-
180.	Eastern Skylark	-	-	-	-	-	-	-	-	-	-	-	-	P
181.	Eurasian Tree Pipit	-	-	-	-	P	-	-	-	-	-	-	-	P

182.	European Calendra Larke	P	-	-	-	-	P	-	-	-	-	-	-	P
183.	Eurasian Reed-Warbler	-	-	P	P	-	-	-	-	-	-	-	-	-
184.	Franklin's Prinia	-	P	-	-	-	-	-	P	-	-	-	-	-
185.	Graceful Prinia	P	-	-	-	-	P	P	P	-	-	-	-	-
186.	Great Grey Shrike	P	-	P	-	P	-	P	P	P	P	-	-	P
187.	Greater Short-Toed Lark	P	-	-	-	-	-	P	-	-	-	-	-	-
188.	Greater Whitethroat	-	-	-	-	-	-	-	-	-	-	-	-	P
189.	Greenish Leaf-warbler	-	-	-	P	-	-	-	-	-	-	P	-	-
190.	Grey Wagtail	-	-	P	P	P	P	P	P	P	P	-	-	-
191.	Grey-headed Flycatcher	-	-	-	-	-	-	-	-	P	-	-	-	-
192.	House Crow	P	P	P	P	P	P	P	P	P	P	-	P	-
193.	House Sparrow	P	-	P	P	P	P	-	P	P	P	-	P	P
194.	Indian Chat	-	-	-	-	P	-	-	P	-	-	-	-	-
195.	Indian Great Reed Warbler	-	-	-	P	-	-	P	P	-	-	-	-	P
196.	Indian Pitta	-	-	-	-	-	-	-	-	-	-	-	-	P
197.	Indian Robin	P	P	P	P	P	-	P	P	P	-	-	P	-
198.	Indian Tree Pie	P	-	P	-	P	-	-	-	P	P	P	-	-
199.	Jungle Babbler	-	-	P	P	P	-	-	-	-	P	-	-	-
200.	Jungle Crow	P	-	P	-	P	-	-	-	-	-	-	-	P
201.	Jungle Myna	-	-	-	-	P	-	-	-	-	-	-	-	P
202.	Large Grey Babbler	P	-	P	-	P	-	P	-	P	P	-	P	-
203.	Large Pied Wagtail	P	-	P	P	-	P	P	P	-	P	-	-	P
204.	Marshall's Iora	-	-	-	-	-	-	P	-	-	-	-	-	-
205.	Northern House Martin	-	-	-	P	-	-	-	-	-	-	-	-	-
206.	Northern Wheatear	P	-	-	-	-	-	-	-	-	-	-	-	-
207.	Oriental Magpie-Robin	P	-	P	-	-	-	P	P	-	-	-	-	-
208.	Oriental White eye	-	-	-	-	-	-	-	-	-	-	-	-	P
209.	Orphean Warbler	P	-	-	-	-	-	-	-	-	-	-	-	-
210.	Paddyfield Pipit	-	-	-	-	-	-	-	P	-	-	-	-	-
211.	Pied Bush Chat	P	-	P	-	-	-	-	-	P	-	P	-	P
212.	Pied tit	P	-	-	-	-	-	-	-	P	-	-	-	-
213.	Plain Leaf-Warbler	-	-	-	-	P	-	-	-	-	-	-	-	-
214.	Plain Martin	-	-	-	-	-	-	-	P	-	-	-	-	-
215.	Plain Prinia	-	-	P	P	-	-	-	-	P	P	-	-	-
216.	Purple Sunbird	P	P	P	P	P	P	-	P	P	P	P	P	P
217.	Red Whiskered Bulbul	-	-	-	-	-	-	-	-	P	-	-	-	-
218.	Red-rumped Swallow	P	-	P	P	P	-	-	P	P	-	-	P	-
219.	Red-throated Flycatcher	P	-	-	-	-	-	-	-	P	-	-	-	-

220.	Red-vented Bulbul	P	P	P	P	P	P	P	P	P	P	P	P	P
221.	Red-winged Bush Lark	-	-	-	-	-	-	-	P	-	-	-	-	-
222.	Rosy Pastor	P	-	-	-	-	-	P	-	-	P	-	-	-
223.	Rufous-backed Shrike	-	-	-	-	P	-	-	P	-	P	-	-	-
224.	Rufous-bellied Babbler	P	-	-	-	-	-	P	-	P	-	-	-	-
225.	Rufous-fronted Prinia	-	-	-	-	P	-	-	-	-	-	-	-	-
226.	Rufoustailed Finch Lark	-	-	-	-	-	-	-	-	-	P	-	-	P
227.	Rufous-tailed Scrub-Robin	-	-	P	-	-	-	-	-	-	-	-	-	-
228.	Rufous-tailed Shrike	-	P	-	-	P	-	-	-	-	-	-	-	-
229.	Rufous-tailed Wheatear	-	-	-	-	-	-	-	-	P	-	-	-	P
230.	Rusty-tailed Flycather	-	P	-	-	-	-	-	-	-	-	P	-	-
231.	Small Minivet	-	-	-	-	-	-	-	P	-	-	-	-	-
232.	Smoky Warbler	-	-	-	-	-	-	-	-	-	P	-	-	-
233.	Spanish Sparrow	-	-	-	P	P	-	-	-	-	-	-	-	-
234.	Southern Grey Shrike	-	-	P	-	-	-	-	-	P	P	-	-	-
235.	Spotted Munia	P	-	P	-	P	-	-	P	P	-	-	-	P
236.	Streak-throated Swallow	-	-	-	-	-	-	-	-	-	P	-	-	-
237.	Striated Babbler	P	-	P	-	P	-	P	P	P	-	-	P	-
238.	Tawny Pipit	-	-	-	-	-	-	-	-	-	P	-	-	P
239.	Thickbilled Warbler	-	-	-	-	-	P	-	-	-	-	-	-	-
240.	Variable Wheatear	-	-	P	-	-	-	P	-	P	P	-	-	-
241.	Warbler	-	-	P	-	-	-	-	-	-	-	P	-	-
242.	Water Pipit	-	-	-	-	-	-	-	-	-	-	P	-	P
243.	White Throated Munia	P	-	-	P	P	P	-	P	P	P	-	P	P
244.	White Wagtail	-	-	P	-	P	-	-	P	-	-	-	-	-
245.	White-browed Fantail Flycatcher	-	-	P	P	-	-	-	-	P	-	P	-	-
246.	White-capped Redstart	-	-	-	-	P	-	-	-	-	P	-	-	-
247.	White-cheeked Bulbul	P	-	P	P	-	-	P	-	P	P	P	-	-
248.	White-eared Bulbul	P	P	P	P	P	P	-	P	-	P	P	P	P
249.	White-headed Babbler	P	-	-	-	-	-	-	-	-	-	P	-	P
250.	White-throated fantail Flytcather	-	-	-	-	-	P	-	-	P	-	-	-	-
251.	Wire-tailed Swallow	P	-	-	-	-	-	-	-	P	-	-	-	-
252.	Yellow Wagtail	-	-	P	-	P	P	-	P	-	P	-	-	-
253.	Yellow-eyed Babbler	-	-	-	-	P	-	-	-	-	-	-	-	-
254.	Dalmatian Pelican	-	-	-	P	-	-	-	-	-	-	-	-	-
255.	Darter	P	-	-	P	-	-	-	-	-	-	-	P	P

256.	Great Cormorant	-	-	-	P	P	P	-	P	-	P	-	-	-
257.	Great White Pelican	-	-	-	-	-	-	-	-	-	-	-	-	P
258.	Indian Shag	-	-	P	-	-	-	-	-	-	P	-	P	-
259.	Little Cormorant	P	-	P	P	P	P	-	P	-	P	-	P	-
260.	Pygmy Cormorant	-	-	-	-	-	-	-	P	-	-	-	-	-
261.	Spot-billed Pelican	-	-	-	-	-	-	-	P	-	-	-	-	-
262.	Greater Flamingo	-	-	-	-	-	-	-	-	-	P	-	-	-
263.	Lesser Flamingo	P	-	-	-	P	-	-	P	-	-	-	-	-
264.	Brown-headed Barbet	P	-	-	-	-	-	-	-	-	-	-	-	-
265.	Coppersmith Barbet	-	-	P	-	-	-	-	P	-	-	-	-	-
266.	Crimson-throated Barbet	P	-	-	-	-	-	-	-	-	-	-	P	-
267.	Lesser Goldenbacked wood pecker	P	-	P	-	P	-	-	P	-	-	-	-	-
268.	Little Grebe	P	-	P	-	P	P	-	P	-	P	-	P	-
269.	Plum-headed Parakeet	-	-	P	-	-	-	-	-	-	-	-	-	-
270.	Rose-ringed Parakeet	P	P	P	P	P	P	P	P	P	P	-	P	P
271.	Spotted Owlet	-	-	P	-	P	-	P	P	-	P	P	-	-

P = Present; (-) = Not recorded

Table 4. Diversity indices of Birds in different orders in the Thar Desert of Rajasthan

Order	Richness		Simpson's Index	Shannon Index	Hill's Number		Evenness	
	R1	R2			N1	N2	E1	E2
Psittaciformes	0.14	0.06	0.95	0.11	1.11	1.05	0.16	0.56
Piciformes	0.85	0.69	0.45	0.94	2.56	2.20	0.68	0.64
Phoenicopteriformes	0.12	0.03	0.85	0.29	1.33	1.18	0.41	0.67
Pelecaniformes	0.89	0.16	0.33	1.42	4.13	3.05	0.68	0.52
Passeriformes	11.37	0.94	0.06	3.36	28.87	17.11	0.72	0.26
Gruiformes	0.84	0.08	0.85	0.38	1.46	1.18	0.17	0.16
Galliformes	0.45	0.14	0.48	0.78	2.19	2.07	0.57	0.55
Falconiformes	3.55	0.70	0.16	2.20	8.99	6.10	0.67	0.33
Cuculiformes	0.44	0.31	0.41	0.94	2.57	2.45	0.86	0.86
Coraciiformes	1.07	0.10	0.39	1.17	3.23	2.54	0.49	0.29
Columbiformes	0.72	0.50	0.40	1.09	2.96	2.50	0.78	0.74
Ciconiformes	2.15	0.29	0.12	2.31	10.03	8.35	0.78	0.53
Charadriiformes	4.47	0.39	0.09	2.79	16.24	11.30	0.74	0.38
Apodiformes	0.18	0.12	0.97	0.08	1.08	1.04	0.12	0.54
Anseriformes	2.29	0.27	0.19	2.11	8.24	5.26	0.69	0.39

Table 5. Diversity indices of Birds in different families in the Thar Desert of Rajasthan

Family	Richness		Simpson's Index	Shannon Index	Hill's Number		Evenness	
	R1	R1			N1	N2	E1	E2
Anatidae	2.21	0.27	0.23	2.03	7.63	4.44	0.68	0.38
Apodidae	0.18	0.12	0.97	0.08	1.08	1.04	0.12	0.54
Burhinidae	0.39	0.23	0.44	0.88	2.41	2.27	0.80	0.80
Charadriidae	0.86	0.13	0.41	1.14	3.13	2.45	0.55	0.39
Glareolidae	0.47	0.35	0.72	0.50	1.64	1.38	0.45	0.55
Jacanidae	0.31	0.41	0.88	0.17	1.19	1.14	0.25	0.59
Laridae	0.93	0.27	0.56	0.85	2.33	1.77	0.44	0.33
Recurvirostridae	0.13	0.04	0.57	0.62	1.85	1.74	0.89	0.93
Scolopacidae	1.88	0.24	0.14	2.28	9.76	7.29	0.80	0.57
Ardeidae	1.16	0.21	0.25	1.58	4.85	3.97	0.69	0.48
Ciconiidae	0.43	0.13	0.57	0.66	1.93	1.75	0.47	0.48
Threskiornithidae	0.44	0.13	0.35	1.09	2.98	2.90	0.79	0.75
Pterocididae	0.26	0.28	0.62	0.55	1.74	1.62	0.80	0.87
Alcedinidae	0.22	0.20	0.64	0.54	1.71	1.57	0.78	0.86
Columbidae	0.33	0.04	0.51	0.79	2.20	1.98	0.57	0.55
Coracidae	0.21	0.19	0.92	0.16	1.17	1.09	0.23	0.59
Meropidae	0.28	0.09	0.85	0.29	1.33	1.17	0.26	0.44
Cuculidae	0.44	0.31	0.41	0.94	2.57	2.45	0.86	0.86
Accipitridae	3.29	0.65	0.17	2.15	8.62	5.98	0.67	0.34
Falconidae	0.37	0.52	0.44	0.69	2.00	2.29	1.00	1.00
Phasiadnidae	0.45	0.14	0.48	0.78	2.19	2.07	0.57	0.55
Gruidae	0.30	0.11	0.81	0.36	1.43	1.23	0.33	0.48
Rallidae	0.32	0.03	0.95	0.13	1.14	1.05	0.09	0.28
Alaudidae	1.54	0.54	0.41	1.26	3.52	2.44	0.55	0.35
Campephagidae	0.51	0.76	0.38	0.68	1.98	2.67	0.99	0.99
Corvidae	0.41	0.10	0.83	0.40	1.49	1.21	0.29	0.37
Dicruridae	0.18	0.13	0.99	0.03	1.03	1.01	0.04	0.51
Estrildidae	0.39	0.22	0.57	0.76	2.14	1.76	0.69	0.71
Hirundinidae	0.78	0.15	0.33	1.22	3.39	3.05	0.63	0.48
Laniidae	1.06	0.56	0.29	1.41	4.11	3.44	0.79	0.69
Motacillidae	1.35	0.27	0.43	1.21	3.37	2.33	0.51	0.31
Muscicapidae	5.59	0.94	0.12	2.59	13.33	8.34	0.68	0.30
Passeridae	0.26	0.06	0.79	0.41	1.51	1.26	0.38	0.50
Pycnonotidae	0.44	0.14	0.42	1.02	2.77	2.37	0.73	0.69
Sturnidae	0.83	0.18	0.29	1.46	4.33	3.46	0.75	0.62
Pelecanidae	0.31	0.12	0.45	0.91	2.50	2.23	0.83	0.83
Phalacrocoracidae	0.53	0.11	0.54	0.83	2.30	1.87	0.52	0.46
Phoenicopteridae	0.12	0.03	0.85	0.29	1.33	1.18	0.41	0.67
Captitonidae	0.80	0.87	0.43	0.82	2.28	2.31	0.75	0.76
Psittacidae	0.14	0.06	0.95	0.11	1.11	1.05	0.16	0.56

Table 6. Diversity Indices of Birds in different Months in the Thar Desert of Rajasthan

Month	Richness		Simpson's Index	Shannon Index	Hill'sNumber		Evenness	
	R1	R2			N1	N2	E1	E2
January	16.92	1.49	0.09	3.28	26.50	11.07	0.65	0.17
February	15.51	1.11	0.06	3.46	31.80	16.61	0.69	0.21
March	16.22	1.75	0.13	3.30	27.24	7.47	0.66	0.19
April	12.72	1.98	0.21	2.45	11.60	4.81	0.53	0.11
May	14.39	2.43	0.03	3.87	48.04	31.13	0.82	0.43
June	10.28	1.19	0.13	2.77	15.99	7.56	0.62	0.18
July	7.08	1.42	0.05	3.43	30.76	21.53	0.87	0.59
September	6.38	1.14	0.22	2.32	10.18	4.53	0.60	0.21
October	13.16	1.60	0.04	3.72	41.18	25.98	0.79	0.36
November	9.19	1.09	0.07	3.16	23.51	13.39	0.72	0.29
December	6.29	1.02	0.23	2.36	10.56	4.37	0.60	0.21

Table 7. Relative abundance and Dominance of Threatened species of Birds in the Thar Desert of Rajasthan

Sl. No.	Species name	Estimated No. of Individual	Dominance
1	Painted Stork	297	35.83
2	Beach Stone Plover	89	10.74
3	Marbled Teal	83	10.01
4	Sarus Crane	74	8.93
5	Spot-billed Pelican	73	8.81
6	Ferruginous Poachard	56	6.76
7	Cinereous Vulture	56	6.76
8	Darter	44	5.31
9	Indian White-backed Vulture	21	2.53
10	Great Indian Bustard	15	1.81
11	Pallid Harrier	10	1.21
12	Red-headed Vulture	6	0.72
13	Eastern Imperial Eagle	4	0.48
14	Black Stork	1	0.12
Total		829	100.00

Table 8. Abundance of Birds in the IGNP Canal area in the Thar Desert of Rajasthan

Sl. No.	Common Name	Total	Abundance
1	Common Coot	1350	30.25
2	Eurasian Collared-Dove	352	7.89
3	Little Cormorant	221	4.95
4	Blue Rock Pigeon	208	4.66
5	Spotted Redshank	200	4.48
6	House Sparrow	162	3.63
7	Common Pochard	140	3.14
8	Common Teal	134	3.00
9	Great Cormorant	129	2.89
10	Bar-headed Goose	100	2.24
11	Red-wattled Lapwing	76	1.70
12	Tufted Duck	74	1.66
13	Striated Babbler	69	1.55
14	Urasian Spoonbill	67	1.50
15	Indain Peafowl	66	1.48
16	Common Babbler	53	1.19
17	Mallard	50	1.12
18	Northern Shoveller	48	1.08
19	Green Bee-eater	46	1.03
20	House Swift	43	0.96
21	Black-winged Stilt	41	0.92
22	Citrine Wagtail	40	0.90
23	Common Swallow	32	0.72
24	Spanish Sparrow	30	0.67
25	Little Brown Dove	29	0.65
26	Grey Francolin	26	0.58
27	Black-throated Munia	26	0.58
28	White Throated Munia	26	0.58
29	Little Grebe	23	0.52
30	Common Starling	23	0.52
31	Large Grey Babbler	22	0.49
32	Little Egret	21	0.47
33	House Crow	20	0.45
34	Yellow-eyed Babbler	18	0.40
35	Marbled Teal	18	0.40
36	Indian Robin	18	0.40

Sl. No.	Common Name	Total	Abundance
37	Rose-ringed Parakeet	18	0.40
38	Wood Sandpiper	18	0.40
39	Black Drango	17	0.38
40	Oriental White Ibis	17	0.38
41	Gray Heron	16	0.36
42	Black Ibis	16	0.36
43	Rosy Paster	16	0.36
44	Egyptian Vulture	13	0.29
45	Red-headed Vulture	13	0.29
46	Spot-billed Duck	12	0.27
47	Painted Stork	12	0.27
48	White-eared Bulbul	12	0.27
49	Jungle Myna	12	0.27
50	Demoiselle Crane	11	0.25
51	Cattle Egret	11	0.25
52	Purple Sunbird	11	0.25
53	Red-vented Bulbul	11	0.25
54	Purple Heron	10	0.22
55	River Tern	10	0.22
56	Bank Myna	9	0.20
57	White Wagtail	9	0.20
58	White-tailed lapwing	9	0.20
59	Common Myna	8	0.18
60	Curlew Sandpiper	8	0.18
61	Eurosian Reed-Warbler	8	0.18
62	Brahminy Starling	8	0.18
63	Northern Pintail	6	0.13
64	Spotted Owlet	6	0.13
65	Little Ringed Plover	6	0.13
66	Black-headed Gull	6	0.13
67	Yellow Wagtail	6	0.13
68	Common Sandpiper	6	0.13
69	Large Egret	5	0.11
70	Rufous-fronted Prinia	5	0.11
71	Tawny Eagle	4	+
72	Common Raven	4	+
73	Medium Egret	4	+
74	Long-billed Vulture	4	+

Sl. No.	Common Name	Total	Abundance
75	Brown Shrike	4	+
76	Blue-cheeked Bee-eater	4	+
77	White-cheeked Bulbul	4	+
78	White-headed Babbler	4	+
79	Black Kite	3	+
80	Cotton Teal	3	+
81	Desert Wheatear	3	+
82	Indian Pond-Heron	2	+
83	Little Stint	2	+
84	Western Marsh-Harrier	2	+
85	Black Winged Kite	2	+
86	Ashycrowned Finch Lark	2	+
87	Black Francolin	2	+
88	Common Crested Lark	2	+
89	Indian White-backed Vulture	2	+
90	White-breasted Kingfisher	2	+
91	Great Grey Shrike	2	+
92	Rufous-backed Shrike	2	+
93	Spotted Munia	2	+
94	Grey Wagtail	2	+
95	Large Pied Wagtail	2	+
96	Eurasian Curlew	2	+
97	Common Tailorbird	2	+
98	Osprey	2	+
99	Indian Shag	2	+
100	Baya Weaver	2	+
101	Green Sandpiper	2	+
102	Common Hoopoe	2	+
103	Plain Leaf-Warbler	2	+
104	Shikra	1	+
105	Eastern Skylark	1	+
106	Small Blue Kingfisher	1	+
107	Black Stork	1	+
108	Indian Roller	1	+
109	Jungle Crow	1	+
110	Asian Koel	1	+
111	Brown-headed Gull	1	+
112	Variable Wheatear	1	+

Sl. No.	Common Name	Total	Abundance
113	Chestnut-bellied Sandgrouse	1	+
114	Painted Sandgrouse	1	+
115	Jungle Babbler	1	+
116	Yellow-legged Gull	1	+
117	Black Redstart	1	+
118	Ashy Prinia	1	+
119	Oriental White eye	1	+

+ = less than 0.1.

ACKNOWLEDGEMENT

The authors thank Dr. J.R.B. Alfred, Director, Zoological Survey of India for providing research facilities during the study period. Thanks are also due to the Ministry of Environment and Forests, Government of India for providing funds to conduct the present study under the project entitled `Studies on the faunal diversity in the Thar Desert of Rajasthan'.

REFERENCES

Adams, R.M. (1873). Notes on the birds of the Sambhar Lake and its vicinity. *Stray Feathers* 1: 361-404.

Adams, R. M. (1874). Additional notes on the birds of the Sambhar Lake and its vicinity. *Stray Feathers* 2: 337- 341.

Ali, S. and S.D. Ripley (1983). *Hand Book of the Birds of India and Pakistan.* Oxford University Press, Oxford. 737 p.

Barnes, H.E. (1886). Birds nesting in Rajpootana. *J. Bombay nat. Hist. Soc.* 1: 38-62.

Bohra, H.C. & P. K. Goyal, S.P. (1992). Checklist of the birds of Machia Safari Desert Park, Jodhpur (Rajasthan). *Pavo* 30 (1&2): 87-97

Burnham, K.P., D.R. Anderson and J.L. Lake (1980). Estimation of density from the transect sampling of biological publications. *Wildlife Monograph* No. 72, Pp. 202

Changani, A. K. (2002). Avifauna in and around Jodhpur City, Rajasthan, India *Newsletters for Birdwatcher* 42(2): 24-26.

Grimmett, R., C. Inskipp and T. Inskipp (1998). *Birds of the Indian Subcontinent,* Christopher Helm, London.

Hoves, J.G. and D. Bakewell (1989). *Shore Birds Studies Manual.* AWB Publications. No.55, Kuala Lumpur. 362 p.

Kumar Anil and Mussavi A. H. (2001). Present Status of Changing Avifauna in IGNP Area of Bikaner and Jaisalmer. In : *Impact of Human Activities on Thar Desert Environment : Symposium Abstracts.* (Eds S. Kathju, Amal Kar, N. L. Joshi, P. C. Pande and Praveen Kumar), CAZRI, Jodhpur, Feb 15-17, 2001. Arid Zone Research Association of India, Jodhpur, India. 158 p.

Kumar, Anil (In Press). Invasion of some new avian species in the Thar desert of Rajasthan. *J. Bombay nat. Hist. Soc.*

Kumar S., Sivaperuman C., Pardesi M and Baqri Q.H. (2002). Comb Duck *Sarkioliornis melanotus* (Pennant, 1769) in the Thar Desert of Rajasthan, India *Newsletters for Birdwatcher* 42 (5): 102.

Lima, S.L. and T.J. Valone (1989). Impact of Brown-headed Cowbird parasitism on the reproductive success of the solitary vireo. *Auk* 106: 476-480.

Ludwig, J.A. and J.R. Reynolds (1988). *Statistical Ecology: A premier on methods and computing.* A Wiley-Interscince Publication. 337 p.

Manakadan, Ranjit and Aasheesh Pittie (2001): Standard common and scientific names of the birds of the Indian subcontinent. *Buceros* 6(1): 1-37.

Mann, H.S. (1988). Future of the Indian Desert. In: *Desert Ecology.* Eds. I. Prakash, Scientific Publisher, Jodhpur. 307-313.

Margalef, R. (1958). Information theory in ecology. *General Systematics,* **3**: 36-71.

Menhinick, E.F. (1964). A comparison of some species-individuals diversity indices applied to samples of field insects. *Ecology,* 45: 859-861.

Milchunas, D.G., W.K. Lauenroth and I.C. Burke (1998). Livestock Grazing: Animal and Plant Biodiversity of Short grass Steppe and the Relationship to Ecosystem. *Oikos,* 83: 64-74.

Prakash, I. (1989). Degradation of desert rangelands due to grazing pressure.. In *Perspectives in Ecology.* Eds., Singh, J.S. and B. Gopal Jagmender; Book Agency, New Delhi; 181-191

Rahmani, A.R. (1994). *Wildlife situation in the Thar Desert.* Report submitted to World Wide Fund for Nature, New Delhi.

Rahmani, A.R. (1996). Changing Avifauna of the Thar Desert. In *Faunal Diversity* In *The Thar Desert of Rajasthan.* Eds. A.K. Ghosh, Q.H. Baqri & I. Prakash. Scientific Publishers, Jodhpur; 306-324.

Rahmani, A.R. (1997): The Effect of Indira Gandhi Nahar Project on the Avifauna of the Thar Desert. *J. Bombay nat. Hist. Soc.,* 94 (2): 233-266.

Rana, B.D., Jain A.P. and Tripathi, R.S. (1996). Avian Diversity. In *Faunal Diversity In The Thar Desert of Rajasthan.* Eds. A. K. Ghosh, Q. H. Baqri & I. Prakash. Scientific Publishers, Jodhpur. 325-331.

Sangha, H.S. (1994). Marbled teal in Rajasthan. *Newsletter for Birdwatchers* 34 (2): 34.

Sangha, H.S. (2002): A supplementary notes on the Avifauna of the Thar desert (Rajasthan). *J. Bombay nat. Hist. Soc.* 99 (1): 120-126.

Sangha and Kulshreshtra (1993): Birds recorded during a visit to the Desert National Park in summer 1993. In: *Bird Conservation, Strategies for the Nineties and Beyond.* (Eds.: Varghese, A.S. Sridhar and A.K. Chakravarthy) OSI, Bangalore.

Sekhar, U.N. (1988). Ecological Status of the Desert National Park past and present. *Tigerpaper* 25 (20): 14-18.

Singh, K.C. (1991). Production of forage grasses with minimal irrigation in Western Rajasthan. Page: 82-86. In: Prospects of Indira Gandhi Canal

Project (Eds Venkateswaralu, J. and I.P. Abrol). *Indian Council of Agricultural Science*, 60(1): 23-28.

Soni, R. G. (1994). Check-list of birds of Indira Gandhi Nahar Project stage –II (Rajasthan) *Indian Forster* 120 (10): 929-935.

Ticehurst, C.B. (1922a). The birds of Sind. *Ibis* 4(3): 526-572.

Ticehurst, C.B. (1922b). The birds of Sind, part 2. *Ibis* 4(4): 605-662.

Ticehurst, C.B. (1923a). The birds of Sind, part 3. *Ibis* 5(1): 1-43.

Ticehurst, C.B. (1923b). The birds of Sind, part 4. *Ibis* 5(2): 235-275.

Ticehurst, C.B. (1923c). The birds of Sind, part 5. *Ibis* 5(3): 438-474.

Ticehurst, C.B. (1923d). The birds of Sind, part 6. *Ibis* 5(4): 645-666.

Ticehurst, C.B. (1924a). The birds of Sind, part 7. *Ibis* 6(1): 110-146.

Ticehurst, C.B. (1924b). The birds of Sind, part 8. *Ibis* 6(3): 495-518.

Whistler, H. (1938): The ornithological survey of Jodhpur state. *J. Bombay nat. Hist. Soc.* 40: 213-235.

Changing Faunal Ecology in Thar Desert, pp. 231-266, 2005.
Eds.: B.K. Tyagi and Q.H. Baqri

12

MAMMALIAN DIVERSITY IN THE THAR DESERT OF RAJASTHAN

S. Chakraborthy, V.C. Agrawal and S.I. Kazmi*

Zoological Survey of India, Kolkata,
*Desert Regional Station, Zoological Survey of India, Jodhpur - 342 005, India

Based on the surveys, existing literature and local reports a total of 66 mammalian species, three of which new, have been recorded along with their geographical and ecological distribution in the Rajasthan desert. New district records for a number of species belonging to the orders Chiroptera, Rodentia and Carnivora have been brought on record. Species like *Panthera pardus, Caracal caracal* and *Melursus ursinus* are feared to have been almost wiped out from the desert biome, while *Hyaena hyaena* appeared to be nearly extinct. Habitat alteration and tranportation through human intervention has caused about introduction of mesic species like *Bandicota indica, B. bengalensis, Rattus norvegicus, Millardia meltada, Hystrix indica* and *Suncus murinus* etc. in remote desert areas. *Antilope cervicapra* and *Gazella bennettii,* though common in the desert ecosystem, are more or less restricted in the protected areas or in and around Bishnoi-inhabited villages, owing mostly to water scarcity and poaching pressure. Afforestation along the canal has attracted many a species to take refuge and prosper in the desert. Identification keys to the extant mammalian species of desert are provided.

INTRODUCTION

On the basis of similar ecology, biome representation, community and species, Rodgers and Panwar (1988) recognized Indian Desert as a distinct biogeographic zone with two provinces, *viz.*, 3A. Desert- Thar and 3B Desert-Katchchh. The province 3A, *i.e.*, Desert-Thar, is extending over the western part of Aravalli Range in Rajasthan (1,96,150 sq.km.), and Punjab and Haryana (20,000 sq.km.). The word `Desert' conjures up the vision of vast, treeless, undulating and buffed expanse of sand with scattered growth of cactus and occasional caravans of heavily-robed nomads on camel-back! However, except for the sand part and certain climatological factors, the Thar Desert does not seem to conform to the popularly held general pattern of desert in many details. It is one of the smallest but most thickly populated (average density of 83 persons per sq. km.) deserts of the world, presenting a spectacular diversity of the landscapes, habitat types and soil formation (Prakash, 1975a;, Rahmani, 1997; Baqri and Kankane, 2001). At the same time extreme temperature, scarce rainfall, acute shortage of water and frequent sand storms make the environment hostile to the living organisms of the Thar desert. Pressures of an enlarged and impoverished human-livestock populations combined with change in the agro-ecosystem have taken a heavy toll of this natural area. Habitat alteration mainly under the impact of the massive Indira Gandhi Nahar Project (IGNP) is also giving pathway to various life forms from mesic areas replacing the indigenous desert's biodiversity (Prakash, 1981; Ghosh *et al.*, 1996). In fact, unless the proper management practices are adopted, the present socio-political exigencies with habitat alteration may break down the natural age-old ecosystem of the Thar desert and its unique biodiversity containing extraordinary ecological adaptability. An understanding of the desert's ecological processes can give an insight into the management of this extraordinary fragile ecosystem.

Mammalian species play a significant role as primary consumer, prey and predator in the food chain and other ecological processes in the ecosystem. Thus, knowledge about the mammalian species composition, their ecological distribution, conservation and identification is the most essential requirement to understand the ecological processes and develop a fruitful management practice. Unfortunately, till the middle of the last century not much attention was paid either for conducting taxonomic survey or carrying expanded ecological research in the Indian desert. To fill up these gaps, Ministry of Environment and Forests, New Delhi, sanctioned the project "Studies on the faunal diversity in the That desert of Rajasthan" in January, 2000. Under the project till March, 2003, five surveys each of 15-20 days duration were conducted by the first author in the different habitats of the Thar Desert of Rajasthan with special reference to mammals. Based on findings of those surveys and existing literature, an attempt has been made to provide a detail account of the mammalian species diversity in Rajasthan, Thar desert, keys to their identification, conservation, status and ecological and geographical distribution.

Table 1. Trapping, netting and scanning sites

District	Locality	Lat./Long.	Habitat
1	2	3	4
Jaisalmar	Sam	26 ° 50'36"N 070 ° 32'42"E	Shifting Sand dunes
	Dabla	26 ° 52'35"N 071 ° 05'33"E	Gravel plains and rocky outcrops
	Lathi	26 ° 14'15"N 072 ° 19'59"E	Commercial and subsistence farming, sandy plains
	Lathi	227 ° 01'54"N 071 ° 28'43"E	Sandy plains, scattered patches of grasses and shrubs
	Chandan, CAZRI Compound	27 ° 01'53"N 071 ° 28'43"E	Gardens, Sandy plains
	Moolsagar	26 ° 55'15"N 070 ° 51'37"E	Rocky outcrops
	Kaknai	26 ° 55'09"N 070 ° 47'33"E	Dry river bed (Kaknoi), sandy
	Ghadisar	26.54'21"N 070 ° 56'09"E	Wetland, fresh water tank
	Sudasari	26 ° 54'38"N 070 ° 35'44"E	Gravel plains with scattered grasses & dry deciduous trees. Protected area
	Barna	26 ° 40'06"N 070 ° 41'37"E	Gravel plains
	Thiyat	26°40'06"N 070 ° 41'37"E	Sandy plains
	Jaisalmer	26 ° 55'08"N 070 ° 47'33"E	Township
	Knod	26 ° 40'06"N 070 ° 41'37"E	Dry river bed (Solandi), Sandy
	IGNP-0-head (1458 RD)	27 ° 18'38"N 071 ° 09'10"E	Canal side plantation
	Mohangarh	27 ° 18'21"N 071 ° 13'23"E	Village, Sandy plains, subsistence cultivation
	Nachna	27 ° 18'21"N 071 ° 13'23"E	Dense canal side Plantation
Bikaner	Bikampur	27 ° 44'49"N 072 ° 11'17"E	Stabilized sand dunes
	Shankala Phanta	27 ° 50'42"N 072 ° 11'17"E	Subsistence farming
	Kalayet	27 ° 50'52"N 072 ° 53'33"E	Sandy plains, freshwater pond (Dry at the time of survey)
	Gajner	27 ° 50'51"N 072 ° 53'33"E	Sandy plains, freshwater pond (Dry at the time of survey)
	Lunkaransar	27 ° 58'24"N 0733 ° 03'22"E	Saline wetland with vast stretches of dry sandy bed
Jodhpur	Kailana	24 ° 44'38"N 072 ° 58'37"E	Fresh water wetland surrounded by rocky outcrops

	Arthi	26° 17'57"N 072° 58'20"E	Shifting sand dunes
	Kui	26° 35'55"N 072° 28'41"E	Rocky outcrops under heavy pressure of stone mining
	Dehra	26° 35'46"N 072° 20'07"E	Sandy plains, scattered subsistence farming
	Near Peelwa	26° 35'43"N 072° 20'07"E	Sandy plains, scattered subsistence farming
	Near Peelwa	26° 48'12"N 072° 25'05"E	Interdunal plains
	Lordi	27° 02'32"N 072° 27'57"E	Commercial farming
	Balsamand	26° 20'01"N 073° 01'25"E	Garden
	Kechan	27° 07'56"N 072° 24'37"E	Gravel Plains
	Khiyan Sarya	26° 40'20"N 072° 19'27"E	Stabilized sand dunes
	Dechu	26° 40'20"N 072° 19'27"E	Village, Sandy plains, subsistence farming
Barmer	Gunga	26° 14'10"N 071° 12'42"E	Shifting sand dunes
	Langera	25° 43'28"N 071° 21'20"E	Rocky outcrops
	Langera	25° 43'46"N 071° 22'59"E	Commercial farming
	Khet Singh Ki Piyan	25° 36'28"N 071° 24'02"E	Stabilised duns with scanty farming
	Dhorimanna	25° 11'59"N 071° 26'41"E	Sandy plains
	Naiyoki Dhani	25° 06'58"N 071° 31'39"E	Dry river bed with high percentage of clay and under subsistence farming
	Luni	27° 50'42"N 072° 11'17"E	Dry sandy river bed with knee deep water at places
Jalore	Pratappura	24° 43'01"N 071° 46'19"E	Commercial farming, tubewell irrigation
	Baragaon	24° 40'17"N 072° 17'03"E	Village, commercial farming
	Ajadar	24° 43'16"N 072° 14'25"E	Subsistence farming depending on rain water
Churu	Dewanii	27° 46'50"N 074° 26'54"E	Stabilised sand dunes with scanty agriculture
	Talchappar	27° 48'03"N 074° 26'16"E	Dry, saline depression; subsistence farming; partly protected
Sikar	Harsh Parbat	27° 30'02"N 075° 10'29"E	Hills with scanty agriculture
Sri-Ganganagar	03 Kms. of Suratgarh	29° 16'47"N 073° 55'58"E	Sandy plains with dens patches of grasses and shrubs

	Badopal	29° 21'24"N 074° 03'33"E	Wetland, Ghaggar depression
Hanumangarh	Kohla	29° 30'34"N 074° 19'55"E	Dry deciduous forest by the side of canal, commercial farming.
	Lakhera	28° 58'54"N 074° 14'58"E	Shifting dunes and interdunal plains
Pali	Sandarao	25° 18'24"N 073° 10'13"E	Township with neighboring commercial farming areas
Jhunjhunu	Indrapura	27° 46'11"N 075° 28'28"E	Village, commercial farming, gardens
	Kotdam	27° 39'03"N 075° 25'13"E	Fresh water wetland, hills, dry deciduous forest
	Shakhambari	27° 37'45"N 075° 24'11"E	Hills, dry deciduous forest
	Manasa Kho	27° 47'34"N 072° 35'42"E	Hills, dry deciduous forest
Nagaur	Merta	27° 09'28"N 073° 47'31"E	Township surrounded by commercial farming areas
	Rian	27° 08'26"N 073° 50'48"E	Clay plains, intensive commercial farming, tub-well irrigation

STUDY AREA

The present study is confined to the Rajasthan part of the Thar Desert. As already stated, about 1,69,150 sq.km. of geographical area of Rajasthan comes under the Thar Desert. The area is bordered by Aravalli Range and Mahendragarh district (Haryana) in the east, Firozpur and Faridkot districts (Punjab), Sirsa and Bhiwani districts (Haryana) in the north, Banaskantha district (Gujarat) in the south and Pakisthan in the west. Entire Sri Ganganagar, Bikancr, Jaisalmer, Barmer, Jodhpur and Churu districts and part of Nagaur, Ajmer, Pali, Jalore, Jhunjhunu and Sikar districts fall under the jurisdiction of the Thar Desert in Rajasthan (Baqri and Kankane, 2001). Extreme variation of temperature from freezing point in winter to about 50°C in summer, low rainfall and humidity are the main characteristics of desert climate. Relative humidity drops down to almost zero during summer and mean annual rainfall is less than 20 cm. in the western section covering Jaisalmer and part of Bikaner, while it is 20-50 cm. in the other part of the Thar. There is also considerable diurnal variation of temperature and relative humidity. Landscape, soil, rivers, vegetation and the most debtable issue of the region, *i.e.*, impact of IGNP have been discussed in great details by Gupta and Prakash (1975a), Prakash (1981), Anon.(1988), Ghosh *et.al.* (1996), Rahmani (1997), Baqri and Kankane (2001).

Relatively a very small percentage of the area of the Thar has been brought under cultivation. Traditional crops are Bajra (*Pennisetum typhoides*), Sorghum or Jwar (*Sorghum vulgare*), Moong (*Phaseolus radiators), Moth (Vigna acontifolia),* Guar (*Cyamopsis tetragonoloba)* and Til (*Sesamun indicum*). However, in the canal irrigated areas these crops are gradually replaced by Groundnut (*Arachis hypogea),* Cotton (*Gossypium* sp.), Paddy (*Oryza sativa),* Wheat (*Triticum sativum) etc.* (Chatterji and Saxena, 1988).

Details of trapping and scanning sites and sections of road covered under census are given below as Table 1 and 2.

Table 2. Road sections covered by census

District	Section of Road	Total distance	Major Habitats
1	2	3	4
Jaisalmer	Forest road within Sudasari enclosure of Desert National Park (On Foot)	About 5 Km.	Gravel plains with dry deciduous forest and patches of grassland
	Barna to Sam through Desert National Park	20 Km.	Gravel plains, sandy plains, sand dunes, grasslands, dry deciduous forest
	Dhauba to Pitchla on Jaisalmer-Maizlar Road	16 Km.	Sandy and gravel plains
	"0" head of IGNP main canal to 42 RD of shaid Birbal Sakha	About 14 Km.	Canal side plantation
	"0" head of IGNP main canal to 1398 RD of main canal	20 Km.	Canal side plantation
Bikaner	Gajner to Bikaner on NH 15	25 Km.	Sandy plains with patches of agriculture
Jalore	Sanchore to Sarnau on State High Way	33 Km.	Sandy plains, farming villages
Jhunjhunu	Indrapura to Todpura on village road	11 Km.	Sandy plains, farming
Jodhpur	Dechu to Lohawat	33 Km.	Sandy plains, villages, subsistence farming, stabilized dunes
Sikar	Harshnath to Harsh village on hill road	10 Km.	Hills with dry deciduous forest and scanty cultivation
Hanumangarh	Kohla farm to ShareKhan on village road	10 Km.	Village, intensive commercial farming, canal
Sri-Ganganagar	Suratgarh to Nirbana	14 Km.	Sand dunes, sandy plains, farming
Barmer	Barmer to Kharin	16 Km.	Gravel and sandy plains, rocky outcrops
Nagaur	Kuchera to Ren on Nagaur-Merta Road	25 Km.	Clay plains with intensive commercial farming

METHODOLOGY

As per Indian Wildlife (Protection) Act, 1972, collection of the majority of the mammalian species is prohibited. Thus, maximum efforts were given for direct sightings, identification of burrows, hideouts, calls, footprints, faecal matter and signs. Interviews of different sections of pcoplc were also taken. A total of 25 plots measuring about 10,000 to 22,500 sq.m. and distributed over the sand dunes, inter-dunal plains, sandy plains, gravelly pediments, forests, hills, river beds, wetlands, IGNP command areas, villages and cities of different districts have been selected during the surveys. Each plot was thoroughly surveyed on foot both during day and night to find out the direct and indirect evidences of the occurrence of the mammalian species and to estimate their relative abundance. Further, 12 plots each of about 10,000 sq.m. in different sub-habitats were selected for trapping of rodent and insectivore specimens with Shermon traps. In each plot, trapping was done for about 2400 trap hours. Specimens thus collected were released at the same plot after identification and noting the relevant data. As locally made Shermon traps are not effective for the very small sized species, such as *Gerbilllus gleadowi, Mus musculus* etc., in another 12 plots of similar nature trapping was carried out with snap-traps for 2400 trap hours. Random trappings with both types of traps were also conducted at the railway stations, market places, and residential complexes. Specimens collected with snap traps have been incorporated in the National Zoological Collection after identification. Roasted coconut, dry fish and peanut butter have been used as bait material in the traps. To reveal the species composition of the bats, their probable roosting sites such as wells, ancient buildings, forts, ruins, huts, tunnels, rock crevices, large trees *etc.*, have been searched for. Whenever, a roost was found, roost size and species involved have been noted with minimum disturbance. However, mist nets were also placed from dusk to early morning in probable roosting and foraging areas of the districts surveyed for collection and identification of bat specimens. Details of scanning, trapping and netting sites given under the heading `Study Area'.

Roadside census (5 to 33 km. stretch) wcrc also carried out by vehicle or on foot along the different types of roads, such as National highways, State highways, District roads, Village roads, forest roads, cross country and canal roads. Details of the roads covered under census are already given under the heading `Study Area'. Same section of the road was covered in the morning (app. 06.00 to 08.00 hr), afternoon (17.00 to 19.00 hr) and night (23.00 to 01.00 hr) For spotting and identification of mammalian specimens field binocular was used during day and spotlight during night. Roads were selected in such a way as to cover all the major desert habitats.

Global Positioning System (GPS) readings for all the scanning and trapping plots have been noted.

For compilation of species composition, geographical and ecological distribution, in addition to the findings of the present study, data from earlier literature (Prakash, 1956; 1959, 1963a, 1963b, 1964, 1975a,b, 1981, 1994; Prakash and Jain, 1967; Prakash *et al.*, 1971; Prakash and Mathur 1979; Gupta and Agrawal 1966; Agrawal 1967, 2000; Biswas and Ghose 1968; Sinha 1975, 1976, 1981, 1996; Gupta and Prakash, 1975; Ghose, 1976; Rana, 1980; Alfred and Agrawal, 1996; Rahmaani,

1997; Bates and Harrison, 1997; Bhargava *et al.*, 2001; Baqri and Kankane, 2001) have also been incorporated. For framing the keys to the identification of the mammalian species of Rajasthan Thar desert, reserve collection present in the National Zoological Collection have been studied. Further identification keys of the different groups of mammals provided by earlier workers (Wroughton, 1918-20; Pocock, 1939, 1941; Roberts, 1977; Sinha, 1981; Corbet and Hill, 1992; Agrawal *et al.*, 1992; Agrawal, 2000) have been taken into consideration.

Classification and nomenclature of mammalian species in the text have been followed after Wilson and Reeder (1993).

RESULTS

Mammalian diversity and significance

Agrawal (1998), on the basis of Wilson and Reeder (1993), listed 13 Orders, 42 Families, 180 Genera and 390 species of Indian living mammals. Out of these 8 Orders, 23 Families, 45 Genera and 66 species have been recorded from Rajasthan desert (Table 3).

Table 3. Mammalian species diversity of Rajasthan Thar desert as compared to that of the country

Living orders of Indian Mammals	No. of Families		No. of Genera		No. of Species	
	India	Rajasthan Desert	India	Rajasthan Desert	India	Rajasthan Desert
Insectivora	3	2	11	2	28	4
Chiroptera	6	7	36	12	110	18
Primates	3	1	6	2	15	2
Pholidota	1	1	1	1	2	1
Carnivora	7	6	33	9	55	14
Artiodactyla	5	2	20	4	32	4
Lagomorpha	2	1	3	1	10	1
Rodentia	4	3	43	14	101	22
Proboscidea	1	-	1	-	1	-
Sirenia	1	-	1	-	1	-
Perissodactyla	2	-	2	-	3	-
Cetacea	6	-	21	-	29	-
Scandentia	1	-	2	-	3	-
Total	42	23	180	45	390	66

Rajasthan part of the Thar Desert is only about 6 per cent of the total land area of the country. However, within this relatively small area with hostile environmental situation, it supports nearly 17 per cent of the mammalian fauna of India. A detailed systematic list of mammalian species of Rajasthan Thar desert along with their conservation status, geographical distribution and relative abundance has been provided in Table 2. Species marked with the asterix in the table could not be observed or collected during the present surveys and included in the list on the basis of literature or local report. For obvious reason, Rajasthan desert is devoid of any truly aquatic or high altitude form.

Among the eight mammalian orders in the Thar, Chiroptera is represented by highest number of the families (7), followed by Carnivora (6). While orders Primates, Pholidota and Lagomorpha have nominees of single family each. The highest generic and species diversity is exhibited by the order Rodentia being 14 and 22, respectively.

A few species such as *Rattus rattus, Mus musculus* have more than one subspecies in the desert and often with overlapping geographical distribution (Table 2). This has happened mainly due to transportation from other parts of country through human agency.

Prakash (1984) recognized eleven species of rodent as major pest species effecting food production in India, out of these as many as ten *viz., Funambulus pennanti, Hystrix indica, Tatera indica, Meriones hurrianae, Rattus rattus, Millardia meltada, Mus musculus, Nesokia indica, Bandicota bengalensis, Bandicota indica* occur in the Rajasthan desert. It is obvious that they inflict severe damage on standing crops, stored food grains and other commodities.

From the analysis of the list of recent species of Indo-malayan mammals (Corbet and Hill 1992), it has been found that the majority of the species occurring in Rajasthan desert are predominantly or extensively Indo-malayan in distribution. However, at least one Afrotropical (*Taphozous perforatus),* six palaearctic/Afrotropical (*Rhinopoma microphyllum, R. hardwickei, Taphozous nudiventris, Tadarida aegyptiaca, Felis silvestris, Caracal caracal),* two palaearctic (*Vulpes vulpes, Gerbillus nanus)*; and two introduced (*Mus musculus, Rattus rattus)* species have been recorded from Rajasthan Thar desert.

No mammalian species has been found endemic strictly to Rajasthan part of Thar desert. However, two endemic species *viz., Millardia gleadowi* and *Gerbillus gleadowi* of desert biogeographic zone is well distributed in its Rajasthan part. Among the Indian endemic species, three *viz., Mus platythrix, Mus phillipsi and Cremnomys cutchicus* are also well distributed in the Thar desert of Rajasthan.

Quite a number of species, such as, *Caracal caracal, Panthera pardus, Melursus ursinus, Vulpes bengalensis, Sus scrofa* which were once found in fairly good numbers are now practically wiped out from the desert scene or, at best, represented by a very few localized populations in certain pockets mainly due to hunting pressure. Depending on their existing populations, genetic features, hunting, trade, habitat loss and also role as vermin, nearly 73 per cent of the mammalian species occurring in the Thar region of Rajasthan have found placed in one or other schedule of the Indian Wildlife (Protection) Act, 1972 (Amended up to 2000) (Table 4). A total of nine species of Rajasthan Thar desert have been included in Schedule 1. Out of these two *viz., Gazea nemmettii* and *Antilope cervicapra* have the strongest possibilities of genetic deterioration owing to free mixing and breeding with village cat (Kankane, 1995). Apart from Schedule I, altogether 8 species (*Felis chaus, Semnopithecus entellus* etc.) are included in schedule II, 2 species (*Boselaphus tragocamelus, sus scrofa*) in Schedule III, 7 species (*Lepus nigricollis, Herpestes* sp.etc) in Schedule IV, 22 species (All fruit bats, rats and mice) in vermin or Schedule V of the said Act (Table 4). Depending on the extent of the threat from international trade of wildlife products, CITES kept four species (*Semnopithecus entellus, Panthera pardus* etc.) of Rajasthan desert in Appendix I, while six (*Pteropus giganteus, Macaca multta* etc.) and eight (*Canis aureus, Herpestes edwardsi* etc.) species are included under Appendix II and III respectively. Among the threatened mammalian species of the country as recognized by National Red Data Book (Anon., 1994), two endangered (*Felis silvestris, Caracal caracal*) and six vulnerable species (*Canis lupus, Vulpes vulpes* etc.) have been recorded from the area. However, on the basis of IUCN criteria, CAMP workshop report (Molur *et.al.*, 1998) considered only two species as vulnerable and another 18 species as on lower risk near threatened category (Table 2).

Table 4. Systematic list of mammalian species from Rajasthan along with their district- wise distribution, relative abundance and conservation status

Order	Family	Species	District-wise distribution	Relative abundance	Conservation status			
					IWPA (1972) as amended upto 2002	CITES as amended upto 2000	Red Data Book Natn, 1994	C.A.M.P Ws. Report, 1998
Insectivora	1. Erinaceidae	Hemiechinus microus	All	++++	-	-	-	-
		H. collaris	8,9,10, 12	++	Schedule IV as *H.auritus*	-	-	LRlc
	2. Soricidae	3a. Suncus murinus sindensis	6,8,9, 10,12	++	-	-	-	LRlc
		3b. S.m. caerulescens	All	++	-	-	-	LRlc
		4a. S. stoliczkanus leucogenys	8,9,10	++	-	-	-	LRlc
		4b. S.s. subfulvus	13	++	-	-	-	LRlc
Chiroptera	3. Pteropodidae	5. Pteroppus giganteus giganteus	4,7,8, 10,11	++	Schedule V	Appendix II	-	LRnt
		6. Rousettus leschenaulti	1,7,10	++	Schedule V	-	-	LRlc
		7. Cynopterus sphinx sphinx	1,8,10, 11	++	Schedule V	-	-	LRlc

	4. Megader-matidae	8. Megaderma lyra	1,2,3,4, 5,10,11, 13	++++	-	-	-	LRlc
	5. Rhinolophi-dae	9. Rhinolophus lepidus lepidus	1,4,5,8, 10	++	-	-	-	LRnt
		10. Hipposiderus fulvus pallidus	7,10,11	+++	-	-	-	LRnt
	6. Rhinopomat-idae	11. Rhinopoma microphyllum kinneari	All	++++	-	-	-	LRnt
		12. R. hardwickei	4,5,6,7, 8,9,10	+++	-	-	-	LRnt
	7. Emballon-uridae	13. Taphozous perforatus perforatus	3,4,5,6, 7,8,9,10 12, 13	+++	-	-	-	LRnt
		14. T. longimanus longimanus	1,7,	++	-	-	-	LRlc
		15. T.nudiventris kachhensis	7,9,10, 11,12	++	-	-	-	LRnt
	8. Molossidae	*16. Tadarida aegyptiaca	7,10	+	-	-	-	LRnt
	9. Vespert-ilionidae	17. Pipistrellus mimus mimus	All	+++	-	-	-	LRlc (as *P. tenuis*)
		18. P. ceylonicus indicus	13	+	-	- .	-	LRlc
		19. P. dormeri caurinus	2,7,10	+	-	-	-	LRnt
		20. Scotophilus heathi heathi	4,5,6,7, 10	++	-	-	-	LRlc
		21. S. kuhlii	1	+	-	-	-	LRnt
		*22. Scotoecus pallidus	10	+	-	-	-	LRnt
Prim-ates	10. Cercopithe-cidae	23. Semnopithecus entellus entellus	All	+++	Schedule II	Appendix I	-	LRlc
		24. Macaca mulatta mulatta	2,4,5,6, 8,10,11	++	Schedule II	Appendix II	-	LRlc
Carni-vora	11. Canidae	25. Canis aureus	1,2,4,5, 6,8,9, 10,11, 12,13	++	Schedule II	Appendix III	-	LRlc
		26. C. lupux pallipes	6,8,10, 11,12,13	++	Schedule I	Appendix		
		27. Vulpes vulpes pusilla	3,6,8,9, 10,12	+++	Schedule I	Appendix III	Volnera ble	LRnt
		28. V. bengalensis	1,2,9, 10,11,13	++	Schedule II	Apppendi x III	-	LRnt

	12. Mustelidae	*29. Lutrogale perspicillta sindica	11,13	+	Schedule II	Appendix II	-	Lrnt
	13. Viverridae	30. Viverricula indica deserti	1,2,4,5, 10,11	+	Schedule II	Appendix III	-	LRnt
	14. Herpestidae	31. Herpestes javanicus pallipes	All	++	Schedule IV	-	-	LRlc
		32a. H. edwardsi nyula	1,2,4,5, 6,7,12	+++	Schedule IV	Appendix III	-	LRlc
		32b. H.e.ferrugineus	8,9,10, 13	+++	Schedule IV	Appendix III	-	LRlc
		33. H. smithi	4,5,7,10 12	+	Schedule IV	Appendix III	-	LRlc
	15. Felidae	34. Felix silvestris ornate	All	++	Schedule I	Appendix II	Enda-ngered	LERnt
		35. F. chaus prateri	All	+	Schedule II	Appendix II	-	LRnt
		*36. Caracal caracal schmitzi	9,12	+-	Schedule I	Appendix I	Enda-ngered LRnt	
		*37. Panthera pardus fusca	4,10,11, 13	+-	Schedule I	Appendix I	Vulner able	Vulner able
	16. Ursidae	*38. Melursus ursinus ursinus	4,11,13	+-	ScheduleI	Appendix I		Vulner able
Artiodac tyla	17. Suidae	39. Sus scrofa cristatus	2,8,9	+	Schedule III	-	-	LRlc
	18. Bovidae	40. Boselaphus tragocamelus	All	+++	Schedule III	-	-	/LRlc
		41. Antilope cervicapra rajputanae	1,3,6,8, 10,11, 12,13	++	Schedule I	Appendix III	Vulner able	LRlc
		42. Gazella bennettii	All	+++	Schedule I	-	Vulner able	LRlc
Phol-idota	19. Manidae	*43. Manis crassicaudata	4,7	+	Schedule I	Appendix II	Vuln-erable LRnt	
Logo-morpha	20. Leporidae	44a. Lepus nigricollis dayanus	8,9,10, 12	+	Schedule IV	-	-	LRlc
		44b. L. n. ruficaudatus	All	+++	Schedule IV	-	-	LRlc
Rode-ntia	21. Sciuridae	*45. Petaurista sp.	4,5	+-	Schedule II	-	-	-
		46. Funambulus penanti	All	++++	Schedule IV	-	-	LRlc
	22. Muridae	47. Vandeleuria oleracea oleracea	12,13	+	Schedule V	-	-	LRlc

		48a. Rattus rattus alexandrinus	1,2	++	Schedule V	-	-	LRlc
		48b. R.r. narbadae	11,13	++	Schedule V	-	-	LRlc
		48c. R.r. arboreus	All	+++	Schedule V	-	-	LRlc
		48d. R.r. rufescens	All	+++	Schedule V	-	-	LRlc
		49. R. norvegicus	7,8,10	+	Schedule V	-	-	Lrlc
		50. Cremnomys cutchicus	All	+++	Schedule V	-	-	LRlc
		51a. Mus musculus praetexus	All	+++	Schedule V	-	-	LRlc
		51b. M.m. homourus	1,2,3,7, 8,10,11	++	Schedule V	-	-	LRlc
		51c. M.m. castaneus	4,5,6,8, 10,11, 12,13	++	Schedule V	-	-	LRlc
		52. M. phillipsi	4,5,10, 12,13	+	Schedule V	-	-	LRlc
		53. M.booduga	1,2,6,8, 9,10,11, 13	++	Schedule V	-	-	LRlc
		54. M. platythrix	5,6,10, 11,12	+	Schedule V	-	-	LRlc
		55. M. saxicola sadhu	5,10,11, 12	++	Schedule V	-	-	LRlc
		56. Bandicota bengalensis bengalensis	1,2,8, 10,12	+	Schedule V	-	-	LRlc
		57. B. indica indica	1,6,11	+	Schedule V	-	-	LRlc
		58. Nesokia indica indica	1,2	+	Schedule V	-	-	LRlc
		59. Golunda ellioti ellioti	4,5,10, 11,13	++	Schedule V	-	-	LRlc
		60. Millardia meltada meltada	All	++	Schedule V	-	-	LRlc
		61. M. gleadowi	8,9,10, 11,12	+	Schedule V	-	-	LRlc
		62. Meriones hurrianae	All	++++	Schedule V	-	-	LRlc
		63. Tatera indica indica	All	++++	Schedule V	-	-	LRlc
		64. Gerbillus gleadowi	All	++++	Schedule V	-	-	LRlc

		65. G. nanus nanus	3,4,5,6, 8,9,10, 12,13	++	Schedule V	-	-	LRlc
	23. Hystricidae	66. Hystrix indica	1,2,4,6, 8,9,10, 16	++	Schedule IV	-	-	LRlc

Explanation of symbols, abbreviations and codes

*Species could not be observed or collected during the present surveys.

Abbreviations: LRlc = Lower Risk least concern; LRnt = Lower Risk near threatened.

District codes: 1 = Sri Ganganagar; 2 = Hanumangarh; 3 = Churu; 4 = Jhunjhunu; 5 = Sikar; 6 = Nagaur; 7 = Ajmer; 8 = Bikaner; 9 = Jaisalmer; 10 = Jodhpur; 11 = Pali; 12 = Barmer; 13 = Jalore.

Relative abundance: ++++ = Most abundant; +++ = Moderately abundant; ++ = Good; + = Poor; +- = Probably wiped outs from the desert biome.

DISTRICT-WISE DISTRIBUTION

Table 4 shows that at least 20 species are found in all the desert districts of Rajasthan. Out of these, only two *viz., Gazella bennettii and Boselaphus tragocamelus* belong to the large mammal category and rest *viz., Pipistrellus mimus, Rattus rattus, Mus musculus, Meriones hurrianae, Tatera indica, Funambulus pennanti* etc. belong to small mammal category. Two large-sized mammalian species, though recorded from all the districts, are mainly concentrated in and around the protected areas or agricultural fields neighbouring Bishnoi villages. Species like *Bandicota bangalensis, B. indica* and *Nesokia indica* are mainly found in the districts with canal irrigation facility. Some of the species like *Cynopterous sphinx, Tadarida aegyptiaca, Scotophilus kuhlii, Vandeleuria oleracea, Manis crassicaudata* are so far restricted to only one or two districts. Species diversity is apparently highest in Jodhpur district and lowest in Churu district containing 53 and 24 species, respectively.

Table 5. Some new district records for chiroptera species

Species	District
Pteropus giganteus	Bikaner
Rousettus leschenaulti	Sri Ganganagar
Megaderma lyra	Sri Ganganagar, Hanumangarh, Churu, Sikar, Pali, Jalore
Rhinolophus lepidus	Sri Ganganagar, Sikar
Hipposideros fulvus	Pali
Rhinopoma microphyllum	Sri Ganganagar, Hanumangarh, Sikar, Churu, Bikaner, Jalore
Taphozous perforatus	Churu, Jhunjhunu, Sikar, Nagaur, Bikaner, Jaisalmer, Jalore
T. longimanus	Sri Ganganagar;
T. nudiventris	Barmer

Out of 66 species listed in the present paper, only three, *viz., Mus phillipsi, Rattus norvegicus and Cynopterus sphinx* may be considered as new records from Rajasthan Thar desert. However, quite a number of species belonging to Chiroptera, Rodentia and Carnivora were found to have much more extensive distribution in the desert than recorded in earlier literature (Prakash, 1959, 1994; Sinha, 1981; Alfred and Agrawal, 1996; Bates and Harrison, 1997). Some new district records for chiroptera species are as follows (Table 5).

Among the carnivores, *Viverricula indica* was observed for the first time in Sri Ganganagar, Hanumangarh, Jhunjhunu and Sikar, while *Herpestes smithi* in Jhunjhunu, Sikar, Jodhpur and Barmer.

Almost all the rodent species, particularly *Hystrix indica, Vandeleuria oleracea, Golunda ellioti, Bandicota indica, B. bengalensis and Mus saxicola* have been well established in various new areas of Rajasthan desert.

Few species, viz., *Panthera pardus, Caracal caracal, Lutrogale perspicillata, Hyaena hyaena, Melursus ursinus, Tadarida aegyptiaca, Scotoecus pallidus, Taphozous melanopogon* and *Manis crassicaudata* through recorded from Rajasthan desert, but during the recent surveys no sign of their occurrence could be found. Further, for *Taphozous melanopogon* and *Hyaena hyaena* even no local report was obtained. As *Hyaena hyaena* is very conspicuous and its occurrence can easily be noted by the local and forest people, its non-sighting indicates that the species is now extinct from Rajasthan desert.

Ecological distribution

On the basis of floral composition, physical and climatological features several ecological units such as sandy plains, sand dunes, interdunal plains, ruderal, riverine, saline flats and depressions, gravelly pediments, hills and rocky outcrops, wetlands have been recognized in the Thar desert by Prakash (1975a, 1981), Anon. (1988), Rahmani (1997), Baqri and Kankane (2001). In connection with the project "Studies on faunal diversity in the Thar desert of Rajasthan", Baqri (2001) recognized altogether six major habitats with a number of sub-habitats (Table 6). During the present surveys ecological distribution of mammalian species was studied in some of the major habitats as recognized by Baqri (2001) and also in cities and villages scattered over different habitats.

Though a number of habitats have been identified but mostly there is no distinct barrier between two habitats or sub-habitats and one often gradually merges with other. Many of the mammalian species are large-ranging species and utilize number of habitats in search of food and shelter. Further, apparently similar habitats in two parts of the desert may differ in mammalian species composition owing to difference in edaphic and climtological factors. Thus, the following account provides a general pattern of ecological distribution of mammalian species without putting any rigid distinction.

Table 6. Habitat types of Thar desert region of Rajasthan (adopted from Baqri, 2001)

Habitats	Sub-habitats			
1. Sand Dunes	Shifting	Stablised	-	-
2. Arable and or sandy	Plains	Gardens	-	-
3. Farming	Subsistence	Commercial	Fallow	Canal side plantation
4. Forests and or Hills	Stony and gravelly with stunted thorny shrubs	Tropical dry deciduous, thorny and mixed tree	-	-
5. River beds	Wind blown sand dunes	Sandy flats	Grassy plains	-
6. Aquatic and or wetlands	Fresh water annual	Fresh water perennial	-	-

(i) ***Sand dunes:*** Sand dunes are mostly met in the western part of the desert and at places they are more than 100 meters high. Shifting sand dunes are compared of fine, loose soil and practically devoid of any plant species. In general, mammalian species avoid shifting sand dunes, but foot marks of *Gazella bennettii, Lepus nigricollis* have been noticed at Sam (Jaisalmer), while of *Hemiechinus* sp. at Lakhera (Hanumangarh) and Kataswar (Bikaner).

The stablised dunes and their slopes have natural growth of *Calligonum polygonoides, Haloxylon salicornicum, Dipterygium glaucum, Acacia senegal, Prosopis cineraria* etc. on the recommendation of CAZRI for raising micro-wind breaks locally available brushwood materials like *Leptadenia pyrotechnica (*Khimp), *Zizyphus nummularia (*Pala), *Crotalaria burhia (*Samia), tree species *Acacia tortilis, Prosopis juliflora, Albizzia lebbek,* grass species *Lasiurus sindicus* and *Cenchrus ciliaris* have also planted in many areas. Crops like Bajara, Guar have also tried in some dune. Rodent species like *Gerbillus gleadowi, G. nanus, Meriones hurrianae* are found here in good numbers particularly in areas where no plantation programme was undertaken. In the low rainfall areas (less than 20 cm annual precipitation) *G. gleadowi* is the most predominant, while in relatively high rainfall areas (more than 20 cm annual precipitation), *M. hurrianae* becomes the most abundant. Another species of Gerbil, *Tatera indica* has been recorded in some dunes but in very smaller number. Dunes with fairly good cover of vegetation also support small population of *Hystrix indica* and *Hemiechinus micropus.* Among the large and medicum sized species, *Boselaphus tragocamelus, Gazella bennettii, Vulpes vulpes, Canis aureus, Lepus nigricollis* frequent the area.

(ii) ***Arable and or sandy:*** Greater parts of the Thar come under this category. Thick grass cover composing of *Lasiurus sindicus, Panicum turgidum, Aristida funiculta* etc. and various tree and shrub species as already mentioned under sand dunes are the characteristic vegetation in many parts of inter-dunal plains. Small herds of *Gazella bennettii* and *Boselaphus tragocamelus* are found scattered throughout the area. Quite a number of carnivore species like *Felis silvestris, Vulpes vulpes, Canis lupus, Herpestes edwardsi* have recorded but in a very patchy fashion. Huge population of *Meriones hurrianae* and *Gerbillus gleadowi* are found throughout, but former in more abundant in Barmer, Jodhpur, Nagaur, Churu, Jhunjhunu and Sikar, while latter in Jaisalmer, Bikaner and parts of Barmer. Of the another two species of Gerbil, *Gerbillus nanus* is mainly found in Jaisalmer, Jalore and Barmer, while *Tatera indica* in Pali, Jodhpur, Sri Ganganagar and Hanumangarh. Clayey flats of Nagaur with moderate growth of *Cyperus* spp. are found as good habitats for *Antilope cervicapra.* Burrows of *Hystrix indica* were also found in and around agricultural field.

Gravel plains of Jaisalmer, Barmer, Sikar, Jodhpur, Jhunjhunu and some other districts mainly support scattered to dense growth of *Prosopis cineraria, P. juliflora, Salvadori persica, Maytenus emarginatus, Calotropis procera* and *Zizyphus nummularia.* Among the four species of Gerbils, *Gerbillus nanus* is practically absent here, while other three are not so abundant as compared to sandy areas. Scattered and small population of *Millardia gleadowi, golunda ellioti, Funambulus pennanti, Suncus stoliczkanus, Herpestes edwardsi, Felis chaus, Canis aureus, Vulpes vulpes, V. bengalensis* have been observed in many areas. Herds of *Gazella bennettii and Boselaphus tragocamelus* frequent the area in search of food and shelter. Very good populations of *Lepus nigricollis* are thriving in the areas where there is no hunting pressure. Report of occurrence of *Manis crassicaudata* is available from the gravel plains of Ajmer and Jhunjhunu.

Sandy plains are naked in many parts, but rich vegetation composing of shrubs and grasses (*Cenchrus biflorus, C.ciliaris, Panicum sp., Lasiurus sindicus, Crotalaria burhia, Aerva tomentosa, Zizyphus nummularia* etc.), trees (*Prosopis cineraria, P. juliflora, Acacia jacquemntii, Cassia siamea* etc.) are not uncommon. Most of the desert species like *Gazella bennettii, Boselaphus tragocamelus, Canis aureus, Vulpes vulpes, Felis silvestris, Herpestes edwardsi, H. javanicus, Suncus stoliczkanus, Hemiechinus* sp. *Lepus nigricollis, Meriones hurrianae, Tatera indica, Millardia gleadowi* are met with in the wild areas of sandy plains,whereas highest concentration of *Antilope cervicapra* has been established in saline and flat depression (Tal Chapper). Study of Prakash (1975b) revealed *M. hurrianae* as the most abundant rodent species of sandy habitats of Barmer, Bikaner, Jodhpur, Churu, Nagaur, Pali and Jalore, whereas in Jaisalmer, Sikar and Jhunjhunu *G. gleadowi* emerge as predominant. However, during present study due to prolonged drought most of the rodent species are replaced by *G. gleadowi* in sandy habitats.

Large number of gardens of vast stretch have been developed and maintained by regular watering even in the driest part of the desert. This gardens are covered with carpet of green grasses and maintained a large number of fruiting and decorative plants including pomegranate, date palm, varities of ber, guava *etc.* Rodent association in these gardens is mainly formed by *Tatera indica, Mus musculus, Mus booduga, Rattus rattus* and *Millardia meltada.* All the species of fruit bats, *viz., Pteropus giganteus, Cynopterus sphinx and Rousettus leschenaulti* used to forage and sometime roost in these gardens particularly in the eastern and northern part of the desert. Some gardens are also regularly visited by *Felis chaus, Lepus nigricollis, Herpestes javanicus, H. edwardsi* and *Canis aureus.*

(iii) ***Farming:*** Traditional crop fields are mainly dependent upon rainwater and Tubewell irrigation. Large number of rodents species such as *Meriones hurrianae, Tatera indica, Mus musculus, Millardia meltada* find permanent shelter in and around fields. It has been found that *Tatera indica* remains mainly along the periphery and *Meriones hurrianae* in the middle of the field after the harvesting of the crop. In the region where percentage of clay is little higher as in Pali. *Tatera indica* populations surpass other rodents. Nests of *Gollunda ellioti* are also made in the dense thorny fence of crop fields. *Lepus nigricollis, Herpestes edwardsi Canis aureus, Vulpes bengalensis* and *Felis silvestris* frequently visit these crop fields particularly after the harvest. In Bishnoi areas, herds of *Antilope cervicapra, Gazella bennettii and Boselaphus tragocamelus* found permanently settled in and around crop fields particularly in those which are near some water bodies. Small roosts of *Pipistrellus mimus* and *Scotophilus heathi* have been observed in the temporary shelters made up of logs of *Calotropis procera, Prosopis cineraria,* dried bushes of *Crotalaria burhia, Leptadenia pyrotechnica* etc. in the fields for the protection of crops.

Vast areas of Sri Ganganagar, Hanumangarh, Bikaner, Jaisalmer have been brought under canal irrigated cropland system and canal side plantation programme, replacing original xeric vegetation including grassland. Original desert rodents have been replaced to some extent from these areas by gradual building up of *Millardia meltada, Bandicota bengalensis, B.indica, Nesokia india, Mus musculus and M. booduga* populations. Among the Bovids *Antilope cervicapra* and *Gazella bennettii* though reported by *Boselaphus tragocamelus* emerging as most dominant. Population of *Hystrix indica* and *Sus scrofa* are also slowly but steadily building up in and around canal irrigated crop fields. Carnivore species in canal irrigated croplands are the worse victim of secondary poisoning due to random application of pesticide. Quite a number of microchiropteran species, *viz., Pipistrellus mimus, Taphozous perforatus, Rhinopoma microphllum, Scotophilus healhi* and *S. kuhlii,* have been caught in the mist nets placed along the canal side plantation. Species of *Hemiechinus* have totally been wiped out from canal irrigated fields, but both the species of *Suncus* have established their populations.

(iv) ***Forest and Hills:*** Rocky outcrops and hills have nil to thin layer of upper soil, but at places huge deposition of wind blown loose soil could be found. Patches of *Euphorbia caducifolia, Acacia Senegal, A.tortilis, Capparis deciduas, Maytenus emarginata, Prosopis juliflora* alongwith several other shrubs and herbs are found throughout. Small to large troops of more than 100 individuals of *Semnopithcus entellus* are met with in the hills which particularly settled in and around some temple. Many of these troops cover more than 100 sq.km for foraging. Caves, crevices, inter-boulder spaces, thick bushes provide numerous microhabitats for a large number of microchiropteran bats (*Rhinopoma* sp., *Taphozous* sp., *Megaderma lyra, Pipistrellus* sp., *Scotophilus* sp., *Tadarida aegyptiaca, Rhimolophus lepidus)* and rodents. As many as eleven species of rodents *viz.,Cremnomys cutchicus, Rattus rattus, Golunda ellioti, Mus saxicola, M. platythrix, M. musculus, M. phillipsi, Millardia meltada, Tatera indica, Funambulus pennanti, Hystrix indica* could be observed in the Harsh Parbat region of Sikar with *C. cutchicus* as most predominant. Fairly good populations of *Lepus nigricollis* and *Vulpes vulpes* were found throughout the hills, whereas *Vulpes bengalensis, Viverricula indica, Canis aureus, Felis chaus* and *Herpestes smithii* also occured but in very low number. Local reports of occurrence of flying squirrel, *Petaurista* sp. at Mansa Kho, Shakhambari of Jhunjhunu, and Harsh Parbat of Sikar are available. Similarly, Indian Pangolin *Manis crasssicaudata* is said to occur in the hills of Jhunjhunu and Ajmer. Reports of *Panthera pardus and Melursus ursinus* were also found from Jodhpur and Pali region, but at present both the species in all probability have been wiped out from desert biome.

Small patches of thorny; tropical dry deciduous type of forests are found in gravel as well as sandy plains. Gravel plains in Sudasari enclosure of Desert National Park has semi-shrub desert vegetation of *Capparis deciduas, Zizyphus nummularia, Lycium* sp., *Prosopis cineraria, Calotropis procera, Lasiurus sindicus, Haloxylon salicornium* etc. Quite a number of threatened species like *Gazella bennettii, Vulpes vulpes, V. bengalensis* are found here. *Gerbillus gleadowi, G. nanus, Lepus nigricollis, Herpestes edwardsi* are also prominent residents of this enclosure. Occasional visits of *Boselaphus tragocamelus* have also been reported. Kohla forest (Hanumangarh) by the side of Pilibanga branch of IG canal has sandy plains with fairly good cover of *Dalbergia sisso, Acacia nilotica, Prosopis cineraria, Morus alba, Cassia siamea* etc. Among the permanent residents *Hemiechinus micropus, Suncus murinus, Meriones hurrianae, Tatera indica, Hystrix indica, Vulpes vulpes, Herpestes edwardsi* are of worth mentioning. Other species like *Gazella bennettii, Boselaphus tragocamelus* and *Semnopithecus entellus* also frequently visit the area. Small population of *Sus scrofa* is also building up in the neighbouring areas.

(v) ***River beds:*** Most of the river beds during our visit were found dry except a section of Luni river (Lat. 24°50'22"N and 71°46'19"E) where knee deep water could be found during September, 2001. Dry river beds are sandy or gravelly and have growth of *Cyperus rotundus,* and grasses like *Desmostachya bipinnata, Eragrostris ciliarius, Dactyloctenium aegyptium, D. sindicum, Lasiurus sindicus* etc. Some species of fruiting creepers are also grown in sandy beds. All the species of Artiodactyles,

Canis lupus, Canis aureus, Felis chaus frequently visit the river beds, while *Herpestes edwardsi, Lepus nigricollis, Meriones hurrianae, Tatera indica* are also found to inhabit the bushes adjacent to river beds. Vegetable crop fields grown on these river beds also support fairly good populations of *Millardia meltada, Mus booduga* and *Tatera indica.*

(vi) ***Aquatic/Wetlands:*** A number of water bodies/ghagars particularly in and around human habitation are located even in the driest part of the desert. Though, no truly aquatic mammal occurs in the desert, but numbers of species are concentrated around these water bodies. Positive records of occurrence of *Lutrogale perpicillata* from Sardarsamand (Pali) and Jawai lake (Pali) in the recent past are available, however, in all probability wiped out from Sardarsamand area (Prakash 1975a). Quite a number of Microchiropteran species belonging to the genera *Pipistrellus, Scotophilus, Rhinopoma, Hipposideros* have been found flying over these water bodies particularly during dusk, indicating their roosts in the neighboring areas. Trees like *Pithecellobium dulce, Azadirachta indica, Saraca asoca, Ficus, Syzygium cumini* in and around water bodies often serve as roosting place for *Pteropus giganteus and Macaca mulatta.* Herds of *Antilope cervicapra, Gazella bennettii and Boselaphus tragocamelus* are also met with around such water bodies.

(vii) ***Village and Cities:*** Village settlements are scattered throughout the desert on almost all landforms depending upon the availability of water. This complex situation is defines as Ruderal habitat by Prakash (1981). Residential premises, sewerage systems, markets, godowns, railway stations, poultry, deserted buildings/huts, ruins, trees, graveyards, temples *etc.* of ruderal habitat and cities provide shelter to a large number of mammalian species. Here species composition has also been modified by the transportation of different species particularly of small mammals from distant areas by availability of train, trucks and other vehicles. Residential premises, railway yards, godowns, market areas often harbour good populations of *Bandicota bengalensis, B. indica, Rattus rattus, R. norvegicus, Mus musculus* and *Suncus murinus.* In or around ruins, graveyards, nullahs movement of *Herpestes javanicus, Herpestes edwardsi, Canis aureus, Vulpes bengalensis, Vulpes vulpes* and, *Felis chaus* could be frequently marked. Troops of *Semnopithecus entellus* and *Macaca mulatta* are found mainly in temple areas. Roosts of all the microchiropteran bats could be seen in the forts, temples, mosques, ruins, uninhabitated buildings, godowns, crevices and false ceiling of houses, under the culverts, tree-holes and many other dark, undisturbed places. In and around poultry and cowsheds, huge colonies of *Tatera indica* and *Rattus rattus* have been established. Ruderal habitats of Bishnoi communities provide shelter to the herds of *Boselaphus tragocamelus, Gazella bennettii* and *Antilope cervicapra.*

RELATIVE ABUNDANCE

Trapping in different habitats revealed preponderance of rodent species. In sandy and interdunal plains with natural herbs and shrubs species, *Gerbillus gleadowi* was the most dominant species. While in the similar habitat with some rain-fed crop fields, *Meriones hurrianae* appeared as prevalent species. In the residential and market areas most species trapped were belong to *Rattus rattus* and *Mus musculus*. In the hills and rocky outcrops a very rich species composition could be found, lead by *Cremnomys cutchicus*. In sandy plains with relatively higher rainfall or having irrigation facilities, *Tatera indica* outnumbered the other rodents. Overall rodent density was found to vary from nil in most of the shifting dunes to over 1000 per hectare particularly in sandy plains and crop fields. Among the insectivores, only the species belonging to the genus *Suncus* could be trapped particularly in the cities, villages and agricultural fields and *Suncus murinus* appeared as predominant. However, burrow counting and census operation in wild sandy plains revealed a fairly good population of *Hemiechinus* sp. In Kohla farm (Hanumangarh), as many as 24 live burrows of *H. micropus* could be found in one hectare area. Further, during road census at night, quite a good number of them could be observed in almost all the habitats except in hills, canal side plantation and commercial farming areas. Roosts of microchiropteran species were located mostly in rock crevices and dark, undisturbed part of various men-made structures. In Mohangarh fort (Jaisalmer) four different roosts of *Rhinopoma microphyllum* containing nearly 50 to 1000 individuals, one roost each of *Taphozous perforatus, T.nudiventris, Pipistrellus mimus* containing 8,7 and 1 individual were found. Roosts of *Rhinopoma hardwickei* were found at various places, of which Harsh Nath temple (Sikar) and Patwa Haveli (Jaisalmer) are worth mentioning containing more than 1000 and 200 individuals, respectively. One deserted cowshed at Sri Ganganagar revealed a colony of nearly 50 animals of *Megaderma lyra*. Huge roosts containing 50 to 500 individuals of *Pteropus* sp. could be found at Balsamand, Rail Sadan (near Jodhpur city), and near Kailana lake (Jodhpur). No roosts of *Cynopterus sphinx* or *Rousettus leschenaulti* could be detected though both the species were caught in mist nets but populations appeared to be few.

Roadside census revealed highest concentration of *Gazella bennettii* in between Dechu and Lohawat (26 per sq. km.), followed by Sundasari enclosure (8 per Sq. Km.), `0' head of IGNP main canal to 1398 RD of main canal (6 per sq. km.) and Gajner to Bikaner (2.5 per sq. km.). However, none could be observed on road section from Kohla farm to Share Khan and Harsh Nath to Harsh village. Highest density of *Antilope cervicapra* found at Tal Chapper, where within about 7 sq. km. of protected area nearly 2000 individuals are thriving. However, entire population is now practically dependent on the fodder supplied by the forest department with the support of NGOs. Roadside census revealed highest concentration of Black Buck in a stretch of from Kuchera to Ren (8 per sq. km.), followed by road section Barmer to Kharim (2 per sq. km.). Another species *Boselaphus tragocamelus* was observed in all the road sections covered by census except in Sundasari, Barna to Sam, Dhauba to Pitchla and Harsh Nath to Harsh village. Highest concentration was found in the stretch between Kohla to Share Khan (9 per sq. km.), followed by Indrapura to Todpara (5.6 per sq. km.), Suratgarh to Nirbana (3.5 per sq. km.), and Kuchera to Ren (3 per sq. km.). It is worthwhile to mention that during the period 2000-2002, this species was mostly observed in herds of 3 to 20 or more individuals. However, in 2003, Nilgai though seen at various places, but there was no herd formation and always found singly.

A herd of 4 individuals of *Sus scrofa* was noticed in the canal side plantation between `0' head of IGNP main canal and 1398 RD. Among carnivores a total 21 specimens of *Canis aureus* (mainly in crop fields in and around cities and villages, and Harsh Parbat), 34 specimens of *Vulpes vulpes* (in sandy plains and near crop fields), 5 specimens of *V. bengalensis* (in DNP areas and canal side plantation), 3 specimens of *Viverricula indica* (1 each at Harsh Parbat, Todpura Village and near Kailana). Mongooses, *Herpestes* spp. were met with frequently at various places particularly in city gardens, deserted man-made structures, agricultural fields, rocky outcrops and many other areas. Among the other species which were found in moderate to fairly good numbers during road census or while passing from one camp to other are *Hemiechinus* sp., *Gerbillus* sp., *Tatera indica, Meriones hurrianae, Hystrix indica* (only one specimens at Manasa Kho hill), *Funambulus pennanti, Lepus nigricollis, Semnopithecus entellus, Macaca mulatta* (only at Kailana, Neem ka Thana and Rain), *Canis lupus* (one near Kharin, other at about 28 km from Merta road on way to Ajmer) and *Felis* spp. Further, it is interesting to note that as compared to the area, canal side plantation particularly in Jaisalmer provides shelter to relatively higher number mammalian species. Thus, canal might be serving as boon for at least some of the mammalian species.

Based on trap success, roadside census results, literature, local reports and indirect evidences, relative abundance of mammalian species in Rajasthan desert have been reflected in Table 4.

KEY TO THE IDENTIFICATION OF MAMMALIAN SPECIES OF RAJASTHAN THAR DESERT

Order: Insectivora

Small sized; long pointed snout projecting far beyond the lower jaw. Limbs short. Orbits open posteriorly. True molars studded with sharp, pointed cusp.

Key to the families

1. Body roundish, thickest; tail short and stumpy. Back and sides covered with stiff spines. Crown of first and second upper molar with a central fifth cusp .... Erinaceidae

2. Elongated body, tail over half the length of head and body. Back and sides covered with soft hairs. Central fifth cusp of upper first and second molar absent .... Soricidae

Key to the species of the family erinaceidae

1. No distinct parting of the spines on the fore crown. Face lacking white fur around ears and upper part of the crown. .... *Hemiechinus collaris*
2. Spines on the fore crown divided by a longitudinal naked furrow. Fur on the upper part of the face and belly white .... *Hemiechinus micropus*

Key to the species of the family soricidae

1. Small in size, head and body length between 60-71 mm .... *Suncus stoliczkanus*
2. Large in size, head and body length over 93 mm .... *Suncus murinus*

Order: Chiroptera

Forearm modified into wings; only group of mammals capable of true flight.

Key to the families

1. Neither nose-leaf nor tragus present. Margin of ear forming a complete ring. Second finger usually having a claw............ **Pteropodidae**

1a. Either nose-leaf or tragus or both present. Margin of ear not forming a complete ring. Second finger without a claw............ 2

2. Both nose-leaf and tragus present .... **Megadermatidae**

2a. Either nose-leaf or tragus present, but not both 3

3. Nose-leaf present but no tragus .... **Rhinolophidae**

3a. Tragus present but no nose-leaf 4

4. Tail entirely enclosed in interfemoral membrane... .... **Vespertilionidae**

4a. Distal portion of tail free from interfemoral membrane 5

5. Tail emerging from upper surface of interfemoral membrane ..**Emballonuridae**

5a. Tail emerging from end of the interfemoral membrane 6

6. Tail like that of a rat, long and slender .... **Rhinopomatidae**

6a. Tail short and stout .... **Molossidae**

Identification of the species of the family Pteropodidae

1. *Rousettus leschenaulti*: Medium-sized, forearm 70-85 mm. Upper part of the body uniformly coloured. Distinct tail with terminal half free from interfemoral membrane. Pollex 24-30 mm., 2nd phalange of third digit less than 40 mm. Five upper cheekteeth.

2. *Cynopterus sphinx*: Small-sized, forearm 67-75. Upper part of the body uniformly coloured but ear with a narrow whitish border. Distinct tail with terminal half free from interfemoral membrane. Four upper cheek teeth.

3. *Pteropus giganteus*: Large-sized, forearm over 120 mm. Hindneck and shoulder markedly paler than lower back. No tail. Pollex 65-70 mm. Five upper cheekteeth.

Identification of the species of the family Megadermatidae

1. *Megaderma lyra*: Medium-sized, forearm 60-70 mm. Ears very large (on average 36 mm. In length) and conjoined for nearly half the length of inner margin. Tragus bifid. No externally visible tail.

Identification of the species of the family Rhinolophidae

1. *Rhinolopus lepidus*: Small-sized, forearm around 40 mm. Nose-leaf horse shoe shaped and posterior nose-leaf divided into transverse hollows. First toe with three joints, remaining with only two. Three premolars in the lower jaw.
2. *Hipposideros fulvus*: Size as in previous species. Nose-leaf square in outline. Posterior nose-leaf divided into longitudinal hollows. Two premolars in lower jaw.

Identification of the species of the family Rhinopomatidae

1. *Rhinopoma microphyllum*: Forearm 57-75 mm., tail relatively shorter being on average less than 80 percent of forearm. Rostrum with ill-defined narial swelling.
2. *Rhinopoma hardwickei*: Forearm 54-64 mm, tail relatively longer being more than 85 per cent of forearm. Rostrum with prominent narial swelling.

Identification of the species of the family Emballonuridae

1. *Taphozous perforatus*: Forearm short, below 66 mm. Body fur extending over part of the wing and interfemoral membrane. Gular sac moderately developed in males but absent in females. Tail tip pointed. Distal margin of tragus not wavy. Calcar weak. Wing span 210-280 mm.
2. *Taphozous longimanus*: Forearm short, below 66 mm. Body fur extending over part of the wing and interfemoral membrane. Gular sac prominent in males, rudimentary in females. Tail tip blunt. Distal margin of tragus wavy. Calcar well developed. Wing span 370-390 mm.

3. *Taphozous nudiventris*: Forearm over 68 mm. Body fur not extending over the wing and interfemoral membrane. Gular sac in males indicated by semi-circular fold of skin. Tail tip blunt.

Identification of the species of the family Molossidae

1. *Tadarida aegyptiaca:* Forearm around 50 mm., upper lip heavily wrinkled. Antitragus present. Ears separated at muzzle. Wing from ankle.

Family: Vespertilionidae

Key to the Genera

1. Size large, wing span 270-370 mm, forearm 45-64 mm. Upper premolars I on each side .... **Scotophilus**

1a. Size small, wing span 140-240 mm, forearm 26-38 mm. Upper premolar 2 on each side ....**Pipistrellus**

Identification of species of the genus Scotophilus

1. *Scotophilus heathi*: Large, forearm more than 56 mm. Dorsum olive to gray brown, venter lemon yellow to orange yellow. Muzzle with swollen glandular pads between the eyes and nostrils.
2. *Scotophilus kuhlii*: Smaller, forearm less than 55 mm. Dorsum reddish to olive brown, venter creamy white with a tinge of red. Muzzle with glandular swellings behind the eyes.

Identification of species of the genus Pipistrellus

1. *Pipistrellus ceylonicus*: Largest of all the pipistrelles occurring in Rajasthan, forearm 36 mm. or more. Dorsum brown, reddish or grayish brown; venter slightly paler. Ear triangular. Outer upper incisor extending distinctly beyond cingulum of inner.
2. *Pipistrellus mimus*: Smallest of all the pipistrelles, forearm 26-31 mm. Dorsum bistre brown, ventral parts paler. Ear scarcely triangular. Outer upper incisors extending distinctly beyond cingulum of inner.
3. *Pipistrellus dormeri*: Medium-sized pipistrelles, forearm 30-37 mm. Dorsum dark brown with ashy tip of fur, venter whitish or lemon yellow. Ear sub-triangular. Outer upper incisor not extending beyond cingulum of inner.

Order: Primates

Arboreal in nature. First digit of hindfoot the largest, opposable and prehensile. Bony ring around the orbit complete.

Identification of species of the order Primates

1. *Semnopithecus entellus*: Face black; check pouch absent. Tail always longer than head and body. Crown of the head little paler than nape and shoulder. Upper side of body gray to bright sepia brown; hands and feet black and brown and strongly contrasted with that of arms and legs. Hairs of the crown behind the brow radiating from a frontal whorl forming a tuft or crest.
2. *Macaca mulatta*: Face red or light pink; hind quarter of the body orange red and brighter than fore quarter. Arms gray, legs washed with bright yellow; undersurface mostly white with slight red in belly. Tail always shorter than head and body. Check pouches present. Hairs of the crown directed backwards from the brows, without a whorl or parting and not forming a definite cap.

Order: Carnivora

Front teeth consisting of a pair of conical tusk like canines and a row of 6 incisors between them, of which outer are larger than inner. Last upper premolar and first lower molar having sharp edged lobes for shearing the flesh.

Key to families of order carnivora

1. Ethmo-turbinals very large, covering the greater part of the nasal chambers and extending anteriorly to the anterior orifice. Auditory bulla composed of two bones. Cowpers gland absent 2

1a. Ethmo-turbinols excluded from the anterior orifice of nasal chamber by large maxillo-turbinols; auditory bulla composed of a single tympanic bone. Cowper's gland present 4

2. Post palatal foramina set far back in the maxillo-palatine suture. Teeth reduced in number and highly sectorial; dental formula being i3/3, c1/1, pm 3 or 2/2, m 1/1. Interramal tuft of vibrissae absent.................... **Felidae**

2a. Post palatal foramina in front of the maxillo-palatine suture. Teeth not reduced in number or so specialized, dental formula being I 3/3, c 1/1, pm 4/4 ¾, m 2/1/2/1. Interramal tuft of vibrissae present 3

3. Ears moderately elongated (over 35 mm.), with well developed bursa and simple supratragus, but devoid of valvular flap over it. Feet compact with short claws. Anus not enclosed in a glandular sac. No bony tube to auditory orifice .... **Viverridae**

3a. Ears small (less than 30 mm.), with vestigial bursa but having valvular supratragus and valvular flap over it. Feet with free digits and fossorial claws. Anus enclosed in a glandular pouch. A well developed bony tube to the auditory orifice ....**Herpestidae**

4. Medium-sized, slim dog like appearance. Legs long, slender, digitigrade; fourth digit of fore foot raised above the planter pad. Baculum deeply channeled below .... **Canidae**

4a. Not dog like in appearance. Legs short, thick, plantigrade or semiplantigrade. First digit of fore foot present close to the second and not raised above the planter pad. Baculum not deeply channeled below 5

5. Large sized with stout body and short tail. Pads of digits forming week curved line. Three lower molars on each side .... **Ursidae**

5a. Small sized, moderately long tail. Pads of digits forming strong curved line. Two lower molars on each side .... **Mustelidae**

Identification of species of the family Felidae

1. *Felis chaus*: Domestic cat like, medium sized, head and body length 560-740 mm., tail less than half of the head and body. In adult, coat unspotted or without any pattern, fore legs with black horizontal stripes. Back of ears reddish chestnut with a black tip. Tail tip black and preceded by 2 or 3 black rings.
2. *Felis silvestris*: Domestic cat like, small sized, head and body length 460-590 mm., tail more than half of the head and body. Back of ears sandy buff with black tip. Pelage with definite pattern, having solid black spot on the body, which in the sides behind the fore legs tend to form a couple of vertical stripes; crown and nape having narrow dark stripes; two stripes on cheek and base of limbs. Tail banded with black and white rings and having blackish tip.
3. *Caracal caral:* Largest of three lesser cats of Rajasthan, head and body length 700-760 mm., tail one-fourth of total length. Ears entirely black or with frosting of white hairs. Coat unspotted, inside of legs without any black stripes. Tail without any bands.
4. *Panthera pardus:* Large cat, head and body length 1040-1420 mm., tail 760-965 mm., height at shoulder about 700 mm. Ground colour variable – pale yellow, warm gray, rich tawny to rufous fawn with dark spots arranged in rosettes on the sides of the body.

Identification of species of the family Viverridae

1. *Viverricula indica:* Small sized, head and body length 500-620 mm., with tail little smaller. Muzzle slender, pointed; ear broad and rounded. Four quarter of body with small black spots; large spots tending to run into longitudinal line on the flank and forming 6-8 stripes down the back. Naked rhinarium reddish brown. Carpel pad single, metatarsal pad absent. Scent glands opening into a highly specialized pouch.

Identification of species of the family Herpestidae

1. *Herpestes javanicus:* Elongated body with sturdy legs and sharp pointed conical face. Head and body less than 350 mm., ears short and about 22 mm. Contour hairs short having maximum of 4 bands. Legs not darker than body. No black tail tip.
2. *Herpestes edwardsi:* Very similar to previous species but larger, head and body over 360 mm. Contour hairs long having more than 4 bands. Legs darker than body. No black tail tip.
3. *Herpestes smithii:* Similar in appearance to the above species, but with black tail tip.

Identification of species of the family Ursidae

1. *Melursus ursinus*: Large sized bear, head and body 1400-1700 mm. Coat rough, deep black, with or without a `V' – shaped breast patch. Claws ivory white. No distinct post-canine space. Last upper molar subequal to the first.

Identification of species of the family Mustelidae

1. *Lutrogale perspicillata:* Thick, fusiform body adapted for semiaquatic life; tail thick, muscular, distal portion more or less triangular in out line. Hind foot wider than fore foot. Web of skins between all the digits of fore and hind foot. Vibrissae thick.

Identification of species of the family Canidae

1. *Canis lupus:* Size of an Alsatian dog, head and body 900-1050 mm., tail less than two third of the head and body. Height at shoulder 650-750 mm. Tail bushy with black tip. Dorsum grayish fulvus, usually with brownish or reddish tinge and sometime mixed with black; face grayer having a mixture of black and white. A crest of longer hairs on neck and shoulder. Arching brows and elevated forehead. First upper molar with distinct outer cingulum.
2. *Canis aureus:* Very similar to previous species but smaller, head and body 600-700 mm., height at shoulder 380-430 mm. Without arching brows and elevated forehead. Body sandy buff but mid dorsum predominantly black. Tail tip black. First upper molar with a prominent outer cingulum.
3. *Vulpes vulpes:* Head and body 480-600 mm., tail (300-460 mm.) more than two thirds of the head and body. Coat colour varies from red to gray; face rufous. Tail bushy with white tip. Back of ears velvety black. Forehead between post-orbital processes flat.

4. *Vulpes bengalensis:* Similar to previous species but smaller, head and body length and tail around 420 mm and 240 mm., respectively. Back of ear sandy brown or dark brown. Tail tip black. A smudge of black hairs around the upper part of the muzzle in front of eyes.

Order: Artiodactyla

Digits provided with hoofs which are equal and symmetrically paired. Ears erect. Medium axis of feet passing between 2nd and 3rd digits.

Key to families of the order Artiodactyla

1. Upper incisors present. Devoid of horns or antlers................................ **Suidae**
2. Upper incisor absent. Unbranched horn present at least in males .... **Bovidae**

Identification of species of the family Suidae

1. *Sus scrofa:* Head and body length in males 1150-1600 mm., females smaller. Short, thick neck; legs slender. Head with sloping muzzle terminating in a specialized flattened disk. Body dark gray to pinkish brown, covered with coarse bristles which form a crest over neck. Canine in adult male developed into tusk.

Identification of species of the family Bovidae

1. *Gazella bennettii*: Slender bodied, height at shoulder about 650 mm. Coat sandy brown to light chestnut. Horns simuous, closely ringed; horns in males relatively longer (250-300 mm.) than in females (100-130 mm.).
2. *Antilope cervicapra:* Slender bodied, height at shoulder about 800 mm. Coat in males blackish brown above, whitish below; females yellowish fawn above. Males with a pair of long (500-600 mm.) spirally twisted horns, female hornless.
3. *Boselaphus tragocamelus:* Horse like cow in appearance. Large sized, height at shoulder about 1200 mm. Stout legs. Males blue gray, females reddish tan. A mane of black-gray and white hairs runs down the hind neck and over the wither. Males with two short, smooth horns.

Order: Pholidota

Body long, tapering, covered above with large overlapping scales; undersurface having coarse bristle like hairs. Head small, pointed. Jaws devoid of teeth.

Identification of species of the order Pholidota

1. *Manis crassicandata:* Head and body 600-750 mm., tail 380-450 mm. Body covered with an armour of 11-13 rows of scales. Eyes small, ear conch rudimentary. Feet with curved, blunt claws.

Order: Lagomorpha

Two pairs of incisors in the upper jaw, placed one behind the other. Cheek teeth separated from incisors by a wide diastema.

Identification of species of the order Lagomorpha

1. *Lepus nigricollis:* Head and body 400-500 mm., tail bushy and short (60-100 mm.). Ear very long (80-90 mm.) and equal to or longer than head. Fur dense, long and soft. Dorsum rufous to sandy brown in colour, with a grayish patch on the nape.

Order: Rodentia

A pair of chisel shaped incisors in each jaw. Distinct diastema between incisors and cheek teeth.

Key to families of order Rodentia

1. Fur modified into quills. Cheek teeth 4/4 in number .... **Hystricidae**

1a. Fur not modified into quills. Cheek teeth never 4/4 in number (except in genus *Ratufa* which does not occur in Rajasthan) 2

2. Infra-orbital foramen not or scarcely open for muscle transmission. Cheek teeth except in genus *Ranufa* 5/4 in number. Skull with post orbital process .... **Sciurdae**

2a. Infra-orbital foramen clearly open for muscle transmission. Cheek teeth 3/3 in number. Skull without post-orbital process .... **Muridae**

Identification of species of the family Hystricidae

1. 1. *Hystrix indica:* Massive in build, head and body length over 600 mm, tail short less than one-fifth of head and body. Hands and feet broad with well developed claws. A long brown crest of bristles present on the head and neck. Shoulder and belly covered with coarse bristles, back with long quills having more than one alternating bands of brown and white. Tail quills white.

Identification of species of the family Sciuridae

1. *Petaurista petaurista:* Size large, head and body more the 300 mm. Flying membrane present on both sides of body. Dorsum

dark brown. Bushy tail, greater part of which black; under parts whitish gray.

2. *Funambulus pennanti:* Size small, head and body on average 145 mm. Bushy tail, creamy or whitish at tip. Dorsum olive gray with five creamy buff or creamy gray longitudinal stripes.

Key to sub families of family Muridae

1. Cusps in cheek teeth arranged in three longitudinal rows. Upper incisors not grooved on front surface except in the genus *Golunda*. .... **Murinae**

1a. Cusps in cheek teeth arranged in two longitudinal rows. Upper incisors grooved on front surface .... **Gerbillinae**

Key to genera of subfamily Murinae

1. Hallux opposable; hallux and fifth toe having a flat nail .... **Vandeleuria**

1a. Hallux not opposable; hallux and fifth toe provided with claw 2

2. Upper incisors grooved on front surface. Molars heavily cuspidate....... **Golunda**

2a. Upper incisors not grooved on front surface. Molars not heavily cuspidate.... 3

3. Upper incisors proodont; condylobasal length of skull exceeds occipitionasal length 4

3a. Upper incisors opisthodont; occipitonasal length of skull exceeds condylobasal 5

4. Anterior palatal foramina very small, less than 6.5 mm. in length, and not extending between maxillary tooth row .... **Nesokia**

4a. Anterior palatal foramina long, more than 6.5 mm. in length, and extending between maxillary tooth row .... **Bandicota**

5. First upper molar more than one-half the length of tooth row, its antero-internal cusp distorted inward to reach the level of second lamina; third upper molar extremely reduced... **Mus**

5a. First upper molar less than one-half the length of tooth row; its antero-internal cusp not distorted inward except in *Millardia gleadowi*. Third upper molar not extremely reduced 6

6. Anterior palatal foramina short, less than one-fifth of occipitonasal length and not extending between maxillary tooth row .... **Rattus**

6a. Anterior palatal foramina long, more than one-fifth of occipitonasal length and extending between maxillary tooth row 7

7. Palate long, exceeding half the length of occipitonasal length; tail sub-equal or shorter than head and body length .... **Millardia**

7a. Palate short, less than half the occipitonasal length; tail longer than head and body length .... **Cremnomys**

Identification of species of the genus Vandeleuria

1. *Vandeleuria oleracea*: Small, arboreal mouse; head and body length less than 100 mm; tail much longer, double the head and

body and not well haired. Dorsum dull to rusty brown or sandy. Tail unicoloured.

2. ***Golunda ellioti:*** Medium sized rat, head and body length 100-170 mm. Tail shorter than head and body and covered with coarse hairs. Upper incisors grooved.

Identification of species of the genus *Nesokia*

1. ***Bandicota bengalensis :*** Size smaller, head and body 130-261 mm. Occcipitonasal length less than 48 mm. Anterior palatal foramina narrower posteriorly; nasals short, less than one-third of occipitonasal length. Tail well shorter than head and body. Dorsum grayish brown to blackish, under surface light to dark gray; tail dark, unicoloured. Mammae 6-8 pairs.
2. ***Bandicota India :*** Robust rat, size large, head and body 205-368 mm. Occipitonasal length more than 48 mm. Anterior palatal foramina equally broad at both ends; nasals long, more than one-third of occipitonasal length. Tail subequal to head and body. Dorsum blackish brown, undersurface gray. Tail dark, unicoloured. Mammae 6 pairs.

Key to the speciees of the genus Mus

1. Tail unicoloured and larger than head and body length. Diastema equal to or less than one-fourth of occipitonasal length ....***M. musculus***

1a. Tail bicoloured, dark above and pale below; equal to or shorter than head and body length. Diastema generally more than one-fourth of occipitonasal length 2

2. Fur soft. Supra-orbital ridges either absent or scarcely developed. Occipito-nasal length less tha 20 mm .... ***M. booduga***

2a. Fur generally spiny. Supra-orbital ridges well developed. Occipitonasal length more than 20 mm 3

3. Smaller form occipitonasal length 20-23 mm., maxillary tooth row less than4 mm .... ***M. phillipsi***

3a. Larger form, occipitonasal length 23-30 mm., maxillary tooth row more han 4 mm 4

4. An anterior accessory cusp present on first lamina of first upper molar .... ***M. saxicola***

4a. No accessory cusp present on first upper molar .... ***M. platythrix***

Identification of species of the genus Rattus

1. *Rattus rattus:* Medium sized rat, head and body less tha 200 mm. Tail generally longer than head and body. Dorsum in various shades of black and brown, venter colour varies from dark, gray, buff to whitish. Tail unicoloured.

2. *Rattus norvegicus:* A bandicoot like large rat, head and body length over 200 mm. Tail shorter than head and body. Dorsum dark brown, venter gray. Tail obscurely bicoloured.

Identification of species of the genus Millardia

1. ***Millardia meltada:*** Larger species, head and body well over 110 mm. Ear shorter than hind foot. Planter pads five in number. Mammae 4 pairs.
2. ***Millardia gleadowi:*** Smaller species, head and body less than 100 mm. Ear subequal to hind foot. Planter pads four in number. Mammae 3 pairs.

Identification of species of the genus Cremnomys

1. *Cremnomys cutchicus:* Soft furred rat. Medium sized, head and body 104-149 mm; tail longer than head and body and well haired. Body dark gray to brown above; white to white mixed with gray below. Tail either wholly dark or faintly bicoloured. Mammae 3 pairs.

Key to the genera of subfamily Gerbillinae

1. Upper molars prismatic in surface pattern, and laminae joined in midline .... **Meriones**

1a. Upper molars biserially cuspidate; laminae not always joined 2

2. Size large, head and body length exceeds 130 mm., and tail less than one and half times of that. Tail dark above and below, paler on sides .... ***Tatera***

2a. Size small, head and body less than 100 mm., tail generally more than one and half times of that. Tail dark above and pale below .... ***Gerbillus***

Identification of species of the genus Meriones

1. *Meriones hurrianae:* Medium sized gerbil, head and body length 100-137 mm. Tail subequal to head and body, well-haired, tufted at tip. Soles partly hairy, ear small, less than one-teeth of head and body. Tail having a blackish streak above.

Identification of species of the genus Tatera

1. *Tatera indica:* Large sized gerbil, head and body length 130-230 mm. Tail longer than head and body, well-haired, tufted at tip. Soles naked. Tail dark above and below with pale sides.

Identification of species of the genus Gerbillus

1. *Gerbillus nanus:* Size small, head and body length less tha 100 mm. Tail very long, more than 140 percent of head and body length, tufted at tip. Soles naked. Body sandy brown above and white below. Tail bicoloured, dark above and pale below. Hind foot length 19-24 mm.

2. *Gerbillus gleadowi:* Very similar to previous species, but hind foot very long 926-30 mm.) and soles of hind foot partly haired.

ACKNOWLEDGEMENTS

The authors are grateful to the Director, Zoological Survey of India for providing kind permission to carry out the work and study the collection present in the National Zoological Collection. Sincere thanks of the authors are due to The Minister of Environment and Forests, Government of India, New Delhi and Dr. Q.H. Baqri, Officer-in-Charge and Principal Investigator of the project, Desert Regional Station, Zoological Survey of India for providing financial support, and facilities to conduct the survey works, and necessary guidelines regarding project work. We are indebted to Dr. P.L. Kankane, Research Scientist, and all the Research Associates and Jr. Research Fellows of the Project for extending all support and cooperation in connection with the field works. We also like to record our sincere gratitude to the authorities of Forest and Public Works Department, Govt. of Rajasthan for rendering cooperation during our field works. Our heartful gratitude and thanks are also due to Late Dr. Ishwar Prakash, for his kind guidance and encouragement for the present project work.

REFERENCES

Agrawal, V.C., 1967. New mammal records from Rajasthan. *LABDEV (J. Sci. & Tech.)* 5:342-344.

Agrawal, V.C., 1998. Faunal diversity in India: Mammalia, Pp. 459-469. In "Faunal Diversity on India" (Eds. J.R.B. Alfred, A.K. Das and A.K. Sanyal). ENVIS Centre, Zoological Survey of India, Calcutta.

Agrawal, V.C., 2000. Taxonomic studies of Indian Muridae and Hystricidae (Mammalia : Rodentia). *Rec. zool. Surv. India, Occ.Paper* No.180, pp. 1- 177.

Agrawal, V.C., P.K. Das, S. Chakraborty, R. K. Ghose, A.K. .Mandal, T.K. Chakraborty, A.K.Poddar, J.P. Lal, T.P. Bhattacharyya and Ghosh, M.K., 1992. Mammalia. *State Fauna Series* 3: *Fauna of West Bengal,* Part I, 27-169.

Alfred, J.R.B. and V.C. Agrawal, 1996. The mammal diversity of the Indian desert. Pp. 335-348. In *"Faunal diversity of the Thar Desert: Gaps in Research".* (Eds. A.K. Ghosh, Q.H. Baqri and I. Prakash). Scientific Publisher, Jodhpur.

Anon., 1988. *Thar Desert Biosphere Reserve.* Project Document. 7. Ministry of Environment and Forests, Govt. of India, New Delhi.

Baqri, Q.H. 2001. Studies on Faunal Diversity in the Thar Desert of Rajasthan. Annual Progress Report (Sept.2000-Aug.2001). Unpublished. Desert Regional Station, Zoological Survey of India, Jodhpur.

Baqri, Q.H. and P.L. Kankanc, 2001. Deserts: Thar. Pp. 94-109. In "*Ecosystems of India*" (Eds. J.R.B. Alfred, A.K. Das and A.K. Sanyal). ENVIS Centre, Zoological Survey of India, Kolkata.

Bates, P.J.J. and D.L. Harrison, 1997. *Bats of the Indian subcontinent.* Harrison Zoological Museum, Kent.

Bhargava, R.N., L.S.Rajpurohit, P. Bissa and S. Madan, 2001. On the porcupine (*Hystrix indica)* in western Thar desert. *Tiger paper* 28(4):1-3.

Biswars, B. and R.K. Ghose, 1968. New records of mammals from Rajasthan, India. *J. Bombay nat. Hist. Soc.* 65: 481-482.

Chatterji, P.S. and Saxena, S.K., 1988. Canal irrigation in arid zone of Rajasthan and its ecological implication. Pp. 223-258. In "*Desert Ecology*" (Ed. I. Prakash). Scientific Publishers, Jodhpur.

Corbet, G.B. and J.E. Hill, 1992. *The Mammals of Indo-Malayan Region: A Systematic Review.* Oxford univ. Press.

Ghose, R.K., 1976. Further new records of mammals from Rajasthan, India. *J. zool. Soc. India* 26: 149-150.

Ghosh, A.K., Baqri, Q.H. and Prakash, I. (Eds). 1996. *Faunal Diversity in the Thar Desert:* Gaps in Research. Scientific Publishers, Jodhpur.

Gupta, P.D. and V.C. Agrawal, 1966. Geographical distribution of Indian Hairy-footed Gerbils, *Gerbillus gleadowi* Murray. Sci.& Cult., 32: 470-471.

Gupta, R.K. and I. Prakash, (Eds.). 1975. *Environmental Analysis of the Thar desert.* English Book Depot, Dehradun.

Kankane, P.L., 1995. Chinkara and Desert Cat in Rajasthan. *Z.S.I. Envis Newsl.*, 3: 6-9.

Molur, S., P.O. Nameer, and S. Walker, 1998. Report of the workshop "Conservation Assessment and Management Plan for Mammals of India" (BCPP – Endangered Species Project). Zoo, C.B.S.G., Coimbatore.

Pocock, R.I., 1939, 1941. *The Fauna of British India including Ceylon and Burma.* Mammalia. Vol. I & II. Taylor & Francis, London.

Prakash, I., 1956. A list of mammals of Rajasthan Desert. *J. Bengal nat. Hist. Soc.* 28: 1-7.

Prakash, I., 1959. Checklist of the mammals in Rajasthan Desert. *Univ. Rajasthan Stud.* (Biol. Sci.) 4: 30-56.

Prakash, I., 1963a. Taxonomical and ecological account of the mammals of Rajasthan Desert. *Ann. Arid Zone* 1: 143-162.

Prakash, I., 1963b. Taxonomic and biological observations on the bats of the Rajasthan Desert. *Rec. Indian Mus.* 59: 149-170.

Prakash, I., 1964. Taxonomical and ecological account of the mammals of Rajasthan Desert. *Ann. Arid. Zone* 2: 150-161.

Prakash, I., 1975a. The ecology and zoogeography of mammals. Pp. 448-467. In "*Environmental Analysis of the Thar Desert*" (Eds. R.K. Gupta and I. Prakash). English Book Depot, Dehradun.

Prakash, I., 1975b. Population ecology of rodents in the Rajasthan Desert, India. pp. 75-186. In "*Rodents in Desert Environments*" (Eds. I. Prakash and P.K. Ghosh). Dr.W. Junk b.v. Publishers, The Hauge.

Prakash, I., 1981. Ecology of the *Indian Desert Gerbil, Meriones hurrianae*. CAZRI, Jodhpur.

Prakash, I., 1994. *Mammals of the Thar Desert*. Scientific Publishers, Jodhpur.

Prakash, I. and A.P. Jain, 1967. Occurrence of *Rattus meltada* and *Gerbillus dasyurus* in Rajasthan desert. *Ann. Arid. Zone* 6: 235.

Prakash, I. and R.P. Mathur, 1979. *Bandicota bengalensis* in Bikaner town. *Rodent Newsl.* 3: 12.

Prakash, I., R.K. Gupta, A.P. Jain, B.D. Rana and B.K. Dutta, 1981. Ecological evaluation of rodent populations in the desert biome of Rajasthan, *Mammalia* 35:384-423.

Rahmani, A.R., 1990. The distribution of Indian Gazelle or Chinakara *Gazella bennetti* (Sykes) in India. *Mammalia,* 54: 605-619.

Rahmani, A.R., 1991. Present distribution of the Black buck *Antilope cervicapra* in India, with special emphasis on lesser known populations. *J. Bombay nat.Hist.Soc.* 88: 35-46.

Rahmani, A.R., 1997, Wildlife in the Thar. WWF-India, New Delhi.

Rana, B.D., 1980. Occurrence of *Bandicota bengalensis and Vandeleuria olecarea* in Western Rajasthan. *J. Bombay nat. Hist.Soc,* 77: 501.

Roberts, T.J., 1977. *The Mammals of Pakisthan.* Ernest Benn Ltd., London and Tonbridge.

Rodgers, W.A. and H.S. Panwar, 1988. *Planning a Wildlife Protected Area Network in India.* Wildlife Institute of India, Dehradun.

Sinha, Y.P., 1975. New records of bats (Chiroptera) from Rajasthan. *Sci., & Cult.,* 41: 608-610.

Sinha, Y.P., 1976. New records of Indian Seath-tailed Bat, *Taphozous longimanus* from Rajasthan with remarks on winter fat deposition in *T.kachhensis. Sci.& Cult.* 42: 168-170.

Sinha, Y.P., 1981. The bats of Rajasthan: Taxonomy and zoogeography. *Rec.zool.Surv. India* 76:7-63.

Sinha, Y.P., 1996. Bats in Indian Thar desert. Pp. 349-352. In "*Faunal Diversity in the Thar Desert: Gaps in Research*" (Eds. A.K. Ghosh, Q.H. Baqri and I. Prakash). Scientific Publishers, Jodhpur.

Wilson, D.E. and D.M. Reeder, (Eds.) 1993. *Mammal Species of the World: A Taxonomic and Geographic Reference.* Smithsonian Inst. Press, Washington and London.

Wroughton, R.C. Summary of the Results from the Indian Mammal Survey of the Bombay Natural History Society. Part I-VII. *J. Bombay nat.Hist.Soc.,* 1918. **25:**547-598; 1918. **26:** 19-58; 1919. **26:** 338-379, 776-802; 1920. **26:** 955-967; 1920. **27:** 57-85.

Changing Faunal Ecology in Thar Desert, pp. 267-288, 2005.
Eds.: B.K. Tyagi and Q.H. Baqri

13

SIGNIFICANCE OF SCENT MARKING GLAND IN DESERT RODENTS

Mohd. Idris

Central Arid Zone Research Institute, Jodhpur-342 003 (Raj.), India

The paper reviews the work done on various aspects of scent marking gland in three commonly occurring rodent species of the Indian desert, viz., desert gerbil, *Meriones hurrianae*, Indian gerbil, *Tatera indica* and soft furred field rat, *Millardia meltada pallidior*. The scent marking gland is present in both the sexes of *M. hurrianae* and *T. indica* whereas it is present in 71.8% males of *M. m. pallidior*. Among the gerbils, the occurrence of the gland varies as per their social organizations. For example, in solitary population of *T. indica*, living in open scrub grass land (low density), 90.4% males and 42.3% females were reported to possess the scent marking gland, whereas the gregarious population of *T. indica* inhabiting the urban locales recorded relatively lower frequency of occurrence of this gland, *i.e.*, 84.4 and 3.8% in males and females, respectively. Likewise, the laboratory-bred population of *M .hurrianae* also showed reduced frequency of the gland, *i.e.*, 46..8 and 48.8% in males and females, respectively. The rolc of scent marking gland in solitary and gregarious population has been discussed in details.

The gland size is positively correlated with body weight and size of the animals of all the three rodent species. The growth of the gland is a continuous process throughout of the life of the individual. Histologically the ventral scent marking gland is composed of complexes of enlarged sebaceous alveoli of holocrine type, which are full of fatty material. All three species exhibit significant monthly variations in the dimension of the ventral scent marking gland. Peaks in gland size occur late in winter and monsoon seasons which generally correspond to the maximum breeding activity during the year. The exudation is of musky odour. Scent marking is performed by the three species following similar pattern, they press the ventrum to ground surface in their home range, on heaps of excavated sand and on vegetation clumps. They perform perineal drag for marking with urine simultaneously. Sexual dimorphism in gland size and scent marking frequency has been observed in male and female rodents of the three species i.e. males have significantly larger gland and the rate of scent marking by males is also of a significantly higher magnitude than in the females.

It has been observed that both sexes prefer sebum odour of their own sex. Scent marking plays an important role in maintaining the dominant status of both sexes as the dominant saturate the immediate environment with their own odour. The function of scent marking in these species are more of a "familiarization nature" to level the habitat for its own use and orientation to signal "home" to the marking animal. The scent marking also denotes the dominant status of an individual in a social hierarchy and "ready to mate" state of a female. It also serves as an innate mechanism of population regulation. The sebum, urine (more so of females in estrous), and their odours bear phago-stimulant and poison shyness masking properties. Thus, these properties of sebum and urine have become good indicators for developing tools for achieving higher control success during rodenticidal baiting for management of rodent pest species.

INTRODUCTION

Most of rodents have keenly developed sense of smell. The phenomenon probably developed during the evolutionary process since the rodents are endowed with scent organs and glands, the odours of which are agents of olfactory communication among them. Moreover, it has been recorded that the sparsely distributed field rodents depend more on olfactory cues than on visual or auditory impulses for inter- and intraspecific communications. Specialized integumentary glands constitute one of the major sources of olfaction among mammals (Ralls, 1971; Johnson, 1973; Mykytowycz, 1970, Muller Schwarze, 1983; Balakrishnan & Alexander, 1985). Sebaceous glands are more common than the other types (Schaffer, 1940). Besides the skin glands, other sources of vertebrate pheromones have also been demonstrated in various mammalian species, such as vagina (Michael *et al.*, 1971; Singer *et al.*, 1976), urine, possibly through the kidney (Vonderburg *et al.*, 1975), faeces (Ewer, 1968; Mykytowycz, 1970; Ralls, 1971), accessory glands of the male reproductive system (Jones & Howell, 1973) and contents of the caecum (Leon, 1975).

The pioneer work on scent-marking behaviour of some desert rodent species was initiated by the Late Dr. Ishwar Prakash at Central Arid Zone Research Institute, Jodhpur under the "Professor of Eminence Scheme" awarded By Indian Council of Agricultural Research, New Delhi. The work was carried out on desert gerbil, *Meriones hurrianae*, Indian gerbil, *Tatera indica* and soft furred field, *Millardia meltada pallidior* which possess a ventral scent marking gland. All the three species have significant differences in size of the gland, occurrence of the gland between the sexes and species though the scent-marking behaviour in the three species is almost similar. Scent-marking plays a significant role in maintaining social hierarchy among the population and also in reproductive manifestation. This article reviews the various salient achievements carried out under this Scheme during 1980-91, under the fruitful and innovative supervision of Professor Dr. Ishwar Prakash.

THE SCENT MARKING GLAND

Morphology

The ventral scent marking gland in the three rodent species are situated on the mid-ventral line of the abdomen. The gland pad can readily be seen by pushing aside the overlying lateral hairs of that region.

The ventral scent marking gland is present in both sexes of desert gerbil, *Meriones hurrianae*, though it is significantly ($P< 0.001$) smaller in females (Kumari, 1982). It is fairly prominent, ovoid in shape, devoid of hair and yellowish in colour which is broader anteriorly in males but is more elongated in females. The yellowish waxy secretion on the gland surface feels oily and has a distinctive musky odour more or less like that of *Kasturi* of the musk deer, *Moschus moschiferus* (Kumari *et al.*, 1981). Unlike *M. hurrianae*, the ventral scent marking gland of *T. indica* is elongated, dirty white in colour but neither as prominent nor as glandular as in *M. hurrianae*. Among females it is a much smaller structure and is more linear (Idris, 1985). In soft furred field rat, *M. meltada pallidior*, the ventral scent marking was reported for the first time by Prakash and Kumari (1979). The gland is present in 71.86% males. The glandular pad is ovoid, broader on the posterior side and is the longest among the three species. It is dirty yellowish in colour and fairly covered with hair. In this species also the growth of the gland was noticed throughout the life of rodent.

Histology

Histological study revealed that the ventral gland is composed of complexes of enlarged sebaceous alveoli of holocrine type, each with its own duct. Outer most layer of skin is composed of stratified epithelium in *M. hurrianae* and in *T. indica* but in case of *M. m. pallidior* the surface of the skin in the mid-ventral gland area is highly papillated. Sebaceous gland alveoli are the only glandular structure present and they can greatly enlarge in comparison to the skin. The holocrine secretion collects in common ducts and is deposited on objects by rodents with a characteristic ventral rubbing action.

Seasonal variation in gland size

Breeding season of animals affects the growth of the ventral scent marking gland. In *M. hurrianae* breeding continues throughout the year with peaks in February, March, July and October, glandular area was also highest during these months of breeding in both the sexes of *M. hurrianae*. The pregnancy rate was lowest during May and June and December-January, glandular area also decreased during these months in both the sexes (Kumari, 1982).

In Indian gerbil *T. indica* the gland is longest during winter and but varies insignificantly during rest of the year. In contrast to male, female *T. indica* possessed longest scent marking gland during summer (May) and the smallest gland size occurred during winter (Idris, 1985).

Like, *M. hurrianae, M. m. pallidior* breeds the year round but peak breeding activity is noticed during months of March-April and July-September. In relation to pregnancy the glandular area also increases during the peak breeding season. Since the pregnancy rate is lowest in the month of January, June and December, glandular area too decreases during these months (Kumari, 1982). Such seasonal variations corresponding with breeding activities have also been reported in many other mammals like *Capreolus capreolus* (Von Schumacher, 1936), *Antilocapra americana* (Moy, 1969), *Dipodomys* sp. (Quay, 1953). In *Arvicola terrestris*, the lateral scent organ was more active during breeding season, as testis's weight during breeding season was higher than that in the non-breeding season, and amount of secretary tissue present is also higher in the breeding season (Stoddart, 1972).

Species-wise occurrence

The ventral scent marking gland occurs in three test species, *viz.*, Indian desert gerbil, *M. hurrianae,* Indian gerbil, *Tatera indica* and soft furred field rat, *M. meltada pallidior.* The ventral scent marking is present in both the sexes of two gerbils, *M. hurrianae* and *T. indica,* but occurs only in the males of *M. m. pallidior.* A number of individuals of *T. indica* inhabit a single burrow system in the village complex or around urban area (gregarious), but they also live individually or in pairs (solitary) in dry scrub grasslands. These two distinct populations exhibit a characteristic difference in the frequency of occurrence of mid-ventral scent marking gland. It was found in 90.4% (N = 344) males and 42.29% females (N = 339) collected from dry scrub grassland whereas, it was observed in 84.39% males (N = 205) and 3.89% females (N = 231) collected from the urban area at Bikaner (Idris, 1985). Similar observations were also reported in lab bred (CFTRI, Mysore) population of *M. hurrianae*. The scent marking gland is clearly visible and functional even in subadult (body weight 30.14 g ± 1.04 males; 35.15 g ± 1.14 female, sexually immature) in wild populations, however, it is absent in animals of even sexually mature gerbils with a 41-60 g body weight (*M. hurrianae* attains sexually maturity at 40 g body weight in the wild) in laboratory bred (LB) population (Prakash *et al.*, 1988).

On the basis of these observations it is conjectured that the laboratory-borne gerbils, which live in a gregarious social organization in contrast to the wild that prefers solitary and sparsely distributed condition, have the scent marking gland weakly and/or belatedly developed in the laboratory-borne *M. hurrianae*. The need for olfactory communication declines since it is replaced by other type of communication cues. Hence, the frequency of occurrence of the gland in the LB population is low, and it develops much later with sexual maturity in both the sexes (Prakash *et al.*, 1988). Such a situation has also been observed in sympatric species, *T. indica*, which is living in two types of social organizations discussed earlier (Idris & Prakash, 1985).

Table 1. Measurement of ventral scent marking gland and body weight of *M. hurrianae* (Mean ± S.E.) (after Kumari *et al.*, 1981)

Body weight (g)	N	Sex	Mean weight (g)	Gland area (mm^2)
20-40	2	M	34.0±3.0	34.0 ± 3.5
	3	F	38.2 ±2.0	35.2±3.8
41-60	15	M	51.0±4.0	58.6±7.1
	24	F	50.0 ± 1.0	39.1±4.1
61-80	24	M	70.0±1.0	97.6±7.7
	35	F	70.0±2.0	57.1±3.3
81-100	26	M	92.0±3.0	152.0±9.9
	12	F	87.0±2.0	66.5±5.4
101-106	10	M	104.0±1.0	176.0±15.3

Gland size

The area of ventral scent marking gland in *M. hurrianae* ranges from 34.0 ± 3.5 mm^2 to 176.0 ± 15.4 mm^2 in male and from 35.2 ± 3.8 to 66.5 ± 5.4 mm^2 in females in minimum and maximum body weight classes, respectively. The gland is significantly ($P < 0.001$) larger in males than females (Kumari, 1982; Kumari & Prakash, 1979) (Table 1). The length, width and area of the gland is found to be positively correlated with body weight as well as head and body length in both the sexes ($r = 0.43$ to 0.91, $P < 0.001$). In adult *T. indica*, the area of ventral scent marking gland varies from 53.78 mm^2 to 65.19 mm^2 in males and 26.54 mm^2 to 54.60 mm^2 (Idris, 1985) (Table 2). The male gland is significantly ($P<0.001$) larger than that of female in both *M. hurrianae* and *T. indica*. The area of the ventral scent marking gland of metads was reported to be

263.79 ± 15.14 mm² (Kumari, 1982) (Table 3). This gland has been noticed in 71.80% (N=135) adult males. Among three different rodent species studied, *M. m. pallidior* males have been observed to possess largest ventral scent marking gland (Table 4).

Table 2. Gland size in relation to g/100g body weight of *T. indica* (Mean ± S.E.) (adopted from Idris, 1985)

Body weight (g)	N	Sex	Gland area (mm²)
40-60	13	M	53.78±4.83
	10	F	42.20±5.73
61-80	47	M	63.62 ± 3.96
	13	F	54.60±6.06
81-100	59	M	55.08±2.11
	45	F	43.46±2.06
101-120	57	M	65.19±2.68
	36	F	39.06±2.57
121-140	50	M	58.35±2.99
	15	F	39.69±4.35
141-160	41	M	63.28±1.87
	11	F	32.78±1.38
161-180	30	M	58.14±2.77
	8	F	30.37±3.50
181-200	14	M	64.59±2.39
	2	F	26.54±3.61

Table 3. Body weight classes and gland size of male *M. meltada pallidior* (After Kumari & Prakash, 1983).

Body weight classes (g)	Body weight (g)	Gland length (mm)	Gland width (mm)	Gland area (mm²)
30-50	41.25±3.28	21.9±2.8	5.95±0.47	139.0±23.4
51-70	62.82±1.09	24.6±1.0	6.05±0.26	155.3±12.2
71-90	79.06±1.2	29.5±3.1	7.06± 0.44	223.40±16.42
91-110	97.50±1.0	30.7±4.0	7.50±0.87	239.50±57.9

In many mammals, males have larger scent marking gland than the females, *viz.*, Mongolian gerbil, (Thiessen, 1968; Thiessen *et al.*, 1969; Glenn & Gray, 1965), Guinea pig (Montagna & Hamilton, 1949; Marton, 1962), rabbit *Oryctolagus cuniculus*, (Mykytowycz, 1962, 1965, 1966, 1968). Like *M. m. pallidior*, in few instances of *Rattus exulans*, too, the scent glands are not present in all the females, except in few large or older ones (Quay & Tomich, 1963).

Table 4. Comparison in gland size of three rodent species: *M. hurrianae*, *T.indica* and *M. m. pallidior.*

Species	Sex	Body weight (g)	Gland length (mm)	Gland width (mm)	Gland area (mm^2)
	Male	69.32±3.39	A 20.81±0.51	5.11+0.29	104.21±8.25
M. hurrianae			R 28.84±0.95	6.79+0.18	151.17±7.19
	Female	64.68±2.41	A 16.44±0.36	3.10+0.26	50.80_1.72
			R 25.65±1.28	5.24+0.25	90.37±4.01
T. indica	Male	123.213±3.0	A 17.34±0.32	3.25+0.11	57.83±2.16
			R 15.02±0.41	2.71+0.10	48.00±1.82
M. m. pallidior	Male	65.39±1.66	A 26.75±0.68	6.51+0.35	180.35±9. 68
			R 39.66±2.58	9.69+0.69	263.79±15.14

A= Absolute value; R= Relative value (mm/100g body weight)

Growth of the gland *vis a vis* body measurements

The ventral scent marking gland of the three rodent species grows throughout their life as evidenced by the relationship of the gland area with their body weights/classes organized at 20 g interval (Kumari, 1982; Kumari & Prakash, 1983; Idris, 1985). In *M. hurrianae*, the gland size showed no significant variation between the subadults of both sexes however, in all the heavier body weight classes, males had larger glands than females, which was more prominent in 81-100 g body weight class, corresponding to 8-12 months of age (Table 1). The gland size has been observed to increase markedly in both males and females upto 80 g body weight and thereafter it increased significantly only in male. Further analysis of growth data revealed that the regression coefficients for each measurement, males differed significantly (P <0.05) from females in slope of regression of gland width with body weight (Male: Y = 1.53 + 0.513x, female: Y = 0.370 + 0.0548 x) but not in gland length (Male : Y = 10.70 + 0.1326 x, Female : Y = 2.59 + 0.2150 x) (Kumari *et al.*, 1981).

In *T. indica,* the 80g body weight correspond with 4 months old animals, whereas, 8 months of old gerbil weigh 140g, and above these weights the growth curve flattens to the extent that age could be from 1-1.5 years (Jain, 1970). The Indian gerbil attain sexual maturity at around a body weight of 70 g indicating that scent marking activity starts prior to attaining sexual maturity. The length, width and area of the scent marking gland in male *T. indica* are significantly and positively correlated with body weight (r = + 0.22, P< 0.05 to r = + 0.84 P<0.05 to 0.001 (Table 2). The correlation coefficient between body weight and gland length (Y = 9.1637 + 0.07x, r = 0.557) and with gland width (y = 0.172 + 0.0264 x, r = 0.8326) are statistically significant.

In *M. m. pallidior*, too, the scent marking gland grows throughout the life of rodent as indicated by its increasing size, along with the body weight (Table 3). The correlation matrix of body weight and gland length (r = 0.4373, P< 0.05), gland width (r = 0.5572, P<0.001) and gland area (r= 0.220, P<0.05) were found to be statistically significant (Kumari & Prakash, 1983). Male metads mature sexually at a body weight of about 45 g (Rana & Prakash, 1984) but the gland can be prominently observed in smaller animals of 30 to 40 g body weight. It is, however, absent in a number of metads of 51-60g body weight category (Kumari & Prakash, 1983 a).

SCENT MARKING BEHAVIOUR

In all the three species of rodents, the marking procedure is more or less similar. The rodents scent-mark the low lying objects as well as flat surfaces in their environment by rubbing them with their sebaceous gland and depositing the sebum exudates simulating musky odour. The *modus operandi* followed by rodents includes lowering their ventrum by flexing the abdomen and by extending the forelimbs, the ventrum is then dragged forward and pressed against the substratum. Scent marking has slight variation in deposition of sebum and urine amongst various different species of rodents. In desert gerbil, *M. hurrianae,* the drag is long and the scent marking activity involves perineal drag, which also indicates that merion gerbils mark with their urine in a single bout. In *T. indica,* however, the drag is short and it involves only sebum marking. However, in urine marking animal lowers its pelvic portion, which is dragged in a forward motion and leaves a distinct smudge of urine (Idris, 1985). *Millardia m. pallidior* seldom marks flat surface and always leaves scent marks on low lying objects on its pathways (Idris & Prakash, 1987).

Sexual dimorphism

The scent gland and marking behaviour generally appear to be sexually dimorphic. Both the sexes usually mark but males do so more frequently. In case of the gerbils, *Meriones hurrianae, T. indica* and *M. unguiculatus* males mark about twice as often as that of females (Kumari & Prakash, 1981; Idris & Prakash, 1982; Thiessen, 1968). While in the rabbit, male marks ten times more than the females do (Mykytowycz, 1962).

Behavioural responses of desert rodents towards odours

Conspecific odours

Behavioural responses of *T. indica* to conspecific sebum odour of male and female ventral scent marking gland were studied in wooden pen and plus maze. All the three categories of *T. indica* (male possessing scent gland, females possessing the gland and female without gland) preferred the part of the pen carrying the odour of strange male. However, the frequency and duration of social acts by male and female *T. indica* clearly indicate that the magnitude of attraction towards strange male odour was much lower among females as compared to males (Idris

& Prakash, 1982, 1985). In this preferential behaviour *T. indica* is similar to *M. tristrami* (Thiessen *et al.*, 1973) and *M. unguiculatus* (Thiessen *et al.*, 1970). *Meriones hurrianae* prefers the gland odour of similar sex given the choice of odour of both sexes. However, in absence of the same sex odour, both the male and female are attracted towards opposite sex (Kumari & Prakash, 1981a). In this behaviour, the females of *T. indica* differ from *M. hurrianae* though they share in similar habitats in the Indian desert and are the two most predominant rodent species of the Indian desert.

The behavioural responses of *M. m. pallidior* towards conspecific sebum odour is similar to that of *T. indica.* Both the male and female show interests in conspecific odour of the ventral scent marking gland of strange male (Prakash & Idris, 1982). The frequency of sniffing, licking ventral scent marking, anal drag and urination by both the sexes of metad was significantly ($P<0.05$ to 0.001) higher order in presence of sebum odour of strange male. Moreover, scent marking in *M .m. pallidior* was performed through two actions, *i.e.*, ventral marking and anal drag (Prakash & Idris, 1982).

The function of scent marking in desert-dwelling rodents is more of a "familiarization" nature or signal "home" so the marking animal or that of labeling the habitat for an animal's own use in orientation (Johnson, 1973). Other functions assigned to the ventral scent marking by the rodents are to increase its own confidence in the environment to provide homeliness (Eibl-Eibesfeldt 1953; Ewer, 1968; Mykytowycz, 1968), territoriality (Thiessen, 1973), individual identification (Daly, 1977) and recognition of pups (Wallace *et al.*, 1973).

Interspecific odours

To see the influence of interspecific odours, an experiment was carried out in the rattery (13 x 7 m). A group of 6 male and 6 female desert gerbils was released in the rattery and their behavioural activities were recorded before introduction of sympatric Indian gerbil, *T. indica* in the rattery (Idris & Prakash, 1986). The observations clearly indicate that the number of intraspecific interactions of desert gerbils was drastically reduced ($P< 0.05$) after the release of *T. indica.* The range of movement was also declined significantly ($P< 0.001$). The average area of their home range was reduced from 36.46 ± 6.47 m^2 to 13.75 ± 6.48 m^2 in males and from 40.45 to 17.087 m^2 in females. The scent marks of sebum exudation of the mid ventral marking gland and urine of the nocturnal *Tatera indica* was perceived by the diurnal *M. hurrianae.* The latter responds to odours of the intruder, *T. indica,* by significantly shrinking its home ranges and by significantly enhancing its sebum scent marking activity in a way to saturate the general area with its own odours. These observations also point out that the olfactory cues of one species signal message to sympatric species and the latter responds to them in a way to establish its own dominance on the former. However, a 'matching of odours' competition sets in between the resident and intruder species in the common space of resource utilization by superimposing their own odours on that of others (Idris & Prakash, 1986).

Exposure of conspecific and interspecific sebum odours in five combinations to *M. hurrianae* indicated that these gerbils perceive not only conspecific odour of the ventral scent marking gland but also that of another species which is not sympatric, *i.e., M. unguiculatus. Meriones hurrianae* preferred ($P<0.05$) uni-sex odour, even *M. unguiculatus* did so, and also enhanced its own scent marking frequency in the presence of scent marking gland odours of the strange gerbils (Idris, 1994).

Hormonal control of scent marking behaviour

Investigations to study the association between hormones and scent marking behaviour in the Indian gerbil, *T. indica* have revealed that castration reduces the frequency of scent as well urine marking potential significantly within two weeks (Idris & Prakash, 1985b). The response of castrated animals to testosterone propionate therapy was gradual and the peak scent marking activity was attained after 8 weeks of administering the injection on a weekly interval. However, this hormone therapy could not enhance the scent marking activity to the level that of normal animals. The marking activity dropped no sooner the weekly injections were stopped. Such clear cut effect of hormonal therapy was, however, not observed in urine marking behaviour in *T. indica* which indicated that control of this activity may be regulated by a number of hormones (Idris & Prakash, 1985). Glandular activity has also been reported to be controlled by sex hormones in rabbit (Mykytowycz, 1970) and in the Mongolian gerbil, *Meriones unguiculatus* (Blum & Thiessen, 1971; Thiessen *et al.*, 1968).

Scent marking behaviour in relation to density

Scent marking frequency is inversely associated with density of the individuals per unit area. The scent marking frequency of individual *M. hurrianae* decreased to about 50% when they were grouped in various densities (Table 5). The number of rodents which scent-marked in cages under varying densities also reduced to about 66.6% in density of three and 33.3% in density of 6, 9 and 12 per cage (90x30x30 cm). Declining trend, similar to that of scent-marking activity, was observed in the acts which was relevant to olfactory communication, such as urination, defecation and in sniffing and exploratory activity (Kumari & Prakash, 1984). Contrarily, the acts concerning agonistic behaviour gradually increased with increased group size.

It is obvious from these observations that with the increase in density, the need for olfactory communication declines. Since the need for olfactory is greater in gregarious mode of living than in an high density situation where intraspecific communication may be carried out by direct contact, visual and/or auditory means. This conjecture finds support from the scent-marking behaviour of another sympatric gerbil, *T. indica* which also occurs in the desert habitat where it lives singly or in pairs (low density like that of *M. hurrianae*) in urban areas where 5-70 of them can be found in a single burrow system (high density). Correspondingly, the occurrence of the ventral scent-marking gland also differs in these two types of social groups (Idris & Prakash, 1982, 1987 a).

Table 5. Scent marking frequency *of M. hurrianae* and *T. indica* per 15 min. in varying density level. (after Kumari & Prakash, 1984 and Idris & Prakash, 1987).

Density in cage (90x30x30cm.)	Scent marking frequency/15 min.	
	M. hurrianae	*T. indica*
Individual	9.75±4.98	9.33±1.08
Group of 3	4.41±2.76*	5.03±1083*
Group of 6	2.41±1.17*	2.77±0.44**
Group of 9/12	1.55±0.78^{NS}	1.06±0.35

Level of significance between the group and the one above (Student's t test): P<0.05, ** P< 0.01, NS= Not significant.

A comparison of scent marking and urine marking behaviour in both the sexes of the two population (Table 6) reveals that the animals of laboratory-bred (LB) population is very poor (P<0.001) in scent marking and urine marking frequency as compared to wild populations (e.g., *M. hurrianae*. The difference in urine marking frequency, however, among females of the two populations was of a lower magnitude (P < 0.05). It is noteworthy that the urine marking frequency by females of both the populations was higher than that of males (wild, P < 0.001, LB (P< 0.05). Kumari & Prakash (1984) reported that since the female gerbils possessed relatively smaller ventral gland, they compensated scent marking activity through a higher urine marking frequency. Similar situation was also found in another species of gerbil, *T. indica* (Idris & Prakash, 1985).

Table 6. Scent marking and urine marking frequency (Mean±S.E.) per 30 min. of *M. hurrianae* in wild and lab bred (LB) population (after Prakash *et al.*, 1987)

Population	Body weight (g)	Sex	Scent marking	Urine marking
Wild	78.81 ± 4.27	Male	12.27 ± 2.03***	1.77 ± 0.27***
LB	88.77 ± 5.56	Male	1.41 ± 0.31	0.33 ± 0.14
Wild	88.33 ± 2.06	Female	7.11 ± 0.72***	3.30 ± 1.26*
LB	98.50 ± 5.66	Female	1.00 ± 0.48	1.81 ± 0.56

Student 't' test between the sexes: *** = P< 0.001, *= P<0.05

Scent marking and reproduction

The frequency of scent marking by females in both the gerbils *M. hurrianae* and *T. indica* increased significantly (P< 0.001) during estrous in comparison of diestrous (Table 7). It indicates that oestrous females advertise their sexual receptivity through this manner of olfactory communication. It is likely that in females *T. indica*, which is devoid of scent marking gland, sexual receptivity might be advertised by the perineal rub, leaving a mark of urine on the surface (Prakash & Idris, 1982a). It was evident from the fact that the frequency of perineal rub was increased manifold (P<0.05) in these females during pro-estrous and estrous. There was a manifold increase of this activity during oestrous in the females possessing the marking gland and was of a higher magnitude than that in the females devoid of the gland (Prakash & Idris, 1982a). Urine odours can act as primer, which affect the reproductive stage of female rodents in a number of ways (Bronson, 1974; Bruce, 1959; Dominic 1969 & Whitten, 1956).

Table 7. Mean frequency (Mean ± S.E.) of scent marking and urine marking by females of *M. hurrianae* and *T.indica* in different phases of estrous (after Kumari & Prakash, 1981 and Idris & Prakash, 1982).

Reproductive phase/ Species	Scent marking			Urine marking		
	Dioestrus	Proestrus	Estrus	Dioestrus	Proestrus	Estrus
M. hurrianae	35.20 ± 7.01	49.10 ± 6.00*	51.10 ± 6.75*	1.70 ± 0.30	2.00 ± 0.30NS	1.30 ± 0.34NS
T. indica with gland	6.72 ± 1.32	12.58 ± 2.05*	11.58 ± 1.95*	2.08 ± 0.31	3.66 ± 0.46**	3.56 ± 0.54**
T. indica without gland	-	-	-	2.58 ± 0.31	3.58 ± 0.52	3.33 ± 0.52

Student 't' test. P values: * = P< 0.05, ** =P< 0.01,
NS = Not significant, as compared to dioestrus

ROLE OF SCENT MARKING

As sex attractant

Olfactory cues may serve to attract the opposite sex, or may act as a stimulant to sexual behaviour. Kumari & Prakash (1981) observed that male as well as female *M. hurrianae* prefer the odour of the ventral gland exudates of their own sex, even in the presence of odour of both the sexes. However, in the absence of their own sex odour, they preferred the sebum odour of the opposite sex. The same has been confirmed in *T. indica* and *M.m. pallidior* also (Idris & Prakash, 1982; Prakash & Idris, 1982).

For social dominance and aggression

The male desert gerbil, *M. hurrianae* was found to be more active and performed all the aggressive acts at a higher frequency than the females (Kumari & Prakash, 1981). While a number of males in a social group perform this acivity however, in females only the dominant one perform this activity. It would, therefore, seem that evolution of sociability in *M. hurrianae* chiefly based on scent marking behaviour, had an adaptive value as a self regulatory mechanism of population control (Wynne Edwards, 1962, Eisenberg, 1967) in this most predominant rodent species of the Indian desert where their numbers are kept below a maximum level by restricting mating to only a few males and the dominant female not allowing through social interactions the dominant male to mate with other females in the social group. It is also possible that the dominant female desert gerbil familiarizes the alpha male with its own odour thus obliterating other females from chance of mating (Kumari & Prakash, 1981 b).

Idris and Prakash, (1988) observed that both male and female members of social groups in desert gerbil, *M hurrianae*, spend more time in association with their sebum scent. Only female members in the group prefer the sebum scent of the dominant male. Identification of the dominant malc by the females through the scent may be for selecting the best mate in social group.

Studies on the role of dominance in a social group of *M. hurrianae* revealed that the frequency of social as well as aggressive activities of the members significantly ($P<0.001$) decline if the dominant male is suddenly removed. However, the subdominant animal enhances its marking activity in that situation. It is further enhanced after the introduction of a strange animal. It shows that the removal of the dominant animal provides an opportunity to the subdominant animal to attain a dominant status by enhancing its scent marking as well as urine marking rates (Idris & Prakash, 1991). It also indicates that chemical communication through scent marking plays an important role in establishing social hierarchy in *M. hurrianae.*

In contrast to *M. hurrianae*, the social organization in *T. indica* is based on multi-dominant animals of both the sexes (Idris & Prakash, 1985a). Though the social organization of *T. indica* is also male-dominated but all females are equally aggressive as the male. The versatility in scent marking and many other behavioural adaptabilities like omnivory may be one of the reasons of its presence in every habitat throughout India. According to the classification of Eisenberg (1963) the social organization of *M. m. pallidior* can be classified under "contact" and pair tolerance category verging towards "group tolerance" (Idris & Prakash, 1987).

Thermo-regulatory mechanism

Desert gerbil builds complex burrow system in which it maintains constant temperature, enabling it to live comfortably in harsh xeric conditions (Prakash, 1981). After exposure to the hot outer environment during the day they develops hyperthermia and it offloads the excessive heat load into the cooler environment of the burrow by intermittently visiting it (Prakash, 1986, 1993). In this context, the glandular region could act as a "heat window", allowing an exchange of heat with the surrounding environment when the sebum exuded out (Kittrell *et al.*, 1982). The scent marking thus assist homeostasis in temperature regulation besides chemo-communication. This function of the scent marking gland may explain the reduction of scent marking frequency when they are in cages and are grouped together since due to huddling the necessity for homeostasis also decline (Prakash *et al.*, 1998).

Familiarizing/territorial nature

All three rodent species deposit two types of scent in their area of movement on the new objects: (i) that of sebum, and (ii) urine (Idris & Prakash, 1986, 1987a; Kumari & Prakash, 1981). It indicates that the function of scent marking in these species is more of a familiarization nature to level the habitat for its own use and orientation to signal 'home' to the marking animal (Idris & Prakash, 1982,1985a). Other works (Ewer, 1973; Mykytowycz, 1970; Ralls, 1971) have also suggested this function in which the scent marks provide a sense of security and increase confidence of animals in the general area of their home range.

Advertisement of sexual receptivity

Communicatory signals that advertise the presence, location and reproductive condition of potential sexual partners are crucial components of the sexual behaviour of most vertebrates. A nearly universal aspect of mammalian behaviour is olfactory "testing" of females by males, usually involving sniffing and licking of the ano-genital region and often sampling urine as it is passed by the females. It was found that when exposed to the urine odours of estrous and anoestrous females simultaneously, the male gerbils *T. indica* and *M. hurrianae* preferred estrous urine in comparison to anoestrous urine. The significant increase ($P<0.001$) in the frequency of sniffing behaviour of male gerbils when exposed to estrous urine indicated that the urine odours send definitive olfactory signals in gerbils (Idris & Prakash, 1983; Kumari & Prakash, 1981b). Estrous female *M. hurrianae* and *T. indica* also scent mark significantly for more times than they do while in diestrous (Table7; Prakash & Idris, 1982a).

Relationship with breeding season

A broad correlation in respect of variations in the size of scent marking gland and the periodicity of breeding activity has been indicated. In *M. hurrianae* breeding continues throughout the year with peaks in February, March, July and October. The glandular area in this animal was also found to be the highest during these months of breeding

(Kumari, 1982). The frequency rate was lowest during May and June, and December and January. The glandular area also decreased during these months in male as well as in female *M. hurrianae.*

Millardia m. pallidior breeds all the year round but the peaks occur during the months of March–April and July–September. The glandular area of scent marking gland is also largest during these months. Likewise, when the pregnancy rate declines during December-January, and June, the glandular area also decreases. In *T. indica* however, the relationship between reproductive activity and the size of the gland area is not conspicuously correlated, rather it indicated an inverse relationship (Kumari & Prakash, 1983a). In all species of *Dipodomys* (except *D. elephantinus, D. magaritae* and *D. mitchelli*), the mid-ventral organ is maximally productive during the breeding season and is comparatively quiescent during the non-breeding season (Quay, 1953). Stoddart (1972) suggested that increase in testicular androgen during the breeding season serves to increase the quantity of odourous secretions produced at a time when the population is diluted. The lateral organs of *Arvicola terrestris,* as reported by Stoddart (1972), were also found to be more active during the breeding season as testis weight was higher than in the non-breeding season. The amount of secreting tissue present was also higher in breeding season than in the non-breeding season in that species. Mykytowycz (1966) observed in the Australian rabbit, *Oryctolagus cuniculus* and in the hare, *Lepus europaeus,* that the weight and secreting activity of the anal gland were highest during the breeding season, *i.e.,* November-December, and lowest during the non-breeding season (May-June). It is apparent, therefore, that scent marking has a definitive role during the breeding season.

Phagostimulant properties of urine/sebum odours

It has been observed that the body odours, which are mainly of four types, *viz.*, body odour, urine odour, faecal' odour and the odour of sebum exudation of the ventral gland, adequately influence food consumption of the Indian gerbil, *T. indica.* When scented millet was drawn daily from the boxes in which male or female gerbils were housed continuously, the scented millet was consumed in significantly higher quantities ($P<0.001$) (Table 8). It is apparent from these results that various body odours of rodents do function as phagostimulants. The ability of the scented food to induce increased consumption by the gerbils can be utilized to ensure that they consume lethal doses of a poison in the bait when bait scented with their own odour is utilized for rodent management. Similarly, conspecific urine of *M. hurrianae* has been observed significantly ($P<0.001$, Kumari & Prakash, 1980) (Table 9).

As observed in the *M. hurrianae,* the addition of conspecific urine of male and female, *T. indica* and *M. m. pallidior* to millet grains, significantly enhanced the food intake of both the sexes, respectively (male $P<0.001$; female $p<0.05$) (Table 9).

Idris & Prakash (1982) reported that sebum exudation possesses a phago- stimulant properties for *T.indica.*

Table 8. Phago-stimulant property of conspecific body odours

Species	Body odours (sex)	Experimental animal	Mean daily food intake g/100g body weight ± S.E.	
			Scented millet with body odours	Ordinary millet
M. hurrinae	Male	Male	2.77 ± 0.48*	1.88 ± 0.21
		Female	2.90 ± 0.43*	1.45 ± 0.32
	Female	Male	2.96 ± 0.30**	0.53 ± 0.18
		Female	2.70 ± 0.18***	0.71 ± 0.13
T. indica	Male	Male	5.03 ± 0.50**	1.81 ± 0.21
		Female	5.66 ± 0.30**	2.91 ± 0.26
	Female	Male	5.23 ± 0.60**	1.28 ± 0.28
		Female	5.31 ± 0.50	1.30 ± 0.25

Student't'test. Level of significant between scented and ordinary millet: *= P<0.5, **= P<0.01, ***=P<0.001

Table 9. Conspecific urine as phago- stimulant property in three rodent species

Species	Urine (Sex)	Experimental animal	Mean daily intake g/100g body weight (M±S.E.)	
			Urine mixed millet	Ordinary millet
M. hurrianae	Male (0.4%)	Male	4.89 ± 0.65**	1.33 ± 0.39
		Female	5.16 ± 0.62**	1.71 ± 0.44
	Female (0.4%)	Male	4.88 ± 0.61**	2.28 ± 0.47
		Female	7.11 ±0.93***	1.28 ± 0.38
T. indica	Male (0.4%)	Male	4.81 ±0.44***	1.23 ± 0.44
		Female	3.62 ± 0.48**	2.05 ± 0.42
	Female (0.4%)	Male	4.11 ±0.86***	1.94 ± 0.37
		Female	4.45±0.78***	1.38 ± 0.45
M. m. pallidior	Male (1.0%)	Male	2.25 ± 0.33*	1.33 ± 0.22
		Female	1.91 ± 0.17*	1.33 ± 0.20
	Female (1.0%)	Male	2.41 ± 0.46	1.83 ± 0.30
		Female	2.00 ± 0.56*	1.18 ± 0.18

Student't'test. Level of significance between urine mixed and ordinary food: *=P<0.5, **=P<0.01, ***=P<0.001

Masking of poison shyness/aversion

A number of rodents develop shyness towards zinc phosphide bait after consuming a single sub lethal dose (Bhardwaj & Prakash, 1979; Prakash & Jain, 1971; Prakash *et. al.*, 1975; Prakash & Ojha, 1978). As a consequence, it becomes very difficult to control poison-shy residual population. To solve such problem, a number of experiments were carried out to mitigate the poison shyness.

Meriones hurrianae develops aversion to poison even after a single exposure to the zinc phosphide and the consumption of poison food on subsequent exposure tends to decline drastically due to induction of poison shyness/aversion behaviour (Prakash & Jain, 1971) which persists for 35 days in *M.hurrianae*. However, addition of urine inhibits the poison aversion in desert gerbils by masking the shyness behaviour (Kumari & Prakash, 1980). Two exposures of zinc phosphide ordinarily develop shyness in merion gerbils towards *bajra* for 35 days and addition of urine (on 6/12 Days) not only masked the shyness amongst them but also enhanced its palatability (Kumari & Prakash, 1980). Prakash and Jain (1971) observed that after a single exposure of sub lethal dose of zinc phosphide to *T. indica*, the bait shyness persisted for 115 days. However, even after one day exposure to urine mixed millet (in which zinc phosphide was added on day 4 and 5) the bait shyness totally vanished and on day 7, 8 and 9 the intake of millet was normalized as observed prior to exposure of poison bait (Kumari & Prakash, 1984).

Above findings opens a new vista for evolving highly compatible rodenticidal baiting technology. The scent marking gland exudates have shown greater promises as phagostimulant for making the poison baits more attractive and palatable besides having the properties of masking the poison shyness /aversion in the pest rodents. These findings require further exploration in laboratory and field for making them more practical. In future, identification and chemical synthesis of the active ingredient of these exudates would certainly pave a new era of rodent pest management in the field.

ACKNOWEDGEMENTS

The author is grateful to Dr. Pratap Narain, Director, Central Arid Zone Research Institute, Jodhpur for providing facilities and encouragement. Thanks are also due to Dr. R.S. Tripathi, Project Co-ordinator, AICRP on Rodent Control, CAZRI, Jodhpur for critical editing and for helpful suggestions in preparation of the manuscript.

REFERENCES

Balakrishnan, M. & Alexander, K.M. 1985. Source of body odour and olfactory communication in some Indian mammals. *Indian Rev.Life Sci.* 5: 277-313.

Bhardwaj, D. & Prakash, I. 1979. Association of bait shyness with the component of poisoned food in *Rattus rattus*. *Z. Angew. Zool.* 66: 329-335.

Blum, S. & Thiessen, D.D. 1971. The effect of different amount of androgen on scent marking in the male Mongolian gerbil. *Horm. Behav.* 2: 93-105.

Bronson, F.H. 1974. Pheromonal influences on reproductive activities in rodents. In "*Pheromone* (Ed. Birch, M.C.) Amsterdam, North Holand Publishing Company. 344-365.

Bruce, H.M. 1959. An exteroreceptive block pregnancy in the mouse. *Nature* (London) 184, 105.

Daly, M. 1977. Some experimental tests of the functional significance of scent marking by gerbils (*Meriones unguiculatus*). *J. Comp. Physiol. Psychol.* 91: 1082-1094.

Dominic, C.J. 1969. Pheromonal mechanisms regulating mammalian reproduction. *Gen. Comp. Endocr.* Suppl. 2: 260-267.

Eibl-Eibesfeldt, T. 1951. Gebangenschaftsbeobachten ander Persischen wustenmaus (*Meriones persicus persicus* Blanford): Ein Beitrag Zur Vergleichenden Ethologie der Nagr. *Z. Tierpsychol.* 8: 400-423.

Eibl-Eibesfeldt, T. 1953. Zur Ethologie des hamsters (*Cricetus cricetus* L.) *Z. Tierpsychol.* 10: 204-254.

Eisenberg, J.F. 1963. The behaviour of Heteromyid rodents. Univ. California Publ. Zool. 69: 1-114.

Eisenberg, J.F. 1967. A comparative study in Rodent Ethology with emphasis on evolution of social behaviour. *I. Proc. U.S. Nat. Mus.* 122: 1050.

Ewer, R.F. 1968. *Ethology of mammals.* New York: Plenum Press.

Ewer, R.F.1973. *The carnivores.* Ithaca: Cornell University Press.

Glenn, M. & Gray, J. 1967. Effect of various hormones on the growth and histology of the gerbil (*Meriones unguiculatus*) abdominal sebaceous gland pad. *Endocrinol.*, 76: 1115-1123.

Idris, M. 1985. Structure and functions of ventral scent marking gland of the Indian gerbil, *Tatera indica indica* Hardwicke, and the role of odours in behavioural communication. Ph.D., Thesis, Jodhpur Univ., Jodhpur. pp. 1-208.

Idris, M. 1994. Behavioural responses of the Indian desert gerbil, *Meriones hurrianae* towards conspecific and interspecific sebum odour. *Annals of Arid Zone.* 33 (2): 137-141.

Idris, M. & Prakash, I. 1982. Behaviour of the Indian gerbil, *Tatera indica* to conspecific sebum of the ventral scent marking gland. *Indian Acad. Sci.* 9 (3): 259-265.

Idris, M. & Prakash, I. 1983. Response of male Indian gerbil, *Tatera indica* to urine from female *Indian J. exptl. Biol.* 21: 97-98.

Idris, M. & Prakash, I. 1985. Scent marking in the Indian gerbil, *Tatera indica* in response to conspecific odours. *Proc. Indian Acad. Sci.* 95 (1): 89-96.

Idris, M. & Prakash, I. 1985a. Social and scent marking behaviour in Indian gerbil, *Tatera indica. Biol. Behav.* 10: 31-39.

Idris, M. & Prakash, I. 1985b. Hormonal control of scent marking behaviour in Indian gerbil. *Acta Theriol.* 30 (16): 267-274.

Idris, M. & Prakash, I. 1986. Influence of odours of Indian gerbil, *Tatera indica* on the social and scent marking behaviour of sympatric desert gerbil, *Meriones hurrianae. Proc. Indian natn. Sci. Acad.* B 52 (3): 333-340.

Idris, M. & Prakash, I. 1987. Scent marking behaviour and social organisation in Indian soft-furred field rat, *Rattus meltada. Acta Theriol.* 32 (19): 315-325.

Idris, M. & Prakash, I. 1987a. Scent marking activity in the Indian gerbil. *T. indica* in relation to population density. *Anim. Behav.* 35: 920-921.

Idris, M. & Prakash, I. 1988.Perception of dominance status through conspecific scent in the desert gerbil, *Meriones hurrianae.* J. *exptl .Biol.* 26: 479-480.

Idris, M. & Prakash, I. 1991. Scent marking behaviour in desert gerbil, *Meriones hurrianae* in relation to social hierarchy .*Proc. Indian natn. Sci. Acad.* 57 (3&4): 263-266.

Jain, A.P. 1970. Body weights, sex ratio, age structure and some aspects of reproduction in the Indian gerbil, *Tatera indica indica* Hardwicke in the Rajasthan desert, India. *Mammalia* 34: 415-432.

Johnson, R.P. 1973. Scent marking in mammals. *Anim. behav.* 21: 521-535.

Jones, R.B. & Howell, H.W. 1973. The coagulating glands as a source of aversive and aggression inhabiting pheromone (s) in the male albino mouse. *Physiol. Behav.* 11: 455-462.

Kittrell, E. Melaine, Greg, B.R. and Thiessen, D.D. 1982. Brood patch function for the ventral scent gland of the female Mongolian gerbil, *Meriones unguiculatus. Dev. Psychobiol.* 15: 197-202.

Kumari, S. 1982. The structure and functional significance of the ventral scent marking gland in the three rodent species (*M. hurrianae, T. indica, and R. meltada*). Ph.D. Thesis, pp. 1-197. Univ. of Agra, India.

Kumari, S., Cowan, P.E. & Prakash, I. 1981. The mid ventral gland of the Indian desert gerbil, *M. hurrianae* Jerdon. *Acta Theriol.* 26 (6): 97-106.

Kumari, S. & Prakash, I. 1979. Conspecific odour as a phagostimulant for the Indian gerbil, *Tatera indica indica* Hardwicke. *Indian J. Exptl. Biol.* 17: 981-982.

Kumari, S. & Prakash, I. 1980. Role of conspecific urine in enhancing food consumption and masking shyness behaviour in the desert gerbil, *Meriones hurrianae* (Jerdon). *Indian J. Exptl. Biol.* 18: 730-732.

Kumari, S. & Prakash, I. 1981. Behavioural responses of *Meriones hurrianae* (Jerdon) to conspecific sebum of ventral sebaceous gland. *Biol. Behav.* 61: 255-263.

Kumari, S. & Prakash, I. 1981a. Observations on the social behaviour of the Indian desert gerbil, *Meriones hurrianae. Indian Acad. Sci. (Anim. Sci.).* 90: 463-471.

Kumari, S. & Prakash, I. 1981b. Scent marking behaviour of *Meriones hurrianae* during oestrus. *Anim. Behav.* 29: 1269-1271.

Kumari, S. & Prakash, I. 1983. Seasonal variation in the dimension of scent marking gland of three desert rodents and its possible relationship of their reproductive performance. *Proc. Indian Acad. Sci.* 92: 299-304.

Kumari, S. & Prakash, I. 1983a. Mid ventral marking gland of *Tatera indica* and *Rattus meltada pallidior. Acta Theriol.* Poland 28: 425-434.

Kumari, S. & Prakash, I. 1984. Association between scent marking density and dominance in the Indian gerbil, *Meriones hurrianae*. *Indian J. Exptl. Biol.* 22: 421-423.

Kumari, S. & Prakash, I. 1984a. Perception of oestrous odours by females. *Anim. Behav.*, London. 32: 927-929.

Leon, M. 1975. Dietary control of maternal pheromone in lactating rat. *Physiol. Behav.* 14: 311-319.

Marton, J. 1962. Effect of castration and androgen replacement on the supracaudal gland of the male guinea pig. J. *Morphol.* 110: 285-298.

Michael, R.P., Keverne, E.B. and Bonsall, R.W. 1971. Pheromones : Isolation of male sex attractants from a female primate. *Science.* 172: 964-966.

Montagna, W. and Hamilton, J.B. 1949. The sebaceous glands of the hamster. II. Some cytochemical studies in normal and experimental animals. *Am. J. Anat.* 84: 365-395.

Moy, R.F. 1969. A study of fine skin glands of the pronghorn (*Antilocapra americana*). M.Sc. Thesis. Univ. Montana, Missoula.

Muller-Schwarze, D. 1983. Scent glands in mammals and their functions. In. *Advances in the study of mammalian behaviour.* (Eds.) Eisenberg, J.F. and Kleiman, D.G. Plenum Press, New York. Spl. Publ. Soc. 7: 150-197.

Mykytowycz, R. 1962. Territorial function of the chin gland secretion in the rabbit, *Oryctolagus cuniculus* (L). *Nature* 193-199.

Mykytowycz, R. 1965. Further observation on the territorial function and histology of the submandibular cutaneous (chin) glands in the rabbit, *Oryctolagus cuniculus. Anim. Behav.* 12: 400-412.

Mykytowycz, R. 1966. Observations on odoriferous and other glands in the Australian wild rabbit, *Oryctolagus cuniculus* (L) and hare *Lepus europaeus* (P), CSIRO *Wildl. Res.* 11: 11-29.

Mykytowycz, R. 1968. Territorial marking by rabbits. *Sci. Am.* 218, 116-126.

Mykytowycz, R. 1970. The role of skin glands in mammalian communication.In : *Advances in chemoreception.*(Eds : J.W. Johnson, D.G. Mourton and A. Turk), New York : Appleton Century Crafts.

Prakash, I. 1981. Ecology of the Indian desert gerbil, *Meriones hurrianae. Monograph* No. 10.CAZRI, Jodhpur. pp. 1-87.

Prakash, I. 1986. Faunal diversity in the arid tracts. Proc. *Indian Acad. Sci.* B. sp no. , 45-58.

Prakash, I. 1993. Food, feeding and shyness behaviour of the gerbils of the Thar desert. *Proc. Indian natl. Sci. Acad.* B59: 31-39.

Prakash, I. & Idris, M. 1982. Scent marking by the female Indian gerbil, *Tatera indica* from two distinct habitat during oestrus. *Indian J. exptl. Biol.* 20: 915-916.

Prakash, I. & Idris, M. 1982a. Response to conspecific odour of the scent marking gland by *Rattus meltada. Z. Angew. Zool.* 69: 481-486.

Prakash, I., Idris, M. & Soni, G.R. 1988. Scent marking behaviour of wild and laboratory bred desert gerbil, *Meriones hurrianae. Proc. Indian natn. Sci. Acad.* 54(1): 31-34.

Prakash, I., Idris, M. & Kumari, S. 1998. Scent marking behaviour of Three desert rodents. *PINSA* B 64(5&6): 319-334.

Prakash, I., & Jain, A.P. 1971. Bait shyness of two gerbils, *Tatera indica* and *Meriones hurrianae. Ann. Applied Biol.* 69 (2): 160-162.

Prakash, I., & Kumari, S. 1979. Occurrence of ventral marking gland in desert rodents. *Saught. Mitteil.* West Germany 27: 315-316.

Prakash, I., & Ojha, P. 1978. The phenomenon of poison aversion and its possible altenuation in Indian gerbil, *Tatera indica indica* (Hardwicke). *Z. angew Zool.* West Germany 65: 385-395.

Prakash, I., Rana, B.D. & Jain, A.P. 1975. Bait shyness in three species of *Rattus. Z. angew. Zool.* 62: 89-97.

Quay, W.B. 1953. Seasonal and sexual differences in the dorsal skin gland of Kangaroo rat (*Dipodomys*). J. *Mammal.* 34: 1-14.

Quary, W.B. & Tomich, P.O. 1963. A specialized mid ventral sebaceous glandular area in *Rattus exulans*. J. *Mammal.* 44: 537-542.

Ralls, K. 1971. Mammalian scent marking. *Science*: 443-449.

Rana, B.D. & Prakash, I. 1984. Reproduction biology of *Rattus meltada pallidior* (Ryley 1914) in the Rajasthan desert. J. *Bombay nat. Hist.* Soc. 81: 59-70.

Schaffer, J. 1940. Die Hautdrusen organe der Saugetiere. *Urban and Schwarz-enberg*, Berlin, 464 pp.

Singer, A.G., Agosta, W.C. O'Connell, R., Pfaffman, C., Bowen, D.V. & Field, F.H. 1976. Dimethyl disulphide : An attractant pheromone in hamaster vaginal secretion. *Science* 191: 948-950.

Soni, G.R. & Prakash, I. 1988. Mitigation of poison shyness in desert gerbil by adding conspecific sebum of ventral scent marking gland and urine in poison baits. *Indian J. exptl. Biol.* 26: 476-478.

Stoddart, D.M. 1972. The lateral scent organs of *Arvicola terrestris* (Rodentia : Microtinae). J. *Zool.* London, 166: 49-54.

Thiessen, D.D. 1968. The root of territorial marking in the Mongolian gerbil : A problem of species – common topography. *Behav. Res. Meth. Instru.* 1, 70-76.

Thiessen, D.D. 1973. Foot holds for survival. *Amer. Scientist* 61: 346-351.

Thiessen, D.D., Blum, S. & Lindzey, G. 1969. A scent marking response associated with the ventral sebaceous gland of the Mongolian gerbil (*Meriones unguiculatus*). *Anim. Behav.* 18: 26-30.

Thiessen, D.D., Friend, H.C. & Lindzey, G. 1968. Androgen control of territorial marking in the Mongolian gerbil. *Science* 160: 432-434.

Thiessen, D.D., Lindzey, G. & Nyby, J. 1970. The effects of olfactory deprivation and hormones on territorial marking in the male Mongolian gerbil (*Meriones unguiculatus*). *Hormones Behav.* 1: 315-325.

Thiessen, D.D., Wallace, P. & Yahr. P. 1973. Comparative studies of glandular scent marking in *Meriones tristrami*, an Israel gerbil. *Horm. Behav.* 4: 143-147.

Vandenbergh, J.G., Whitsett, J.M. & Lombardi, J.R. 1975. Partial isolation of a pheromone accelerating puberty in female mice. J. *Reprod. Fert.* 43: 515-523.

Von-Schumacher, S. 1936. Das stirnorgun des Rh (*Capreolus capreolus* L) ein bisher unbekanntes Dufforgan. *Z. Mikrosk. Anat. Forsch.* 39: 215-230.

Wallace, P., Owen, K. & Thiessen, D.D. 1973. The control and function of maternal scent marking in the Mongolian gerbil (*Meriones unguiculatus*). *Physiol. Behav.* 10: 463-466.

Whitten, W.K. 1956. Modifications of the oestrous cycle of the mouse by external stimuli associated with male. J. *Endocrinol.* 13: 399-404.

Wynne Edwards, V.C. 1962. *Animal Dispersion in relation to social behaviour.* (Edinburg : Oliver and Boyd) pp. 653.

Changing Faunal Ecology in Thar Desert, pp. 289-304, 2005.
Eds.: B.K. Tyagi and Q.H. Baqri

14

REPRODUCTION IN DESERT RODENTS

R.S. TRIPATHI

Central Arid Zone Research Institute, Jodhpur-342 003 (Rajasthan), India

INTRODUCTION

The Indian desert exhibits an admixture of peninsular, extra-peninsular and Indo-Gangetic features. Therefore, like other great deserts of the world, Thar is not an endless stretch of sand dune bereft of life or vegetation. The hot arid region in north India extends from Punjab to Gujarat through Rajasthan. The sandy tract is well interspersed with hills and the dominating Aravalli ranges are a distinct landmark of the Indian desert. A few large lakes, small *nadis* and a number of ephemeral and perennial streams occurring in the desert constitute the invaluable surface water resources of the region. The climate of the Indian desert is characterized by high atmospheric temperature and low and erratic rainfall. More than 90% of annual precipitation occurs during the monsoon months of July-September. The annual rainfall varies from 80-500mm. The lowest and highest temperatures recorded are -4.4^0 C and 50.5^0 C, respectively. The mean annual relative humidity of the air varies from 26-60 per cent. Winds are strong and for most parts of the year wind direction is southwest to northwest. It is very rich in floral and faunal biodiversity. Biogeographically, the majority of plants and animals found in the Indian desert exhibit 'western' affinities. The animal life in Thar Desert is quite exciting. Ghosh (1996) has reported that more than 50 mammals inhabit the region. Among these about one third (18 species) belonging to order Rodentia occur in this region (Tripathi *et al.*, 2002). Similarly other mammalian groups are also well represented in the desert. For example, Chiroptera, Carnivora, Artiodactyla and Insectivora are represented by 18, 16, 5 and 4 species, respectively (Alfred and Agrawal, 1996).

Table 1. Rodent species occurring in arid zone (adapted from Tripathi *et al.*, 1992)

Family / sub family	Species
(a) Family Hystricidae :	(1) *Hystrix indica* Kerr
(b) Family Sciuridae :	(2) *Funambulus pennanti* Wroughton
(c) Family Muridae	
(i) Sub-family Gerbillinae	(3) *Tatera indica* (Hardwicke)
	(4) *Meriones hurrianae* (Jerdon)
	(5) *Gerbillus nanus indus* Thomas
	(6) *Gerbillus gleadowi* Murray
(ii) Sub-family-Murinae	(7) *Vandeleuria oleracea* (Bennett)
	(8) *Rattus rattus* (Linn.)
	(9) *R. cutchicus* (Wroughton)
	(10) *R.* (=*Millardia*) *meltada pallidior* (Gray)
	(11) *R. gleadowi* Gray
	(12) *Mus musculus* Linn.
	(13) *M. cervicolor* Hodgson
	(14) *M. platythrix* Bennett
	(15) *M. booduga* Gray
	(16) *Golunda ellioti* Gray
	(17) *Nesokia indica* (Gray and Hardwicke)
	(18) *Bandicota bengalensis* (Gray)

RODENT DIVERSITY

Rodents constitute one of the largest mammalian groups in the Indian desert. In the first exhaustive field study on ecological evaluation of desert rodents conducted in 11 arid districts of western Rajasthan, Prakash *et al.* (1971a) reported occurrence of 17 species in the region. The authors trapped 14 species and sub species belonging to families, scuiridae and muridae (subfamily murinae and gerbillinae). Besides, they recorded observations on three more species also, *viz.*, a hystricide (*Hystrix indica*) and two murids (*Rattus rattus* and *Nesokia indica*). In later years, Prakash & Mathur (1979) and Idris & Rana (2001) recorded occurrence of lesser bandicoot rat, *Bandicota bengalensis* inhabiting the irrigated crop fields in Sri Ganganagar and Pali districts and urbane locales of Bikaner and Jodhpur cities. Tripathi *et al.* (1992) had listed 18 rodent species occurring in Indian arid zone (Table 1). Among these 4 species viz., *Gerbillus nanus indus, G. gleadowi, Meriones hurrianae* and *Rattus gleadowi* are essentially denizens of arid zone and can be referred as true xcric fauna. Rodent species like, *Rattus cutchicus cutchicus* and *Mus cervicolor phillipsi* inhabit the rocky areas in the desert. Similarly the species viz., *Funambulus pennanti, Tatera indica indica, Rattus meltada pallidior, Mus booduga booduga , Golunda ellioti gujerati, Mus platythrix, Bandicota bengalensis, N. indica and H. indica,* reported from arid zones are also widely distributed in humid, sub humid and semi arid regions of the country. Similarly, the two most common commensal rodents, *Rattus rattus* and *Mus musculus* are also well distributed in the similar habitats of this region also. *M musculus*, no doubt have been reported from cotton fields in Sri Ganganagar district in recent years showing atavism in its habitat.

Table 2. Habitat preference of different rodent species occurring in Indian desert

Habitats	No. of species	Name of species
(a) Exclusive to habitat		
Sandy	One	*G. nanus indus*
Rockey	Two	*R. cutchicus* and *M. cervicolor phillipsi*
Ruderal	Three	*M. musculus, M. booduga* and *B. bengalensis*
(b) Habitat sharing		
Rockey and Ruderal	Four	*F. pennanti, T. indica indica, R. meltada pallidior,* and *M. platythrix sadhu*
Sandy and Gravel	Six	*T. indica indica, M. hurrianae, R. meltada pallidior, M. platythrix sadhu, R. gleadowi* and *G. ellioti gujerati*
Gravel and Ruderal	Five	*T. indica indica, M. hurrianae, R. meltada pallidior, M. platythrix sadhu* and *G. ellioti gujerati*
Rockey and Ruderal	Four	*F.pennanti, T. indica indica, R. meltada pallidior* and *M. platythrix sadhu,*
Sandy and Ruderal	Seven	*F.pennanti,G. gleadowi, T. indica indica, R. meltada pallidior, M. platythrix sadhu, M. platythrix sadhu* and *N.indica*

Most of these rodents are fossorial in habit and make extensive burrows in the soil, except the squirrels, *F. pennanti* and a tree mouse, *Vandeleuria oleracea,* which are exclusively arboreal animals. Cutch rock rat prefers to live in the cracks and crevices in the rocky habitats. About 70 per cent of the species are nocturnal in their activity pattern, whereas about 18 per cent are diurnal or crepuscular in nature. *R. rattus* and *M. musculus* are basically nocturnal species but are also active in the day time. On the basis of ecological distribution of rodents, Prakash *et al.* (1971a) have classified the desert biome in four habitats, viz., sandy, rocky, gravel and ruderal (village complex, residential areas, hutments and crop fields) habitats. Rodents occur in all these habitats in varying numbers. Sandy and ruderal habitat supports more number of rodents than the gravel or rocky habitats, however in ruderal habitats maximum numbers of rodent species have been reported. The desert rodents exhibit some degree of habitat specificity (Prakash *et al.*, 1971a, Prakash & Rana, 1973, Prakash, 1975 and Tripathi *et al.*, 1992). For example, the species like, *M. booduga, M. musculus, R. rattus, and B.. bengalensis* are found exclusively in ruderal habitats; *R. cutchicus* and *M. cervicolor phillipsi* in rocky habitats and *Gerbillus* species in sandy habitats. Occurrence of *N. indica* have been reported from afforestation sites and irrigated crop lands in arid zones (Jain and Tripathi, 1988; Jain *et al.* 1995 and Idris *et al.* (2003) and *H. indica* from rocky habitats in Jodhpur district (Idris & Rana, 2001). Other rodents are found in more than one habitat. On the basis of their percent occurrence (preference) these rodents may be assigned a particular habitat (Table 2). An overall analysis of habitat preference of rodents indicated that four species share rocky and sandy habitats, six species sandy and gravel habitats, five species gravel and ruderal and four species share rocky and ruderal and seven species share sandy and ruderal habitats together (Table 2).

The faunal diversity of rodents in Indian arid zone indicates that *M. hurrianae, T. indica, R. meltada* and *F. pennanti* are the most predominant species in wild/field situation and *R. rattus* and *M. musculus* are predominant under commensal conditions. In fact these species have acquired pest status, because they cause colossal losses to all major production systems of arid zone (agriculture, horticulture, forestry, grasslands etc.). With the incoming of canal irrigation through Gang canal, Ghagghar canal and Indira Gandhi Canal, the land use pattern of the arid zone is undergoing a quick transformation. With the result vast stretches of undulating sand dunes and degraded sandy scrublands are being converted in croplands. These changes are leading to shifts in faunal diversity of rodents also. In the extensively irrigated crop fields in Sri Ganganagar district, xeric rodents, like *Gerbillus* and *Meriones* spp. have been driven away to drier areas and are replaced first by submesic species, *R. meltada* and latter by mesic species, viz., *B. bengalensis* and *N. indica.* Thus, diversity of rodents represented by 12 murid and one scuirid fauna recorded from true desert biome in sandy and ruderal habitats by Prakash *et al.* (1971a) have been reduced to nine in irrigated crop fields of Sri Ganganagar district (Tripathi *et al.*, 2002). Besides decrease in the biodiversity, the new intruders are more destructive and may pose serious public health hazards (Prakash, 1991; Jain *et al.*, 1996).

REPRODUCTION

Reproduction is the main reason for increase in size of population, which is typically true for rodent species that are capable of rapid population growth especially when the conditions are favourable (Fig. 1). It has been observed that most of the tropical mammals breed during winter season, whereas in this desert, monsoon is considered as most conducive season for reproductive activities. A rich faunal diversity of 18 species in this region reflects their sustenance capabilities under various natural stresses of the desert biome. These animals are morphologically, physiologically and behaviourally well adapted to face the abiotic and biotic adversaries of desert and maintain their population in fairly good numbers. Increase in population of desert rodents is generally governed by a combination of two factors, *viz.*, high reproductive potential and a shorter period of sexual maturation. Most of the rodent species of the region breed round the year with certain peak periods in breeding, commonly refered as breeding seasons, which largely depends on availability of nutritious food in plenty. The desert region show rich floral diversity, particularly in monsoon season (July-September) with lush green vegetation all over containing an inexhaustible source of foraging material for the very large number of herbivores, domestic as well as wild ones including rodents. Availability of green food in plenty during this season triggers the breeding activity of native rodents. After winters, again with the advent of spring emerges a fresh crop of sprouts. This period also records increased breeding activity in certain rodent species. During summers, which are really a tough time for desert fauna, the breeding in majority of desert rodents attains lowest ebb.

Fig. 1. High breeding potential of rodents.

Detailed information on various aspects of reproduction are available for only a few species, viz., *M. hurrianae, T. indica* and *R. meltada* (Ghosh & Taneja, 1968; Prakash, 1969; Kaul & Ramaswamy, 1969; Jain, 1970; Prakash *et al.*, 1971b; Prakash, 1981; Rana & Prakash, 1984). Besides this, reports on breeding season and litter size are available for a number of rodent species inhabiting the arid zone (Prakash, 1971, 1974; Purohit *et al.*, 1966; Prakash & Jain, 1971; Prakash & Purohit, 1967; Prakash *et al.* 1973; Prakash & Mathur, 1987; Tripathi *et al.* 1992). The Information available on reproduction and breeding biology of desert rodents are being detailed in the following pages.

(i) Oestrous cycle

On the basis of average period of interval between vaginal cornification, Ghosh and Taneja (1968) worked out the oestrous cycle for *M. hurrianae* and *T. indica* to be 6.22 and 3.82 days. The authors, after adjusting the maximum probable duration of cornification (1 day), said that average oestrous cycle for *M. hurrianae* and *T. indica* be regarded as 7.22 and 4.82 days. In another study at Jaipur, Kaul and Ramaswamy (1969) reported that in *M. hurrianae* the duration of proestrous varies from 0.5 to 3.0 days (Av. 1.4 days); oestrous 0.5 to 3.0 days (Av. 1.7 days); metoestrous 0.35 – 3.0 days (Av. 1.1 days) and dioestrous: 0.5-10 days (Av. 1.9 days). Male fecundity in *T. indica indica* is mostly exhibited when the animals reach a body weight of about 105 g, however, a few animals weighing as low as 84 g possessed sperms in epididymis (Jain, 1970). In case of *R. meltada pallidior*, the males acquire sexual maturity at a body weight above 45 g. At higher body weights of 174g (in *T. indica*) and 70-73 g (in *R. meltada*), the animals exhibited no sperms in epididymis due to regression of testis, which is quite common in mammals (Rana & Prakash, 1984).

(ii) Mating behaviour

Information on mating behaviour of rodents is available for two most common diurnal species, viz., *M. hurrianae* and *F. pennanti.* According to the reports of Prakash (1981) the female and male desert gerbils are engaged in mating for about 10-40 minutes with actual coitus lasting for 5-10 minutes only. The female assumes the lordosis position during mating and allows the male to mount. The mating consists of multiple mounts (13-21 times) in a series. Both the sexes at times lick their genitals after coitus. *F. pennanti* also exhibits almost similar coitus behaviour (Banerjee, 1955). Here extensive chasing is major part of courtship and the coitus lasts for 10-20 minutes Tails of both the partners twine and untwine several times during copulation.

Table 3. Prevalence of pregnancy in some rodent species occurring in Indian desert

Rodent Sp.	Jan	Feb	Mar	Apr	May	Jun	Jul	Aug	Sep	Oct	Nov	Dec
F. pennanti	0.0	38.7	70.0	31.2	32.5	16.2	42.5	17.5	10.0	0.0	0.0	0.0
T. i. indica	26.6	47.4	30.0	12.5	16.6	41.1	46.6	61.0	9.7	10.5	38.8	15.7
M. hurrianae	7.6	24.2	20.8	20.0	12.0	11.4	21.0	16.6	18.2	26.6	20.0	9.3
R. cutchicus	-	-	50.0	-	45.5	-	5.9	88.9	-	77.0	-	0.0
R. m. pallidior	31.0	51.5	90.0	75.0	62.5	55.0	100.0	100.0	100.0	83.0	33.0	31.0

(iii) Gestation period

Data on gestation period is available for only a few species of desert rodents. The squirrel, *F. pennanti* exhibits a very long gestation period of 42.0 days (Banerjee, 1957). This period for the two most predominant gerbil species, *T. indica* and *M. hurrianae* was found to be 27-30 days (Av. 28.27 days) and 28-30 days (Av. 29 days), respectively (Prakash, 1964; Prakash *et al.*, 1971c). *R. meltada pallidior* recorded a gestation period of 20 days (Bindra & Sagar, 1968). On the basis of the period between two deliveries in zoo, Prakash (1968) made a rough estimate for the gestation period of Indian crested porcupine, *H. indica* as 109 days.

(iv) Breeding season

A majority of desert mammals litter between July to September, while the number of species giving birth to young ones is low during extreme winters. The species, which breed year round, too show maximum prevalence of pregnancy during rainy season (Table 3). These peaks coincide with the nutritional status of green food and moderate temperature with higher relative humidity during this period The overall health of the mother in this season remains better to sustain the stress of lactation and hence survival of young ones is enhanced greatly. Moreover, when the young ones are weaned, the green food is also available for new recruits in the population. Species wise account of breeding season is described as under:

(a) Funambulus pennanti: The squirrels are reflex ovulators. A mature female may produce two litters in one breeding season. The period of suckling their young ones is six weeks and the young ones grow to full size in four months. It breeds from March to September in arid and semi arid areas of Rajasthan (Purohit *et al.*, 1966). In another study, Prakash (1971, 1975) reported two breeding peaks, one during March-April and other during July-September (Table 4). Agarwal (1965) noted a case of pregnancy with five young ones during December also. Seth and Prasad (1969) also reported similar breeding patterns in palm squirrels from Delhi.

(b) Gerbillus species: Both the species occurring in Rajasthan desert, viz., *G. nanus indus* and *G. gleadowi* have two breeding seasons in a year i.e. winter and summer months (Prakash & Jain, 1971; Prakash & Purohit, 1967). The authors recorded high prevalence of pregnancy during April, June and December for *G. nanus Indus* and during May-June and October-January for *G. gleadowi* (Table 4). The breeding season of this species was quite paradoxical as it recorded peak breeding in summer months, when most other species maintain a low reproductive profile. It indicates that both the truly xeric gerbil species are well acclimatized for the desert conditions and the young ones borne during summers are initially suckled by mothers inside the burrows for

4-6 weeks and by the time monsoon set in they are fully grown and ready to move out for foraging.

Table 4. Peak breeding season and litter size of rodent pests (adapted from Tripathi *et al.*, 1992)

Rodent species	Peak breeding season	Litter size	References
H. indica	Monsoon and in December (zoo)	1-3	Prakash (1975)
F. pennanti	March to September	1-5	Purohit *et al.*, (1966)
	March-April and July-September		Prakash (1975)
G. gleadowi	May, June and October to January	2-5 (summer)	Prakash and Prakash (1967)
		5-6 (winter)	
G. nanus indus	April, June and December	2-3	Prakash and Jain (1971)
T. indica	February, July- August and November	1-9	Jain (1970)
M. hurrianae	(i) February to April	2-7	Prakash (1964)
	(ii) July and September to November	2-7	Kaul and Ramaswamy (1969)
R. gleadowi	August-October	2-3	Prakash (1975)
G. ellioti gujerati	March-August	5-10	Prakash (1975)
R. rattus	April-September	1-9	Prakash (1975)
R. meltada pallidor	Monsoon and Spring	3-9	Rana & Prakash (1984)
M. musculus	All the year round	1-8	Mann (1969)
M. booduga booduga	September-October and February and June	6-13	Srivastava (1968)
N. indica	January-March and August-October	2-5	Gariyali and Saxena (1975)

(c) Meriones hurrianae: Females were reported to litter round the year, however, two breeding peaks were noticed by Prakash (1964) during February and July. Kaul and Ramaswamy (1969) observed three peaks in pregnancy during February to April, July and September to November at Jaipur. Prakash (1981) observed that the gerbil passes the cold spell during December and January in a state of torpidity, which breaks in February and sexual activity is initiated leading to peak in littering during February (Table 4). Similarly peak in July was correlated with the availability of green food and mean monthly day length. The reproductive activity is low during winter, when day length is shortest, apparently due to the ovarian, refractoriness to photo periods. It has also been mentioned by Prakash (1981) that during drought years the breeding potential of *M. hurrianae* is decreased considerably.

(d) Tatera indica: Indian gerbils inhabiting the crop fields, grasslands etc. did not show year round breeding (Prakash, 1975). Jain (1970) has studied the breeding biology of this gerbil species in detail. He found pregnant fcmalcs in thc population round the year. The percentage of pregnancy ranged from 9.7-61.0 per cent (Av. 29.70). Maximum reproductive activity was noticed during February, July-August and November (Table 4). These observations were recorded on gerbils occurring in Bikaner town, where the animals get ample food and safe shelter round the year and the impact of climatic changes are minimal in comparison to open fields. The males in population residing in urban locales of Bikaner too recorded fecundity round the year. It tends to indicate that physiological conditions of males may determine the breeding intensity of this species.

(e) Rattus meltada pallidior: It is a sub-mesic species and inhabits irrigated crop fields and grasslands in western Rajasthan. Two year long study by Rana and Prakash (1984) revealed that this species also breeds throughout the year. The period of maximum breeding activity was reported during March-September, which coincides with larger litter size. The winter months recorded least pregnancy rates, which attain a second peak during spring season. Thus like most of the species, this species also show maximum breeding activity during monsoon and spring season (Table 4). This species have been reported to show maximum breeding during March to October is U.P. and Punjab (Srivastava, 1968; Bindra & Sagar, 1968; Guraya & Gupta, 1975).

(f) R. cutichicus cutchicus: Monthly collections from rocky habitats revealed that the pregnant rock rats were collected maximum during monsoon (August) which declined to zero level during extreme winters (December). Interestingly a peak was observed in March (Table 4) and then the female fertility decreased again to 5.9 per cent in July, which was quite paradoxical as most of the desert rodents record maximum breeding in this month. The male fecundity was reported round the year, though at a reduced rate from October-December, however, the male fecundity index was higher (83.8-100%) from March to August (Prakash *et al.*, 1973).

(g) Other rodents: Data on breeding season of other rodents occurring in Rajasthan desert are quite scanty. *Rattus gleadowi* and *Golunda ellioti gujerati* were reported to breed from August-October and March to August (Prakash, 1975). Among the *Mus* species, prevalence of pregnancy in *M. cervicolor phillipsi* was observed to be higher during July (50%) and December (25%). Similarly *M. booduga* was found to record maximum breeding in October in Rajasthan desert, whereas it breeds during September-October, February and June in U.P. (Srivastava, 1968). This species was however reported to breed all through the year with low reproductive activity during March to July in South India. The commensal, *Mus musculus* also breeds round the year, however in Rajasthan it is least during summers and winters, thus recording a bimodal pattern of breeding season (Rao, 1992). Data from Jodhpur and Bikaner Zoo, indicated that the porcupine *H. indica* show peak breeding during monsoon followed by another peak during December (Prakash, 1974). Breeding biology of the two mole rat species, *N. indica* and *B.*

bengalensis which are relatively new inhabitants of Rajasthan desert is not yet studied in detail except a few observation of breeding in *N. indica* where littering was recorded during October-November months. Gariyali and Saxena (1975) reported that *N. indica* breeds from January to March and August to October (Table 4). Similarly *B. bengalensis* recently established in urban localities of Jodhpur city recorded year round breeding. Reports from other parts of the country too indicate considerably high percentage of pregnant bandicoot females in the population all through the year. In fields, two distinct breeding peaks have been observed during September – October and January- March.

Prakash (1975) while reviewing the population ecology of desert rodents concluded that, most of the rodent species of the region breed from March-September, although a few species breed all through the year. Generally minimum births are reported to occur during the winter season, and not during summers. This was quite contrary to the mammals of other desert regions of the world, where breeding season was reported in late winters in Sahara-Sind region; in late summers in Sudano-Deccan region and in spring in Irano-Turanian region (Bodenheimer, 1957). This indicates that breeding pattern of rodents of Indian desert is an admixture of breeding pattern observed in East Africa, Sudano-Deccanian and Irano-Turamain region (Prakash, 1975; Tripathi *et al.*, 1992).

(v) Litter size

Rodents are prolific breeders and in one delivery they give birth to many young ones. The litter size is highly variable, may go up to 19 young ones depending upon the species, animal concerned, population density and several other factors. In desert region the rodents regulate their population size depending on the future availability of food and shelter. During drought years the gerbils decrease their breeding activity considerably. Some species like *M. hurrianae, T. indica, G. gleadowi* etc. practice cannibalism when mothers feed on its own young ones in case of any possible natural calamity leading to paucity of food to their off springs. The truly xeric species viz., *G. nanus indus, G. gleadowi* and *M. hurrianae* were reported to show a litter size of 2-3, 2-6 and 1-9 respectively (Prakash, 1964, 1981; Prakash & Purohit, 1967; Prakash & Jain 1971). The litter size of *T. indica indica* ranged from 1-9 (Jain, 1970). Minimum litter size of 1-3 and 2-3 was noticed in case of *H. indica* and *R. gleadowi*, respectively (Prakash, 1975). *G ellioti gujerati* inhabiting the thorny bushes in arid zone recorded highest litter size of 5-10. *Nesokia indica* trapped from irrigated crop fields near Bikaner delivered 1-6 young ones per litter. The younger females produce smaller litter size. In laboratory bred *N. indica* the litter size was 1-2. In case of *R. meltada*, the litter size varied from 3-9 (Av. 5.69). The litter size of various rodents is detailed in Table 4.

(vi) Superfoetation

This phenomenon is quite common in rats, mice, rabbits, cattle and sheep, however, Rana and Prakash (1979), Jain (1970), and Rana and Prakash (1984) reported superfoetation in *Suncus murinus sindensis* (an insectivore) *and* two rodents, viz., *T. indica indica* and *R. meltada pallidior*, respectively. These workers reported repeat conception within 9-12 and and 12-15 days of pregnancy by *R. meltada* and *T. indica*, respectively. Cases of pregnancy in lactating mothers have also been reported in *T. indica*, which reveals that the gerbils may conceive just after delivery indicating a possibility of post-partum oestrus.

(vii) Embryonic mortality

Like other mammals embryonic mortality during pre as well as post implantation periods have been reported in desert rodents also. It has been observed that number of corpora lutea almost corresponds with number of embryos in the respective uterine horns in rodents. Pre-implantation losses may be judged by comparing the number of corpora lutea and the implanted embryos, whereas post implantation losses are judged by identifying the resorbed and mummified embryos in the uterus. The pre- and post-implantation losses, in *R. meltada* was to the extent of 13.3-36.8 and 1.55 per cent, respectively (Rana & Prakash, 1984) and Embryonic mortality in case of *T. indica* was around 20 per cent (Jain, 1970).

ANNUAL PRODUCTIVITY

The productivity is rarely achieved to its potential levels in wild populations due to intrinsic regulatory mechanisms and lower carrying capacity of the environment. On the basis of various reproduction characteristics, viz., breeding season, gestation period, prevalence of pregnancy, number of embryos, and pre and post implantation mortality etc., annual productivity of rodents (no. of young ones/female/year) have been worked out for some desert rodent species. Among three field rodent species, *R. meltada pallidor* recorded highest annual productivity of 52.52 young ones/female/year whereas it was 19.96 and 17.72 for *R. cutichicus* and *T. indica* (Rana & Prakash, 1984; Prakash *et al.*, 1973; Jain, 1970). The reason for higher productivity of *R. meltada* may be due to better carrying capacity of its habitat (irrigated crop fields and grassland) where the green food is available almost throughout the year. The commensal rodents, *R. rattus* and *M. musculus* had an annual productivity rates of 27.03 and 31 young ones/female/year (Rana *et al.*,1983; Rao, 1992).

POPULATION DYNAMICS

The seasonal fluctuations of rodent population in Indian desert have not been extensively investigated on a long-term, year-to-year basis, but some information on this aspect is available (Table 5) for the predominant rodent species (Tripathi *et al.*, 1992). The trapping data on *F.. pennanti* indicated a gradual decrease of its population from April to October (Prakash & Kametkar, 1969). But in case of *T. indica,* Jain (1970) noticed a broad pattern of monthly fluctuation with two minor peaks in the population, the first one during March-April and the other during August-September. The merion gerbil, *M. hurrianae* has been reported to show a population buildup during winters, which continues till spring and then their number declines during summer (Prakash *et al.* 1971b). The data on other rodent species did not record any particular trend but bimonthly collections revealed uniform catches for *R. m. pallidior* and maximum catches of *M. cervicolor phillipsi* and *M.latythrix sadhu* in July and March respectively (Prakash & Rana, 1970).

An overall comparison of seasonal fluctuations of predominant species revealed that (i) their number is lowest during the hot summer months due to hostile living conditions in the desert, and (ii) the population build up is during spring and soon after the monsoon. These two population build up periods coincide with that of highest breeding activity among desert rodents (Prakash & Mathur, 1987). Bimodal breeding patterns and corresponding increase in rodent population coincides with the cropping patterns of the region and exhibits serious economic concerns. The young ones borne during both the peak breeding periods (monsoon and spring seasons) become mature when the respective *kharif* and *rabi* crops are at flowering/maturity stage which is considered as highly vulnerable stage to rodent attack in large numbers.

Table 5. Monthly fluctuations in the number of various desert rodent species (after Prakash, 1975)

Rodent Sp.	Jan	Feb	Mar	Apr	May	Jun	Jul	Aug	Sep	Oct	Nov	Dec
F. pennanti	-	-	-	102	54	61	36	47	28	30	10	6
T. i. indica	30	39	51	53	63	69	55	74	78	42	40	43
M. hurrianae	-	-	300	-	-	131	-	170	-	-	-	385
R. cutchicus	-	-	49	-	26	-	47	15	-	32	-	23
R. m. pallidir	-	-	12	-	10	-	12	0	-	10	-	-
M. c. phillipsi	-	-	15	-	11	-	21	0	-	2	-	-
M.p. sadhu	-	-	21	-	9	-	11	1	-	1	-	-
G. e. gujerati	-	-	4	-	4	-	5	1	-	1	-	-

CONCLUSIONS

Success of a species in any particular habitat depends on its breeding potentiality *vis-a-vis* carrying capacity oh the habitat. The breeding potential of any species is affected by several factors viz., gestation period, liter size, period between delivery and next conception and the reproductive life of females. As explained above, all these factors are in favour of rodents indicating their vast breeding potentiality. Some rodent species attain sexual maturity at a very early age, due to rapid growth mainly during first 2-3 weeks after birth. Thus with a life span of 1-2 years a typical rodent residing in Rajasthan desert may attain puberty at an age of 6-16 weeks. The oestrous cycle lasts for 4-7 days and the females remain in heat up to 24 hours. They observe bimodal breeding pattern (two breeding peaks during monsoon and spring season). The litters containing 1-13 young ones are delivered after a gestation period of 20-30 days. The annual productivity too is quite high, *i.e.*, 17-52 young ones/female/year. After weaning the new recruits in the population face the challenges of the hostile climate and other adversaries of the desert region, including the predators. These small mammals have well adapted themselves for better survival in the desert. Fossorial and mainly nocturnal life protects them from high temperatures and natural predators. Strong and well-developed tympanic bulla helps in improved audio reception and the animal can discriminate the sounds of wing beats of a predatory/non predatory birds. The list of predators of desert rodents is quite long and includes reptiles (snakes, monitor lizards *etc.*), birds (shikra, eagle, owls, kestrel *etc.*) and mammals (cats, foxes, hedgehogs *etc.*). All these predators have a wide range of food habit and none is specific rodent feeder.

High breeding potentialities coupled with several ecological, behavioural and morphological peculiarities have made the desert rodents most successful mammal (perhaps next to *Homo sapiens?*).

REFERENCES

Agarwal, V.C., 1965. Observation on habits of five striped squirrel, *Funambulus pennanti* in Rajasthan. *J. Bengal nat. Hist. Soc.* 34: 76-83.

Alfred, J.R.B. and V.C. Agarwal, 1996. The Mammal Diversity of the Indian desert. In: *Faunal Diversity in the Thar Desert: Gaps in Research* (Eds. Ghosh, A.K., Baqri, Q.H. and Prakash, I.). Scientific Publishers, Jodhpur. Pp. 335-348.

Banerjee, A., 1955. The family life of five striped squirrel, *Funambulus pennanti* Wroughton. *J. Bombay nat. Hist. Soc.* 53: 261-265.

Banerjee, A., 1957. Further observations on family life of the five striped squirrel, *Funambulus pennanti* Wroughton. *J. Bombay nat. Hist. Soc.* 55: 336-343.

Bindra, O.S. and P. Sagar, 1968. Breeding habits of field rat, *Rattus meltada. J. Bombay nat. Hist. Soc.* 65: 477-481.

Bodenheimer, F.S., 1957. The ecology of mammals in arid zones. In Human and Animal Ecology. *Rev. Res. Arid Zone Research* 8: 100-137, UNESCO, Paris,

Gariyali, V. and R.N. Saxena, 1975. Some aspects of physiology of reproduction and control of fertility in male Indian bandicoot rat, *Nesokia indica. Proc. All India Rodent Seminar,* Ahmedabad, India. 133 p.

Ghosh, A.K., 1996. The Thar Desert Ecosystem. In *Faunal Diversity in the Thar Desert : Gaps in Research.* (Eds. Ghosh, A.K., Baqri, Q.H. and Prakash, I.). Scientific Publishers, Jodhpur. Pp. 1-18.

Ghosh, P.K. and G.C. Taneja, 1968. Oestrous cycle in the desert rodents, *Tatera indica* and *Meriones hurrianae. Ind. J. Exptl. Biol.* 6 : 54-55.

Guraya, S.S. and S. Gupta, 1975. Observation on the reproductive biology of soft furred field rat (*Millardia meltada*). *Proc. All India Rodent Seminar,* Ahmedabad. India. Pp. 70-75.

Idris, M. and B.D. Rana, 2001. Occurrence of the lesser bandicoot rat, *B. bengalensis* in Jodhpur town. *Rodent Newsletter.* 25(3-4): 9.

Idris, M. and B.D. Rana, 2001. Some observations on infestation of porcupine, *Hystrix indica* Kerr in forest nursery of arid region. *Rodent Newsletter.*25 (1-2): 5.

Idris, M., R.S. Tripathi, V. Chaudhary and B.D. Rana, 2003. Incidence of Indian field mice, *Mus booduga* Gray in Thar Desert. *In: Human Impact on Desert Environment* (Eds. Narain, P., Kathju, S., Kar, A., Singh, M.P. and Parveen-Kumar). Scientific Publishers, Jodhpur. Pp. 343-347.

Jain, A.P. and R.S. Tripathi, 1988. Occurrence and incidence of short tailed mole rat, *Nesokia indica* in the arid zone afforestation ecosystem. *Proc. Natl. Symp. Adv. Ecol. Zool.,* Jodhpur.

Jain, A.P., 1970. Body weight, sex ratio, age structure and some aspects of reproduction in the Indian gerbil, *Tatera indica indica* Hardwicke in Rajasthan desert, India. *Mammalia* 34: 415-432.

Jain, A.P., R.S. Tripathi and N. Patel, 1995. Influence of aridity on burrowing and other behavioural traits of *Nesokia indica* Gray. *Ann. Arid Zone* 34 (1): 67-69.

Jain, A.P., R.S. Tripathi, and B.D. Rana, 1996. Impact of ecological transformation on the rodent fauna of the Thar desert. In *Faunal Diversity in the Thar desert : Gaps in Research* (Eds. Ghosh, A.K., Baqri, Q.H. and Prakash, I.). Scientific Publishers, Jodhpur. Pp. 389-402.

Kaul, D.K. and L.S. Ramaswamy, 1969. Reproduction in the Indian desert gerbil, *Meriones hurrianae* Jerdon. *Acta Zologica.* 50: 233-248.

Prakash, I., 1964. Ecotoxicology and control of the Indian desert gerbil, *Meriones hurrianae* (Jerdon). II. Breeding season, litter size and postnatal development. *J. Bombay nat. Hist. Soc.* 61 : 142-149.

Prakash, I., 1968. Biology of the rodents of Rajasthan desert. *Proc. Symp. National Resources of Rajasthan,* Jaipur. Pp. 337-352.

Prakash, I., 1969. Ecotoxicology and control of Indian desert gerbil, *Meriones hurrianae* (Jerdon) V. Food preference in fields during monsoon. *J. Bombay nat. Hist. Soc.* 65: 581-589.

Prakash, I., 1971. Breeding season and litter size of Indian desert rodents. *Z. Angew. Zool.* 58: 442-454.

Prakash, I., 1974. Ecology of vertebrates of the Desert. In *Biogeography and ecology in India* (Ed. Mani, M.S.) Dr. Junk b.v. Publishers, The Hague. Pp. 369-420.

Prakash, I., 1975. The population ecology of the rodents of Rajasthan desert, India. In Rodents in desert environment (Eds. Prakash, I. and Ghosh, P.K.) Dr. W. Junk b.v. Publishers, The Hague. Pp. 75-116.

Prakash, I., 1981. Ecology of the Indian desert gerbil, *Meriones hurrianae*. CAZRI Monograph No. 10 (CAZRI, Jodhpur). 1-86 p.

Prakash, I., 1991. Rodent communities, epidimeological and economic considerations. In *Prospects of Indira Gandhi Canal Project*. (Eds. Abrol, I.P. and Venkateswarlu, J.). ICAR, New Delhi. 110-115.

Prakash, I., A.K., Gupta, A.P. Jain, B.D. Rana, and B.K. Dutta, 1971a. Ecological evaluation of rodent population in the desert biome of Rajasthan. *Mammalia*. 35: 384-423.

Prakash, I. and A.P. Jain, 1971. Some observation on Wagner's gerbil, *Gerbillus dasyrus indus* (Thomas). *Mammalia* 35: 614-628.

Prakash, I. and L.R. Kametkar, 1969. Body weight, sex and age factor in population of the Northern Palm Squirrel, *Funambulus pennanti* Wroughton. *J. Bombay nat. Hist. Soc.* 66: 99-115.

Prakash, I. and R.P. Mathur, 1979. *Bandicota bengalensis* in Bikaner town. *Rodent Newsletter* 3(2) : 12-13.

Prakash, I. and R.P. Mathur, 1987. *Management of rodent pests*. Indian Council of Agricultural Research, New Delhi. 133 p.

Prakash, I. and K.G. Purohit, 1967. Some observation on hairy footed gerbil, *Gerbillus gleadowi* Murray in the Rajasthan desert. *J. Bombay nat. Hist. Soc.* 63: 431-434.

Prakash, I. and B.D. Rana, 1970. A study of field population of rodents in the Indian desert. *Z. Angew. Zool.* 57: 129-136.

Prakash, I. and B.D. Rana, 1973. A study on field population of rodents in the Indian desert. III. Sand dunes in 100 mm rainfall zone. *Z. Angew. Zool*. 66: 31-41.

Prakash, I., A.P. Jain and K.G. Purohit, 1971c. Breeding and postnatal development of the Indian gerbil, *Tatera indica indica* Hardwicke in the Rajasthan desert. *Saug. Mittel*, 19 (4) : 375-380.

Prakash, I., B.D. Rana and A.P. Jain, 1973. Reproduction in cutch rock rat, *Rattus cutichicus cutichicus* in the *Indian* desert. *Mammalia* 37 : 457-467.

Prakash, I., G.C. Taneja and K.G. Purohit, 1971b. Ecotoxicology and control of Indian desert gerbil, *Meriones hurrianae* (Jerdon). VII. Relative numbers in relation to ecological factors. *J. Bombay nat. Hist. Soc.* 68 : 86-93.

Purohit, K.G., L.R. Kametkar and I. Prakash, 1966. Reproductive biology and postnatal development in the Northern palm squirrel, *Funambulus pennanti* Woughton. *Mammalia* 30: 538-546.

Rana, B.D. and I. Prakash, 1979. Reproductive biology and population structure of the house shrew, Suncus murinus sindensis in western Rajasthan. *Saug. Kunde* 44 : 333-343.

Rana, B.D. and I. Prakash, 1984. Reproduction biology of the soft furred field rat, *Rattus meltada pallidior* (Ryley, 1914) in the Rajasthan desert. *J. Bombay nat. Hist. Soc.* 81: 59-70.

Rana, B.D., R. Advani and B.K. Soni, 1983. Reproductive biology of *Rattus rattus rufescens* in Indian desert. *Zeit. Angew. Zool.* 60: 207-216.

Rao, AMKM, 1992. The mice, *Mus* spp. In *Rodents in Indian Agriculture, Vol. I* (Eds. Prakash, I. and Ghosh, P.K.). Scientific Publishers, Jodhpur. Pp. 147-164.

Seth, P. and M.R.N. Prasad, 1969. Reproduction cycle of the female five striped Indian palm squirrel, *Funambulus pennanti* Wroughton. *J. Reprod. Fert.* 20: 211-222.

Srivastava, A.S. 1968. Rodent control for increased food production. Rotary Club (West), Kanpur. 152 p.

Tripathi, R.S., M. Idris and B.D. Rana, 2002. Ecological distribution of rodents in desert environment. In *Modern trends in environmental biology* (Ed. G. Tripathi) CBS Publishers & Distributors, New Delhi. pp. 115-123.

Tripathi, R.S., A.P. Jain N. Kashyap, B.D. Rana and I. Prakash, 1992. North-western Desert. In *Rodents in Indian Agriculture, Vol. I.* (Eds. Prakash, I. and Ghosh, P.K.) Scientific Publishers (India), Jodhpur. Pp. 357-395.

Changing Faunal Ecology in Thar Desert, pp. 305-322, 2005.
Eds.: B.K. Tyagi and Q.H. Baqri

15

SCENARIO OF CLINICAL STUDIES OF CUTANEOUS LEISHMANIASIS OF HUMAN AND CANINES

Anil Ahuja

Department of Clinical Veterinary Medicine,
College of Veterinary and Animal Science, Bikaner - 334 001 (Rajasthan), India

Leishmaniasis, caused by parasites of the genus *Leishmania* - originally parasites of rodents in which they cause a mild cutaneous disease – is regarded as one of the world's most complex diseases. The disease, endemic in the tropical and subtropical regions of 88 countries in five continents, has prevailed insidiously through centuries affecting people in both old and new worlds. There is an overall prevalence of 12 million cases worldwide, and estimatedly there occur 1.5 to 2 million new cases of cutaneous leishmaniasis (CL) annually. Leishmaniasis is with several forms, some of which are self-curing (e.g., cutaneous leishmaniasis), others hideously disfiguring (e.g., muco-cutaneous and/or post-kala azar dermal leishmaniasis), still others fatal (e.g., visceral leishmaniasis). There are four species of *Leishmania* pathogenic to man. They are transmitted by sandflies and have become adapted to canines and man in whom they cause three main clinical types of the disease, of which cutaneous leishmaniasis is endemic in the Thar Desert, western India.

INTRODUCTION

Cutaneous leishmaniasis is a protozoan disease caused by various species of *Leishmania*. The main hosts of the disease are man and dog but the infection is also found in rodents such as rats, mice and the Indian desert gerbil (*Meriones hurrianae*). Disease is transmitted by the sand fly (*Phlebotomus* spp.) among the hosts (Adler and Theodor, 1957). The parasite occurs in two forms, *viz.*, the amastigote form and the promastigote form. The amastigote form is found in cutaneous lesions while promastigote form is found in the sand fly and culture (Ahuja *et al.*, 2000 a; Ahuja *et al.*, 2002).

Attempts to find the etiology of cutaneous leishmaniasis (oriental sore) were started by Cunnigham (1985) in India. The organisms isolated from the oriental sore were described as a "Peculiar Parasitic Organism" in the tissue of specimen of Delhi boil. Firth (1891) identified the parasite causing cutaneous leishmaniasis as *Sporozoa furnuculosa*. Marchand (1904) reported the leishmania species from the spleen of Chinese subjects. Wright (1903) reported the first complete description of parasite. He observed the parasite in an Armenian child who was brought to Massachusetts General Hospital and named it *Helcosoma tropica*. Marzinowsky and Bogroff (1904) discovered the parasite in a sore of boy in Russia and named the organism as *Ovoplasma orientale*. Luhe (1906) changed the name *Helcosome tropica* given by Wright (1903) to *Leishmania tropica*. Thus, the parasite was ultimately named as *L. tropica* (Wright,1903 and Luhe,1906).

PREVALENCE OF CUTANEOUS LEISHMANIASIS

The disease is widespread in human beings and occurs in various parts of the world, and it is present in the old as well as in the new world (Fergusson & Richards, 1910; Thomson & Balfour, 1910; Carter, 1911; Archibald, 1911; Martoglia, 1913; Christopherson, 1914; Kojevnikov, 1941a; Kovziakine, 1941; Beltran, 1944; Craig & Faust, 1953; Cahill, 1964; Slappendel, 1988; Hruzik *et al.*, 1977; Egoz & Michaeli, 1978; Klaus *et al.*, 1994; Ridley, 1983; Koulmann *et al.*, 1985; Lin *et al.* 1986; Kibbi *et al.* 1987; Kubba *et al.*, 1987; Zahaf *et al.* 1986; Dereure *et al.* 1989; Hayatee *et al.* 1990; Pereira & Castro, 1991; Opitz, 1992; Mebrahtu *et al.*, 1992; Momeni & Aminjavaheri, 1994). The cutaneous leishmaniasis of dogs has also been prevalent in the whole world (Neligan, 1913; Pedroso, 1913; Yakimoff, 1914; Gachet, 1915; Kirk & Drew, 1938; Issayev, 1941; Dedet *et al.*, 1979; Pozio *et al.*, 1981; Garcia *et al.*, 1983; Ferrer *et al.*, 1988; Marty *et al.*, 1988; Bourdeau & Groulade, 1988; Hermeto *et al.*, 1993; Dereure *et al.*, 1991 and Khiami *et al.*, 1991).

In India, the distribution of the disease is not uniform, as in some of the places, only sporadic cases of leishmaniasis have been reported while in some other parts of the country, it is so prevalent that these endemic areas are known by their local names given to the disease, *i.e.*, Delhi boil, Ambala sore, Karnal sore, Gujarat sore, Bombay sore *etc.* (Sinton, 1925).

Avari and Mechie (1924) found ulcers on the ears of a dog positive for *Leishmania* species in Bombay. A large-scale epidemic of oriental sore was recorded in Delhi during 1940 in which approximately 20,000 people suffered in Karol Bagh area (Shah, 1941). Several reports about the incidence of this disease have been recorded from the district of Aurangabad (Maharashtra) and its adjoining areas (Young, 1937; Daner and Ahmed, 1943; Farooq & Cutubuddin, 1945).

Cutaneous leishmaniasis occurred in the central, western and north-western parts of the Indian sub-continent and the noteworthy centres of infection were Lahore, Multan, Delhi, Dera-Ismilekhan etc. (Manson-Bahr, 1951). Ray (1945) referred to the reports of cutaneous leishmaniasis in dogs from North-West Frontier province and United Provinces in India. Faust and Russel (1957) mentioned its incidence in dogs from Bombay. Omran (1961) and Lysenko (1971) suggested that cutaneous leishmaniasis or oriental sore was restricted to dry western half of Indo-Gangetic plains, including arid zones of Rajasthan and Gujarat states. Saxena *et al.* (1970) reported occurrence of this disease in Bikaner. Its endemicity has also been reported form Bikaner (Bharadwaj & Joshi, 1971; Sharma *et al.*, 1973a). Cases were detected in and around Bikaner, *i.e.*, in Shivbari, Gangashahar and Nokha in Bikaner district. Lodha *et al.* (1971) were the first to report the endemicity of cutaneous leishmaniasis in dogs and men in Bikaner. Thereafter, Mishra (1971) also reported cutaneous leishmaniasis in dogs and men in Bikaner. In India, the evidence of the possibility of existence of a zoonotic infection based upon clinico-epidemiological analysis of the cases was collected in 1973 during the course of an investigation of an outbreak of cutaneous leishmaniasis in Bikaner as reported by Sharma *et al.* (1973c). Later, the presence of sporadic cases in the villages in the vicinity of Rajasthan canal area at Tibbi and Hanumangarh were detected during March, 1973 (Sharma *et al.*, 1973 b) The prevalence rate for cutaneous leishmaniasis in human beings and dogs in Bikaner is reported by many workers and their results are presented in tables 1 and 2. Purohit *et al.* (1982) also reported the human cases of cutaneous leishmaniasis in Bikaner. A case of 18 month old boy suffering from cutaneous leishmaniasis in Bikaner was reported by Pathak *et al.* (1990).

Arora *et al.* (1990) reported cutaneous leishmaniasis due to *L. tropica* infection in 18 patients during July, 1986 through June, 1989. Later on, Ahuja *et al.* (1993) reported cutaneous leishmaniasis in domestic dogs of Bikaner. Further, Chhangani *et al.* (1995), Sharma (1998), Ahuja (1999), Ahuja *et al.* (2000 b, 2001), Kumar (2002) and Joshi (2003) diagnosed cutaneous leishmaniasis in dogs and human beings in Bikaner.

Present review showed that cutaneous leishmaniasis shared by both human beings and canines of this zone (Table 1 and 2), indicating that the disease has great zoonotic importance in the interest of public health. In particular, areas where high population of stray dogs make a close association with human beings, sand fly transmits the disease among them, with an obligatory zoonosis in dogs. Lack of knowledge and awareness of the disease amongst the local folks, obfuscated further by an stupendous population of a mélange of rats, mice and gerbils as well as sand flies both in the canal irrigated command and non-command areas, together with lack of treatment of stray dogs, are the important factors responsible for the endemicity of the disease in the particular zone.

Table 1. Prevalence of cutaneous leshmaniasis in human beings in Bikaner

S.No.	Name of workers	No. of suspected cases	No. of confirmed cases	Percentage
1.	Bharadwaj & Joshi (1971)	88	31	35.20
2.	Agarwal (1980)	86	18	20.93
3.	Nirban (1985)	84	16	19.04
4.	Chhangani (1993)	74	36	48.64
5.	Sharma (1998)	95	64	68.04
6.	Ahuja (1999)	100	90	90.00
7.	Kumar (2002)	204	189	92.6
8.	Joshi (2003)	104	38	36.5

Table 2. Prevalence of cutaneous leshmaniasis in dogs reported by various workers

S.No.	Name of workers	No. of suspected cases in dogs		Confirmed cases in dogs		Percentage	
		Stray	Pet	Stray	Pet	Stray	Pet
1.	Bharadwaj & Joshi (1971)	20	-	8	-	40	-
2.	Lodha *et al.* (1971)	15	-	8	-	53.3	-
3.	Agarwal (1980)	64	-	14	-	21.86	-
4.	Nirban (1985)	17	-	4	-	23.52	-
5.	Chhangani (1993)	16	10	11	6	68.75	60
6.	Sharma (1998)	79	25	17	6	21.5	24
7.	Ahuja *et al.* (2001)	18	30	12	14	66.6	46.6
8	Kumar (2002)	-	20	-	13	-	65
9.	Joshi (2003)	39	11	11	6	28.2	54.54

Sex-wise prevalence

Sex-wise prevalence of cutaneous leishmaniasis in human beings and dogs are studied by many workers and results are presented in Table 3 and 4.

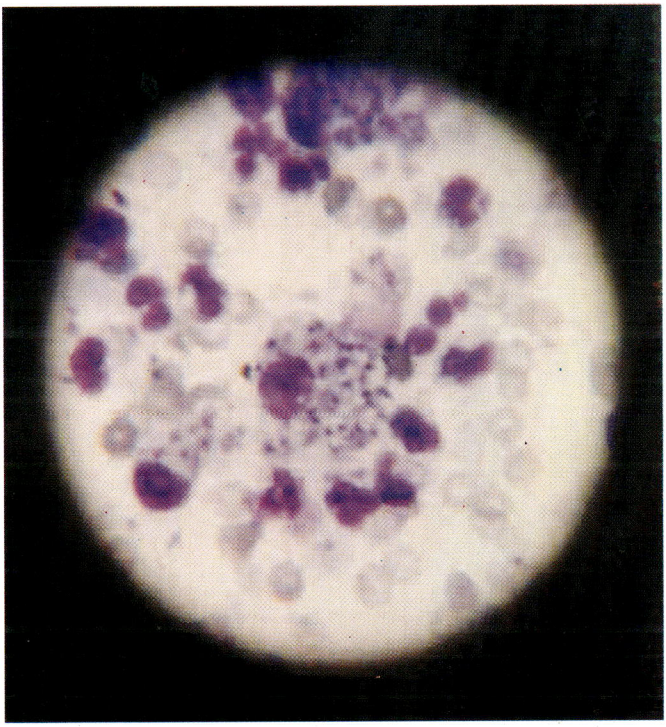

Figure 1- Mononuclear cells packed with *Leishmania tropica* amastigote form [L. T. bodies] from a contact smear of skin lesion of human.[1000 x Leishman stain]

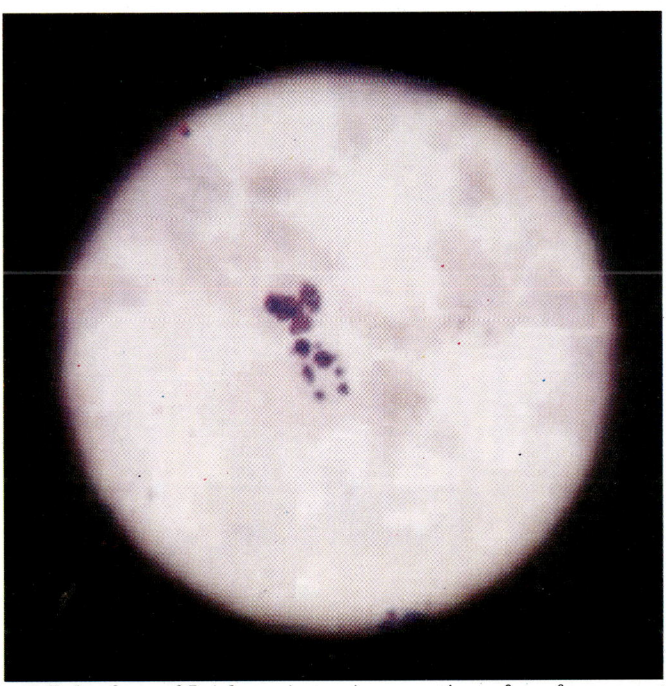

Figure 2- Extracellular form of *Leishmania tropica* amastigote form from a contact smear of skin lesion of a dog (1000 x Leishman Stain)

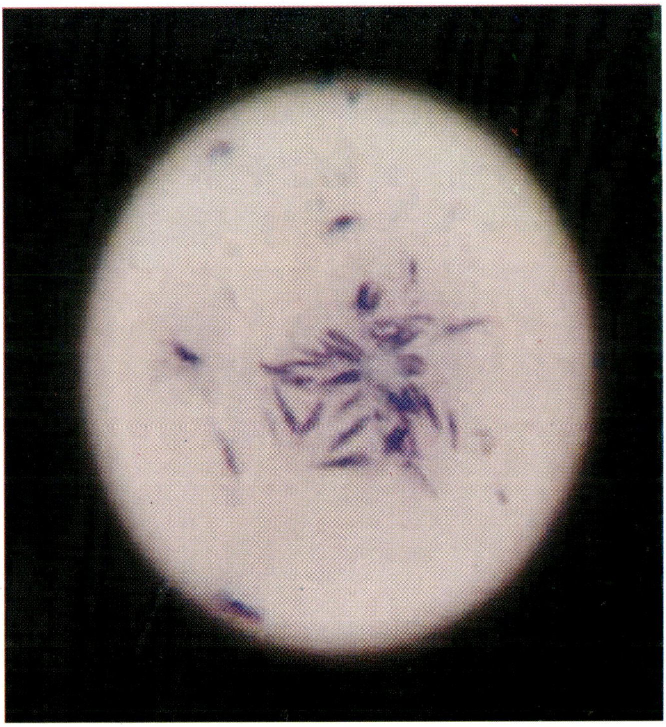

Figuro 3 Promastigotes form of *Leishmania tropica* from culture on tobie medium. (1000 x Leishman Stain)

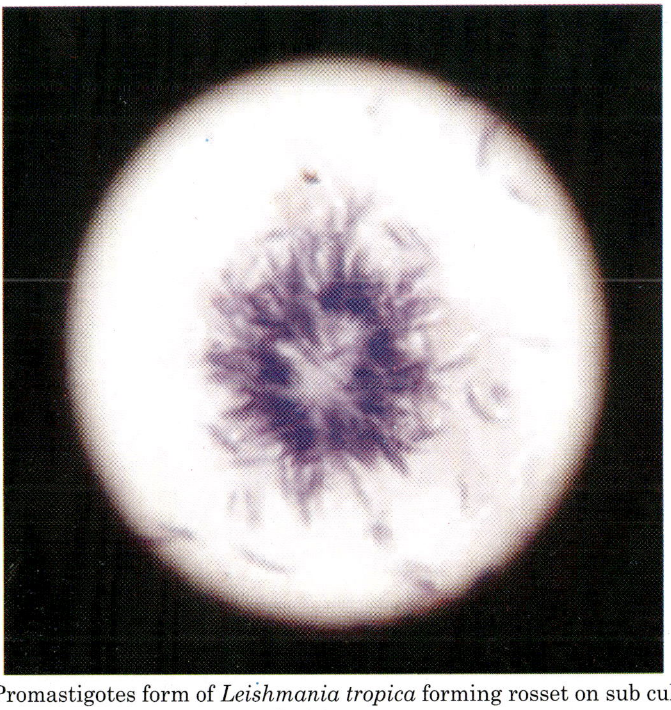

Figure-4 - Promastigotes form of *Leishmania tropica* forming rosset on sub culture on tobie medium (1000 x Leishman Stain)

Table 3. Sex-wise prevalence of human cutaneous leshmaniasis

S.No.	Name of Scientists	No. of male	No. of female	Male/female ratio
1.	Sharma *et al.* (1973a)	—	—	1.8 : 1
2.	Agarwal (1980)	16	2	8 : 1
3.	Singh (1980)	46	29	1.5 : 1
4.	Nirban (1985)	14	2	7 : 1
5.	Chhangani (1993)	19	17	1.1 : 1
6.	Sharma (1998)	43	21	2.09 : 1
7.	Ahuja (1999)	62	28	2.2 : 1
8.	Kumar (2002)	99	90	1.1 : 1
9.	Joshi (2003)	25	13	1.9 : 1

The Table 3 indicates that cutaneous leishmaniasis in human beings is gender biased and it is more prevalent in males than females. The possible explanation is that since females are adequately clothed for cultural reasons and are therefore protected from the bite of sand flies who transmit cutaneous leishmaniasis.

Table 4. Sex-wise prevalence of canine cutaneous leishmaniasis

S.No.	Name of scientists	No. of male dogs	No. of female dogs	Male/Female ratio
1.	Mishra (1971)	5	3	1.6 : 1
2.	Nirban (1985)	4	3	1.3 : 1
3.	Chhangani (1993)	10	7	1.4 : 1
4.	Sharma (1998)	16	7	2.2 : 1
5.	Ahuja (2001)	10	16	1 : 1.6
6.	Kumar (2002)	8	5	1.6 : 1
7.	Joshi (2003)	9	8	1.1 : 1

In case of dogs, cutaneous leishmaniasis is a sporadic disease without any scientific explanation of gender bias.

Age-wise prevalence

Age-wise prevalence of cutaneous leishmaniasis in human beings is reported by various scientists as presented below in Table 5.

In the present review, it was observed that cutaneous leishmaniasis occurred as early as in an infant of 6 months of age to as old as 60 years old man (Ahuja, 1999). However, Lodha *et al.* (1971) reported 18 human patients suffering from cutaneous leishmaniasis who were between 12 to 17 years of age.

Majority of scientists reported that persons between 11 to 20 years of age group showed highest prevalence of the disease. This is explained by the fact that those having ulcers in the earlier life, rarely suffered from this disease in the later years; immunity appeared to last a life time (Farooq & Cutubuddin, 1945).

Table 5. Age-wise prevalence of human cutaneous leishmaniasis (in per cent)

S.No.	Age group (in years)	Name of Scientists							
		Agarwal (1980)	Singh (1980)	Nirban (1985)	Chhangani (1993)	Sharma (1998)	Ahuja (1999)	Kumar (2002)	Joshi (2003)
1.	Up to 1	25				4.6	3.3	1.0	
2.	1-5					23.4	14.4		
3.	6-10	12.5	18.7	37.5	63.8	21.8	14.4	24.4	21.0
4.	11-20	43.7	38.7	43.7		20.3	27.8	19.1	23.7
5.	21-30	12.5	22.7	12.5		15.6	22.4	23.3	13.2
6.	31-40	6.2	13.3	6.2	36.2	9.3	11.1	10.0	15.8
7.	41-50	-	4.0	-		4.6	3.3	12.2	10.5
8.	51-60	-	2.6	-			3.3	10.0	15.8

Age-wise prevalence of cutaneous leishmaniasis in domestic dogs has been reported by Ahuja *et al.* (2001), Kumar (2002) and Joshi (2003), which is presented in Table 6.

Table 6. Age-wise prevalence of cutaneous leishmaniasis in domestic dogs

Name of Scientists	No. of dogs in different age groups						
	Up-to ½	½ - 1	1-2	2-3	3-4	Above 4	Total
Ahuja *et al.* (2001)	1	2	3	5	2	1	14
Kumar (2002)	-	3	5	5	-	-	13
Joshi (2003)	1	-	3	-	1	1	6

According to present observations, cutaneous leishmaniasis was more prevalent between 1-3 years of age group. At present there is no specific scientific explanation available of this observation.

Prevalence of type of lesions

The skin lesions of cutaneous leishmaniasis in human beings are usually classified as cutaneous ulcerative, nodular, mucocutaneous-cum-cutaneous ulcerative, mucocutaneous ulcerative and diffuse infiltrative types. Types of lesions reported by a number of scientists are presented in Table 7.

Table 7. Types of lesions reported in human beings (in per cent)

Sl. No.	Type of lesions	Agarwal (1980)	Singh (1980)	Nirban (1985)	Sharma (1998)	Ahuja (1999)	Kumar (2002)	Joshi (2003)
1.	Cutaneous ulcerative	16.6	22.6	12.5	70.3	46.6	36.5	31.6
2.	Nodular	77.7	77.4	81.2	23.4	45.5	55.1	55.3
3.	Mucocutaneous-cum- cutaneous ulcerative	—	—	—	—	5.5	—	—
4.	Mucocutaneous ulcerative	—	—	—	—	1.1	—	—
5.	Diffuse infiltrative	5.5	—	6.2	6.2	1.1	8.4	13.1

Some of the scientists reported more than 75 per cent prevalence of nodular lesions of cutaneous leishmaniasis while Sharma (1998), Ahuja (1999), Kumar (2002) and Joshi (2003) reported cutaneous ulcerative lesions in human beings for 70.3 per cent, 46.6 per cent, 36.5 per cent and 31.6 per cent, respectively. Ahuja (1999) also reported 5.5 per cent prevalence of mucocutaneous-cum-cutaneous ulcerative lesions. Present review also indicates increasing trend of diffuse infiltrative lesions in human beings (Kumar, 2002; Joshi, 2003).

In case of dogs various scientists reported the type of lesions which are shown in Table 8.

Table 8. Type of lesions in dogs presented in per cent.

S.No.	Type of lesions	Name of Scientists				
		Sharma (1998)	Ahuja *et al.* (2001)	Kumar (2002)	Joshi (2003)	
					Stray dogs	Pet dogs
1.	Nodular	17.3	—	39.1	27.3	—
2.	Cutaneous/ mucocutaneous uncreative	52.1	100	34.8	54.5	90.9
3.	Diffuse infiltrative	30.4	—	26.1	18.2	9.1

Sharma (1998) reported cutaneous/mucocutaneous ulcerative lesions in 52.1 per cent dogs while Ahuja *et al.* (2001) reported 100 per cent, however, the lesions included cutaneous ulcerative, mucocutaneous ulcerative and mucocutaneous cum cutaneous ulcerative 75 per cent, 16.66 per cent and 8.34 per cent, respectively.

Table 9. Comparative distribution of cutaneous lesions (in per cent) on various parts of the body in human beings reported by various scientists

S.No.	Parts of the body	Name of workers						
		Farooq and Cutubuddin (1945)	Sharma *et al.* (1973a)	Agarwal (1980)	Nirban (1985)	Chhangani (1993)	Ahuja (1999)	Joshi (2003)
1.	Face	38.09	18.6	44.44	31.25	22.22	22.22	15.7
2.	Neck	01.13	—	—	—	—	02.22	2.7
3.	Shoulder	—	—	—	—	—	02.80	—
4.	Arms	—	—	16.67	18.75	33.33	20.00	15.7
5.	Hand	28.40	38.6	22.22	12.50	16.69	25.55	36.8
6.	Legs	—	—	11.12	12.50	13.88	15.55	13.2
7.	Feet	21.59	22.8	05.55	18.75	08.33	11.11	13.2
8.	Abdomen	—	07.1	—	06.25	05.55	00.55	—
9.	Trunk	10.79	—	—	—	—	—	2.7
10.	Back	—	12.9	—	—	—	—	—

Distribution of lesions on the body

Generally lessons of cutaneous leshmaniasis are reported on the exposed parts of the body (Ahuja, 1999). Various scientists observed distribution of cutaneous lesions on different parts of the body in human beings which is presented in Table 9.

Most of the workers reported that lesions of cutaneous leshmaniasis were present maximum on arms and hands while Farooq & Cutubuddin (1945) and Agarwal (1980) reported on face. Joshi (2003) reported 36.8 per cent lesions on hand which is the highest percentage in human. Farooq & Cutubuddin (1945) and Ahuja (1999) also reported lesions on neck 1.13 and 2.22 per cent, respectively. Farooq and Cutubuddin (1945) reported 10.79 per cent lesions on the trunk while Sharma *et al.* (1973a) observed 12.9 per cent lesions on back of human beings. Ahuja (1999) reported 2.80 per cent lesions on the shoulder of human beings.

Like human beings, lesions of canine cutaneous leishmaniasis are prevalent on various body parts (Table 10). However, majority of the lesions were observed on face and extremities (Mishra, 1971; Agarwal, 1980; Nirban, 1985; Sharma, 1998 and Ahuja *et al.*, 2001). Ahuja, *et al.* (2001) reported 43.24 per cent on the face of dogs while Sharma (1998), Kumar (2002) and Joshi (2003) observed 69.5 per cent, 64.7 per cent and 47.8 per cent lesions, respectively. Further Nirban (1985) reported 75 per cent lesions on the extremities while Mishra (1971) reported 77.2 per cent.

Table 10. Comparative distribution of cutaneous lesions (in per cent) on various parts of the body in dogs reported by different scientists

S.No.	Parts of body	Name of Scientists						
		Joshi (2003)	Kumar (2002)	Ahuja *et al.* (2001)	Sharma (1998)	Nirban (1985)	Agarwal (1980)	Mishra (1971)
1.	Face and Nostrils	47.8	64.7	43.24	69.5	25.0	26.6	6.0
2.	Neck	4.3	—	2.70	—	—	—	0.8
3.	Fore limbs	21.8	11.8	18.92	30.5	50.0	13.3	28.6
4.	Hind limbs	21.8	23.5	32.44	—	25.0	30.0	48.6
5.	Trunk	4.3	—	2.70	—	—	10	2.6
6.	Vulvular lips and tail	—	—	—	—	—	20	17.3

Single and multiple lesions have been reported by many workers, which are presented in Table 11.

Table 11. Single and multiple lesions on the body of human beings suffering from cutaneous leishmaniasis (in per cent)

S.No	Type of lesions	Joshi (2003)	Kumar (2002)	Ahuja (1999)	Sharma (1998)	Chhangani (1993)	Nirban (1985)	Singh (1980)	Agarwal (1980)
1.	Single	55.3	65.6	56.6	54.6	69.4	43.7	54.6	38.8
2.	Multiple	44.7	34.4	43.4	45.4	30.6	56.3	45.3	61.2

Most of the workers reported single lesion in more than 50 per cent human cases, while Agarwal (1980) and Nirban (1985) reported 38.8 and 43.7 per cent, respectively. Agarwal (1980) reported 61.2 per cent multiple lesions while Chhangani (1993) reported 69.4 per cent single lesions in human beings suffering from cutaneous leishmaniasis.

Ahuja (1999) reported that 69.21 per cent dogs suffering from cutaneous leishmaniasis had single lesion, 19.3 per cent dogs had two lesions and 11.5 per cent dogs had three lesions. Similarly, in human beings, Chhangani (1993) observed that 69.4 per cent human had single lesion.

Size of lesions

Lesions were observed to vary in size from a few millimeters to 80 mm in cases of human cutaneous leishmaniasis. Per cent-wise size of lesions reported by different scientists is presented in Table 12.

Table 12. Size of lesions in human beings reported by different scientists (in per cent)

S.No.	Diagonal/ diameter of lesions (in mm)	Name of Scientists						
		Agarwal (1980)	Nirban (1985)	Chhangani (1993)	Sharma (1998)	Ahuja (1999)	Kumar (2002)	Joshi (2003)
1.	Up to 20	77.7	75.0	69.3	85.9	65.0	48.1	63.2
2.	21 to 40	—	—	16.8	—	29.4	42.4	—
3.	41 to 60	16.8	18.7	11.1	14.1	4.4	—	36.8
4.	Over 60	5.5	6.2	2.7	—	1.1	9.5	—

Abdalla (1973) reported size of lesions of human cutaneous leishmaniasis ranging from 10 to 60 mm. Presently, all the scientists have reported higher percentage of lesions up to 20 mm; while Sharma (1998) has reported as high as 85.9 per cent Ahuja (1999) has reported 65 per cent. Ahuja (1999) also reported 80 mm maximum size of lesions of cutaneous leishmaniasis.

Similar to human beings, in dogs suffering from cutaneous leishmaniasis, the size of lesions was reported to vary from a few mm to more than 60 mm. Per cent-wise size of lesions reported by scientists is presented in Table 13.

Table 13. Size of lesions in dogs reported by different scientists (in per cent)

S.No.	Size of lesions in mm diameter/diagonal length	Name of scientists				
		Mishra (1971)	Nirban (1985)	Ahuja *et al.* (2001)	Joshi (2003) Stray	Joshi (2003) Pet
1.	Up to 20	48.7	50.0	62.1	60.9	45.4
2.	21 to 40	34.8	25.0	27.0	30.5	27.3
3.	41 to 60	15.7	12.5	8.1	4.3	9.1
4.	Over 60	0.8	12.5	2.7	4.3	18.2

Most of the scientists observed that more than 50 per cent lesions of dogs measured less than 20 mm in size. Nirban (1985) reported 12.5 per cent lesions are of over 60 mm, while Mishra (1971) and Ahuja *et al.* (2001) reported same observation 0.8 and 2.7 per cent, respectively. Similarly, Lodha, *et al.* (1971), Agarwal (1980), Pathak *et al.* (1990), Ahuja *et al.* (1993) and Chhangani (1993) found skin lesions in dogs to vary between 5 mm and 65 mm.

Duration of lesions

Duration of lesions was observed from 20 days to maximum 3 years in case of human cutaneous leshmaniasis (Ahuja, 1999). Duration of lesions reported by various workers is presented in Table 14.

Table 14. Duration of lesions of cutaneous leshmaniasis in human beings (in per cent)

S.No.	Duration in months	Name of Scientists							
		Agarwal (1980)	Singh (1980)	Nirban (1985)	Chhangani (1993)	Sharma (1998)	Ahuja (1999)	Kumar (2002)	Joshi (2003)
1.	Less than 1	16.6	26.7	18.7	8.3	7.8	12.2	12.3	7.9
2.	1 to 2	72.3	—	68.7	61.2	35.9	40.0	21.2	—
									50.0
3.	3 to 4	11.1	45.4	12.6	19.4	21.8	24.4	35.9	—
4.	5 to 6	—	6.6	—	11.1	7.8	13.5	29.1	—
5.	7 to 8	—	14.7	—	—	—	2.2	—	23.7
6.	9 to 12	—	6.6	—	—	—	4.4	—	—
7.	12 to 24	—	—	—	—	—	2.2	1.5	—
8.	36	—	—	—	—	26.7	1.1	—	18.4

Agarwal (1980) observed that 72.2 per cent lesions were between 1 to 2 months duration while Sharma (1998) reported 35.9 per cent for the eanes duration. Ahuja (1999) reported that 1.1 per cent lesions were 36 months (3 years) old. Agarwal (1980), Singh (1980), Nirban (1985), Chhangani (1993), Sharma (1998), Ahuja (1999), Kumar (2002) and Joshi (2003) recorded largest group of patients (100 per cent, 72.2 per cent, 100 per cent, 88.9 per cent, 65.7 per cent, 76.6 per cent, 69.4 per cent and 57.9 per cent, respectively) with the lesions less than 4 months.

Duration of lesions in pet dogs reported by Ahuja *et al.* (2001) in the group of less than 1, 1-2, 2-3 and 4 months are 28.5 per cent, 35.7 per cent, 28.5 per cent and 7.3 per cent, respectively. However, Joshi (2003) reported that 66.6 per cent dogs had a duration of lesions less than one month and rest 33.4 per cent had 2 months old lesions.

Largest percentage of duration of lesions in human beings and dogs are less than four months. This is probably due to the fact that the lesions can heal spontaneously after 6 months (Domoncos, 1971).

Breed-wise prevalence in dogs

Breed-wise prevalence of cutaneous leishmaniasis in dogs is reported by various workers and the same is presented in Table 15.

Table 15. Breed-wise prevalence of cutaneous leshmaniasis in dogs

Scientist	Breed								
	Spitz	German shepherd	Dober-mann	Dachs hound	Labrador	Great Dane	Dalm-atian	Non-descript	Total
Ahuja *et al.* (2001)	1	3	4	3	1	1	1	12	26
Kumar (2002)	7	1	2	—		2	1		13
Joshi (2003)	1	3				1	1	11	17

According to these observations prevalence of cutaneous leshmaniasis in domestic dogs is high in short haired breed of dogs as compared to long haired breeds. High prevalence in short haired breed is justified because of easy accessibility of skin by the vector sand fly as compared to long haired breeds.

REFERENCES

Abdalla, R.E., M. Ali, A.I. Wasfi and A.M.El. Hassan, 1973. Cutaneous leishmaniasis in Sudan. *Trans. Roy. Trop. Med. Hyg.* 67 (4) : 544-59.

Adler, S. and M. Ber, 1941, The transmission of *Leishmania tropica* by the bite of *Phlebotomus papatasi. Indian J. Med. Res.* 803-9.

Adler, S. and O. Theodor, 1925. The experimental transmission of cutaneous leishmaniasis to man from *Phlebotomus papatasii. Ann. Trop. Med. and Parasit.* 19 : 365-71.

Adler, S. and O. Theodor, 1927. The transmission of *Leishmania tropica* from artificially infected sand flies to man. *Ann. Trop. Med. and Parasit.* 21: 89-110.

Adler, S. and Theodor, O., 1957. Transmission of disease agents by Phlebotomine sandflies. *Ann. Rev. Entomol.* 2: 203-26.

Agarwal, S.K., 1980. A study of prevalence of human cutaneous leishmaniasis in Bikaner city. A thesis submitted to the Rajasthan University, Jaipur, Rajasthan for the Degree of Doctor of Medicine.

Ahuja, A.K, S.K. Purohit, J.S. Yadav and P.R. Netra, 1993. Cutaneous leishmaniasis in domestic dogs. *Ind. J. Pub. Hlth.* 37 (1): 29-31.

Ahuja, A.K, R.K. Tanwar, A.K. Gahlot, and S.K. Purohit, 2000a. Cutaneous leishmaniasis. *Ind. J. Vety. Pract.* 1(2): 19-24.

Ahuja, A.K., A.K. Gahlot, S.K. Purohit and D. Rana, 2000b. Treatment trials of cutaneous leishmaniasis in dogs. *Veterinary Practitioners* 1(1): 53-55.

Ahuja, A.K., 1999. Diagnosis and treatment trials of cutaneous leishmaniasis in dogs and antigenic characterization of fractions of *Leishmania tropica*. A thesis submitted to Rajasthan Agricultural University for the Degree of Doctor of Philosophy in the subject of Veterinary Medicine.

Ahuja, A.K., A.K. Gahlot, S.K. Purohit, D.K. Bihani and Dinesh K. Rana, 2002. Transmission trials of *Leishmania tropica* infection. *J. Canine Dev. Res.* 2: 17-21.

Ahuja, A.K., A.K Gahlot, S.K. Purohit and Dinesh, K. Rana, 2001. Some clinical studies on canine cutaneous leishmaniasis. *J. Canine Dev. Res.* 1: 35-40.

Archibald, R.G., 1911. Fourth Report Wellcome Tropical Research Lab. Gordon Memorial College, Khartoum, p. 207-11. (Cited by Abdalla *et al.*1973. Trans. *Roy. Soc. Trop. Med. Hyg.* 67 : 549-57).

Arora, P.N., K.R. Ramakrishnan, J.R. Bhardwaj and K.V. Singh, 1990. Cutaners [cutaneous] leishmaniasis. *Ind. J. Dermatol.* 35 : 123-18.

Avari and Machie, 1924. Cited by Wenyon, C.M. (1965) : Protozoology (1) Bailliere, Tindall and Cassel Ltd., London. pp. 430.

Barsky, S., W. Storino and K. Salgea, 1978. *Arch. Dermatol.* 114 : 1354-55. (Cited by Singh, 1980).

Beltran, E., 1944). Cited by Chandler, A.C. and Read, C.P. (1961), Introduction to Parasitology. Toppan Co. Ltd. Tokyo, Japan, 10th ed. pp. 129.

Berberian, D.A., 1966. Mechanical transmission of Leishmania. *Trans. Roy. Soc. Trop. Med. Hyg.* 60 : 277-8.

Bharadwaj, T.P. and K.R. Joshi, 1971. Preliminary report on the study of endemic focus of tropica leishmaniasis with a note on the existence of canine reservoir. *J. Comm. Dis.* 3 : 92-102.

Bjorvatn, Bjarne and Franklin A. Neva, 1979 b. Experimental therapy of mice infected with *Leishmania tropica. Am. J. Trop. Med. Hyg.* 28 (3) : 480-5.

Bjorvatn, Bjarne and Franklin A. Neva, 1979 a. A model in mice for experimental leishmaniasis with a west African strain of *Leishmania tropica. Am. J. Trop. Med. Hyg.* 28 (3) : 472-9.

Bourdeau, P. and P. Groulade, 1988. Results of an inquiry into leishmaniasis. *Pratique Medicale and Chirurgicale de l' Animal de Compagnie.* 23: Supplement (5) 5-10.

Cahill, K.M., 1964. Leishmaniasis in the Sudan Republic. XXI. Infection in American personal. *Am. J. Trop. Med. Hyg.* 13 : 794.

Carter, R.M., 1911. Non-ulcerating oriental sore. *Ann. Trop. Med. Parasit.* 5 : 15-35.

Chhangani, S.K., 1993. *Investigation on cutaneous leishmaniasis in men and in Bikaner, Rajasthan.* M.V.Sc. thesis, Rajasthan Agricultural University, Bikaner.

Chhangani, S.K., S.K Purohit, G. Kalla, and A. Ahuja, 1995. Diagnosis and treatment of canine and human cases of cutaneous leishmaniasis. Paper presented in second annual conference of IAAVR, Jan. 24-25., Dept. of Vet. Pub. Hlth. & Epid., HAU, Hisar. Compendium, PS-11. pp. 119.

Christopherson, J.B., 1914. On a case of naso oral leishmaniasis and on a case of oriental sore, both originating in the Anglo-Egyptian Sudan. *Ann. Trop. Med. Parasit.* 8 : 485-96.

Craig, C.F. and E.C. Faust, 1953. Clinical Parasitology. Lea and Febiger, Philadelphia, 6th ed. pp. 114.

Cunningham, 1985. Cited by Wenyon, C.M., 1965. Protozoology (1), Bailliere Tindall and Cassel Ltd., London. pp. 397.

Daner, M.B. and S.S. Ahmed, 1943. The occurrence of oriental sore in the Hyderabad state. *Ind. Med. Gaz.* 73 : 296.

Das Gupta, B.M. and B.B. Dikshit, 1929. *Ind. Med. Gaz.*, 64 : 67-70. (Cited by Singh, 1980).

Dedet, J.P., F. Derouin, and B. Hubert, 1979. Ecology of a focus of cutaneous leishmaniasis in the Thies region (Senegal, West Africa). I. The status of cutaneous leishmaniasis in Bulletin in de la Societe de Pathologie Exotique et de ses Filiales. 73 (2) : 124-31.

Dereure, J., H.A. Rageh, W. Daoud and J.A. Rioux, 1989. Eco-epidemiology of visceral and cutaneous leishmaniasis in the Arab Republic of Yemen. *Bull. Soc. Pathol. Exot. Filiales.* 82 (5) : 665-8.

Dereure, J., J.A. Rioux, M. Gallego, J. Perieres, F. Pratlong, J. Mahjour and H. Saddiki, 1991. *Leishmania tropica* in Morocco infection in dogs. *Trans. Roy. Soc. Trop. Med. Hyg.* 85 : 595.

Domoncos, A.N. (1971) : Cited by Asher, W. and Cohen, C. 1977. Cutaneous leishmaniasis. In efficiency of treatment with metronidazole. *Arch. Dermatol.* 113 : 1299.

Egoz, N. and D. Michaeli, 1978. Epidemiological, Immunological and clinical problems of cutaneous leishmaniasis in Israel. Revue Internationale des services de sante des Armees de Terre deMeretde - 1 51 (2) : 151-7.

Farooq. M. and M. Qutubuddin, 1945. Oriental sore in the Nizams dominions, some epidemiological factors. *Ind. Med. Gaz.* 80 : 85-89.

Faust, E.C. and P.F. Russel, 1957. Clinical Parasitology, Lea and Febiger, Philadelphia. 6th ed. pp. 114.

Fergusson, A.R. and O. Richards, 1910. Parasitic granuloma a condition allied to oriental sore occurring in Egypt. *Ann. Trop. Med. Parasit.* 4 : 151-166.

Ferrer, L., R. Rabanal, D. Fondevila, J.A. Ramos and M. Domingo, 1988. Skin lesions in canine leishmaniasis. *J. Small. Anim. Pract.* 29 : 381-88.

Firth, 1891. Cited by Wenyou, C.M., 1965. Protozoology (1), Bailliere Tindall and Cassel Ltd. London, pp. 397.

Gachet, 1915. Cited by Wenyon, C.M., 1965. Protozoology (1) : Bailliere, Tindall and Cassel Ltd. London. pp. 430.

Garcia de Lomas, J., F. Marin and A. Altuna, 1983. Antigenic study of leishmania : immunization, absorption of sera, immunofluorescence and adlers test. Revista Iberica de Parasitologia., 43 (1): 68-77.

Gerald, D. Dorman, 1974. Treatment of cutaneous leishmaniasis by freezing. J.A.M.A. 27 : 658-9.

Hayatee, Z.G., Z.A. Hanoon, and Z.H. Jadex, 1990. Oriental sore in northern part of Iraq. *Indian Journal of Parasitology* 14 (2) : 231-34.

Hermeto, M.V., D. Vieira, O. Genaro, C.A. Costa, V.P.C.P. Toledo, M.S.M. Michalick, W.L Tafuri, K.M. Chaves, W. Mayrink, and C.A. Costa, 1993. Delayed hypersensitivity skin-test using Leishvacin R for epidemiological survey of canine cutaneous leishmaniasis in a rural area of Minas Gerais State, Brazil. *Memorias do Instituto Oswoldo Cruz* 88(4) : 635-6.

Hruzik, J., Z. Lukacovicova, E. Urbancokova and K. Sobota, 1977. Parasitic infections in subjects coming from the tropics. *Bratislavske-Lekarske Listy* 68(1) : 29-36.

Issayev, L.M., 1941. Cutaneous leishmaniasis in dogs. Ashkhabad, p. 263-4. (Cited by Hoare, C.A., 1944; 41(5) : 331-45).

Joshi, V., 2003. Epidemiological survey on cutaneous leishmaniasis in dogs and human beings in and around Biknaer. M.V. Sc. thesis submitted to the Rajasthan Agricultural University, Bikaner in the subject of Veterinary Public Health.

Kaul, H.N., V. Dhanda, and G.B. Modi, 1973. The Phlebotomine Sandflies (Diptera : Psychodidae) from Rajasthan, India, with description of *Sergentomyia* (Sintonius) *sirohi sp. nov. Indian J. Med. Res.* 61 : 528-39.

Khiami, A., J. Dereure, F. Pratlong, A. Martini, and J.A. Rioux, 1991. Human cutaneous leishmaniasis caused by *Leishmania major* in the region of Damascus (Syria). *Bull. Soc. Pathol.Exot.* 84 (4) : 340-4.

Kibbi, A.G., P.G. Karam and A.K. Kurban, 1987. Sporotrichoid leishmaniasis in patients from Saudi Arabia - clinical and histologic features. *J. Am. Acad. Dermatol.* 17 (5): 759-64.

Kirk, R. and C.B. Drew, 1938. Cited by Hoare, C.A., 1944. *Trop. Dis. Bul.* 41 : 265.

Klaus, S., O. Axelrod, F. Jonas and S. Frankenburg, 1994. Changing patterns of cutaneous leishmaniasis in Israel and neighbouring territories. *Trans. Roy. Soc. Trop. Med. Hyg.* 88 (6) : 649-50.

Kojevnikov, P.V., 1941a. Cutaneous leishmaniasis in Turkmenia and the flight against it. In problems of cutaneous leishmaniasis Ashkhabad, p. 93-100. (Cited by Hoare, C.A. (1944): *Trop. Dis. Bull.* 41 (5) : 331-45).

Kojevnikov, P.V., N.V.; Dobrotrorskaya, A.N. Sokolova and A.A. Schakhova, 1941. Experimental cutaneous leishmaniasis in white mice. In problems of cutaneous leishmaniasis, Ashkhabad, p. 249-62 (cited by Hoare, C.A. (1944). *Trop. Dis. Bull.* 41 : 331-45).

Koulmann, M., J. Maria, P. Tagliana, M. Verdier and J.C. Roche, 1985. A case of oriental sore. *Revuse de stomatologie et de chirurgie maxillo faciale.* 86 (2) : 117-21.

Kovziakine, A.I., 1941. Foci of cutaneous leishmaniasis in Eastern Tanscaucasia. Ashkhabad. p. 113-119. (Cited by Hoare, C.A. (1944). *Trop. Dis. Bull.* 41 (5) : 331-45).

Krjukova, A.P., 1941 b. Experimental cutaneous leishmaniasis in wild rodents of Turkmenia. Ashkhabad, 241-8. (Cited by Hoare, C.A. (1944): *Trop. Dis. Bull.* 41 (5) : 331-45).

Kubba, R., A.I.; Gindan, A.M. Hassan and A.H.S. Omer, 1987. Clinical diagnosis of cutaneous leishmaniasis (oriental sore)., *J. Am. Acad. Dermatol.*, 16 (6) : 1183-9.

Kumar, S., 2002. Studies on diagnosis and treatment of oriental sore in human beings and dogs in and around Bikaner city. A thesis, submitted to the Rajasthan Agricultural University, Bikaner for the Degree of M.V.Sc. in the subject of Veterinary Public Health.

Lainson, R. and J.J. Shaw, 1968. Leishmaniasis in Brazil. I. Observations on enzootic rodent leishmaniasis. *Trans. Roy. Soc. Trop. Med. Hyg.* 62 : 385-95.

Latyshev, N.I. and A.P. Krjukova, 1941. Epidemiology of cutaneous leishmaniasis in the sandy deserts. Ashkhabad. p. 55-73. (Cited by Hoare, C.A. (1944), 41 (5) : 331-345).

Lin, C.S., W.J.; Wang, C.K. Wang and D. Chao, 1986. Cutaneous leishmaniasis - clinical histopathologic and electron microscopic studies. *Int. J. Dermatol.* 25 (8) : 511-5.

Lodha, K.R., B.B. Singh and P.R. Jatkar, 1971. Cutaneous leishmaniasis in Bikaner, Rajasthan. *Ind. Vet. J.* 48 : 121-3.

Luhe, 1906. Cited by Craig, C.F. and Faust, E.C., 1953. Clinical Parasitology Lea and Febiger, Philadelphia. 6th ed. pp. 114.

Lysenko, A.J.A., 1971. Distribution of leishmaniasis in the old world. *Bull. Wld. Hlth. Org.* 44 : 515-20.

Manson Bahr, P.E.C., 1951. Manson's Tropical Diseases. Cassel and Co. Ltd., London.

Marchand, 1904. Cited by Wenyon, C.M., 1965. Protozoology (1) Bailliere Tindall and Cassel Ltd., London. pp. 429.

Martoglia, 1913. Cited by Lysenko, A. J.A. (1971).

Marty, P., Y. Le-Fichoux and Y. le. Fichoux, 1988. Epidemiology of Leishmania infection in the South of France. Pratique Medicinale and Chirurgicale de l' Animal de compagnie. 23. Supplement (5) : 11-15.

Marzinowsky and Bogroff, 1904. Cited by Wenyon, C.M. (1965): Protozoology (1), Bailliere Tindall and Cassel Ltd. pp. 429.

Mebrahtu, Y.R., P.G. Lawyer, P.M. Ngumbi, G. Kirigi, J. Mbugua, G. Gachihi, K. Wasunna, H. Pamba, J.A. Sherewood and D.K. Koech, 1992. A new rural focus of cutaneous leishmaniasis caused by *Leishmania tropica* in Kenya. *Trans. Roy. Soc. Trop. Med. & Hyg.* 86 (4) : 381-7.

Mirzyam, A.A., 1941. Cutaneous leishmaniasis in Armenian, S.S.R. in problems of cutaneous leishmaniasis, Ashkhabad., 120-24. (Cited by Agarwal, 1980).

Mishra, A.K., 1971. Studies on *Leishmania tropica* (Wright, 1903) Luhe, (1906) infection in dogs. A thesis submitted to the University of Udaipur, Rajasthan for the degree of M.V.Sc.

Mohan, K. and J.C. Suri, 1975. Studies on cutaneous leishmaniasis in India V. Isolation of *L. tropica* from gerbils, Sandflies andhuman. *J.Com.Dis.* 7 (4) : 353-7.

Mohan, K., J.C. Suri and M. Madan, 1973. Studies on cutaneous leishmaniasis in India II. Behaviours of *L. tropica* strain isolated from dogs in various experimental animals. *J. Com. Dis.* 5 (3) : 143-8.

Momeni, A.Z. and M. Aminjavaheri, 1994. Clinical picture of cutaneous leishmaniasis in Tsfahan, Iran, Int. J. Dermatol., 33(4) : 260-5.

Neligan, 1913. Cited by Wenyon, C.M. (1965). Protozoology (1). Bailliere, Tindall and Cassel Ltd., London. pp. 330.

New, R.R.C. and M.L. Chance, 1980. Treatment of experimental cutaneous leishmaniasis by liposome-entrapped pentostam. *Acta Tropica* 37 (3): 253-6.

Nirban, V.K., 1985. Incidence of *Leishmania tropica* infection in and around Bikaner and preliminary trials of its cultivation on artificial media. A Thesis submitted to the Sukhadia University, Udaipur, Rajasthan for the Degree of M.V.Sc.

Omran, A.R., 1961. The ecology of leishmaniasis in studies in disease ecology edit by May, J. M., Jafner Publ. Co., Newyork (Cited by Sharma *et al.* (1973 a). *J. Comm. Dis.* 5 (1) : 54-72.

Opitz, M., 1992. Cutaneous manifestations of leishmaniasis in dogs. Jahrestagung Fachgruppe Kleintierkrankheiten, 31. Oktober bis 2. November 1991, Fellbach bei Stuttgart., 119-27.

Parrot, D. and Begne, 1921. Cited by Craig, C.F. and Faust, E.C., 1953. Clinical Parasitology. Lea and Febiger, Philadelphia, 6th ed. pp. 114.

Pathak, K.M.L., D. Kochar, and M. Kapoor, 1990: Cutaneous leishmaniasis in Rajasthan, India. *Ind. J. Anim. Hlth.* 29 (2) : 187.

Patton, 1922. Cited by Wenyon, C.M., 1965. Protozoology (1). Bailliere, Tindall and Cassel Ltd., London. pp. 431.

Pawlowsky, 1937-39. Cited by Craig, C.F. and Faust, E.C., 1953. Clinical Protozoology, Lea and Febiger, Philadelphia, 6 th ed. pp. 117.

Pedroso, 1913. Cited by Wenyon, C.M., 1965. Protozoology (1). Bailliere, Tindall and Cassel Ltd., London. pp. 430.

Pereira, A.C. and A.C.L. Castro, 1991. Oriental sore. *Andis-brasileiros de Dermatologia* 66 (4) : 211-2.

Pirmez, C., M.C.A. Marzochi, and S.G. Coutinho, 1988. Experimental canine mucocutaneous leishmaniasis. (*Leishmaniasis braziliensis braziliensis*). *Mem. Inst. Oswaldo Cruz, Rid de Janeriro.* 83 (2) : 145-51.

Pozio, E., L. Gradoni, S. Betini, and M. Gramiccia, 1981. Leishmaniasis in Tuscany (Italy) : VI Canine leishmaniasis in the focus of Monte Argentario (Grosseto). *Acta. Tropica* 38 (4) : 383-93.

Pressat, 1905. Cited by Wenyon, C.M., 1965. Protozoology (1). Bailliere, Tindall and Cassel Ltd., London. pp. 433.

Purohit, S.K., D.K. Kochar, B.B. Lal, and P.R. Jatkar, 1982. Cultivation of *Leishmania tropica* from untreated and treated cases of oriental sore. *Ind. J. Pub. Hlth.* 26 (1) : 34-37.

Ray, H.N., 1945. Presidential address, section of Zoology and Entomology, 32nd Indian Science Congress.

Ridley, D.S. and M.J. Ridley, 1983. The evolution of the lesion in cutaneous leishmaniasis. *Journal of Pathology* 141 (1) : 83-96.

Rioux, J.A., J. K. Dereure, F. Pratlongh, K. Sirdar, and M. Lambert, 1990. Eco-epidemiology of leishmaniasis in Syria. 1. Infection of Psammomys obesurs Cretzschmar to *Leishmania major* Yakimoff and Schokhor. *Ann. Parasitol. Hum. Comp.* 65 (5-6): 203-7.

Satoskar, R.S., D. Kale and S.D. Bhandarkar, 1976. Pharmacology and Pharmacotherapeutics, 5th ed. Popular Prakashan (Pvt) Ltd., Bombay. pp. 570.

Saxena, H.C., V.N.S Tomar and M.R. Songra, 1970. Efficacy of a new salt of berberine (Uni. Berberine) in Oriental sore. *Current Med. Practice.* 247-52.

Scott, P.A. and J.P. Farrell, 1981. Experimental cutaneous leishmaniasis. I. Non-specific immunodepression of mice infected with *Leishmania tropica. Journal of Immunology* 127 (6) : 2395-400.

Sergent, Ed and Et., 1905. Cited by Wenyon, C.M. (1965). Protozoology (1). Bailliere, Tindall and Cassel Ltd., London. pp. 433.

Shah, M.H., 1941, Report on the epidemic of oriental sore in Delhi. *Ind. Med. Gaz.* 76 : 449-57.

Sharma, M.I.D., J.C. Suri, N.L. Kalra, K. Mohan, and P.N. Swami, 1973a. Epidemiological and entomology features of an outbreak of cutaneous leishmaniasis in Bikaner (Raj) during 1971. *J. Com. Dis.* 5 : 54-72.

Sharma, M.I.D., J.C. Suri, N.L. Kalra and Mohan Krishna, 1973b. Studies on cutaneous leishmaniasis in India. I. A note on the current status of cutaneous leishmaniasis in north-western India as determined during 1973. *Ibid,* 5 : 73-79.

Sharma, M.I.D., J.C. Suri, N.L. Kalra and Mohan Krishna, 1973c. Studies on cutaneous leishmaniasis in India. III. Detection of a zoonotic focus of cutaneous leishmaniasis in Rajasthan. *Ibid,* 5(3) : 149-53.

Sharma, Vikas, 1998. Studies on cutaneous leishmaniasis in pets, stray dogs and human beings in and around Bikaner. M.V.Sc Thesis, Rajasthan Agricultural University, Bikaner.

Singh, Pritpal, 1980. A comparative trial of Metronidazole, Co-trimoxazole and Beberine disodium citrate in the treatment of cutaneous leishmaniasis - A clinical study conducted in Bikaner. MD Thesis, Rajasthan University, Jaipur.

Sinton, J.A., 1925. Notes on some Indian species of the genus *Phlebotomus. Ind. J. Med. Res.* 12 : 701-29.

Slappendel, R.J., 1988. Canine leishmaniasis. A review based on 95 cases in the Netherlands. *Veterinary Quarterly* 10 (1) : 1-10.

Strong, 1945. Cited by Wenyon, C.M., 1965. Protozoology (1). Bailliere, Tindall and Cassel Ltd., London. pp. 432.

Thomson, D.B. and A. Balfour, 1910. Two cases of non-ulcerating "Oriental sore" better termed-Leishman Nodules. *J. Roy. ArmyMed. Corps.* 14 : 1-22.

Vavilova, M.P., 1960. Experimental infection of dogs with both types of cutaneous leishmaniasis. Med. Parasit., Moscow, 38 : 660-65.

Wenyon, C.M., 1911, 1926. Cited by Wenyon, C.M. (1965). Protozoology (1). Bailliere, Tindall and Cassel Ltd., London. pp. 432.

Wright, J.H., 1903. Protozoan in case of tropical ulcers (Delhi boil). *J. Med. Res.* 10 : 472-82.

Yakimoff, 1914. Cited by Wenyon, C.M., 1965. Protozoology (1). Bailliere, Tindall and Cassel Ltd., London. pp. 430.

Young, A.G., 1937. Oriental sore simulating Leprosy. *Ind. Med. Gaz.* 72 : 421-2.

Zahaf, A., A. Jarraya, S. Sarhani, G. Lanotte and J. Perieres, 1986. *Leishmania major* Yakimoff et Schokhor, 1914. Leishmania Taxonomie et phylogenese. Applications eco-epidemiologiques. IMEEE, Montpellier, 455-57.

Changing Faunal Ecology in Thar Desert, pp. 323-367, 2005.
Eds.: B.K. Tyagi and Q.H. Baqri

16

THAR DESERT'S CHANGING CLIMATE AND THE EMERGENCE/EXACERBATION OF VECTOR-BORNE DISEASES: *PAST, PRESENT AND FUTURE*

B.K. Tyagi

Centre for Research in Medical Entomology (ICMR), 4-Sarojini Street, Chinna Chokkikulam, Madurai - 625 002 (TN), India

With the Thar's changed ecology, largely because of the triad of the Gang canal, Bhakra-Seerhind feeder canal and the Indira Gandhi canal, the behavioural and physiological ecology of both its flora and fauna, as well as their associated impacts such as the transmission of diseases like malaria and dengue through mosquitoes, have also changed a great deal. Malaria has conflagrated with *P. falciparum* exacerbated manifold, cutaneous leishmaniasis has spread over wider areas than earlier, dengue too has ushered in newer areas with more intensity and lymphatic filariasis has indicated its presence. In far older days, plague was a menace in the Thar Desert, while Guineaworm disease has but recently been wiped of from the face of the Thar and the country. The fast changing Thar poses many challenges and the future only seems grim, particularly from the point of view of aggravation of many a vector-borne infection.

INTRODUCTION

Following introduction of the three major canal systems, the triad of Gang canal (1927), Bhakra-Sirhind feeder canal (1955) and the Indira Gandhi canal (1958), particularly the last one with respect of the Indira Gandhi Nahar Pariyojana (IGNP) – one of the world's largest canal systems in a xeric ecosystem, the physiograpgy of the Thar Desert has changed a very great deal especially with reference to high rise in ground water table, more green vegetation including irrigation-intensive crops maintaining lower surface temperatures, changed water retention properties of soil as a result of transformation from sandy type to alluvial type, more surface water bodies and settlements of large clusters of human habitations nearby and higher quantum of cattle – all a mélange of inevitably indispensable constituents to exacerbate or conflagrate most of the vector-borne diseases particularly malaria, dengue, Japanese encephalitis and filariasis (Blanford, 1876; Gupta, 1975; Heron, 1938; Singhvi and Kar, 1992; Tyagi, 1992, 2002). The Thar Desert, which was earlier known for endemic cutaneous leshmaniasis and high incidence of Guineaworm disease (now eradicated from India), soon became set for exacerbating a nearly benign malaria scenario and ushering the hitherto uninvited break-bone 'dengue' fever. In recent years, the Thar Desert has virtually turned into a hotbed for several emerging and/or re-emerging vector-borne diseases, which not only exact heavy toll of human lives annually but also inflict severe and prolonged physical incapacitation to scores of local folks causing irreparable economic losses.

Before dealing with the vector-borne diseases prevalent in the Thar Desert, it is however considered opportune to describe eco-physiological differences between the IGNP command and the IGNP non-command/ non-IGNP areas, *i.e.*, the Thar Desert ecostem under the imposing impact of extensive canalization.

Irrigation and transformation in the Thar's ecology

As shown in Fig. 1, three major canal sysytems exist in the Thar desert region today (Anon., 1994a). The first was the Gang canal project which became fully operative in 1928, with its 1251 km distribution system in Sri Ganganagar district only. Along with the Bikaner canal feeder, it provides water to about 65% of the 300,000 ha cultivatable land. The Bhakra-Sirhind canal project, with its 1219 km long distribution system, was started in 1955 to utilize a further 300,000 ha of cultivable land. The most important of all the canal systems is the Indira Gandhi Nahar Pariyojana Project (IGNP) which is one of the world's largest project of its kind in the desert. It started in 1958 and irrigation of part of the area began in 1961. The project, when completed will include a 649 km long main canal (of which 445 km is in the Thar Desert) and over 9000 km of distributaries. Although only two of the three stages envisaged are completed by now, the main canal including feeder and 5500 km of distributaries is already irrigating 950,000 ha of land (approx. 50% of the planned irrigated area). The extent of irrigation of all these canal systems, particularly the IGNP, in the Thar Desert is shown in Fig. 2. This Herculian task, bringing Himalaya's water to the Thar desert, aims at transforming vast stretches of western desert terrain into verdure. As a result, the annual agriculture production from the newly irrigated land has by early 1990s exceeded US$ 2.6 million (Anon., 1994a,b). While several irrigation intensive crops, such as sugarcane, mustard and groundnut have been established in some of the canalized desert districts, the production of rice in Sri Gnaganagar district, the most irrigated among all desert districts, has topped the list in the whole State. Rice cultivation has often been associated with malaria transmission (Bradly, 1988; Russell and Rao, 1940; Rao, 1984; Dossou-yovo *et al.*, 1994; Hill and Cambournac, 1941; Tyagi and Yadav, 1994, 1998), although Sharma *et al.*, (1994) have doubted a clear correlation between rice cultivation and rise in malaria incidence.

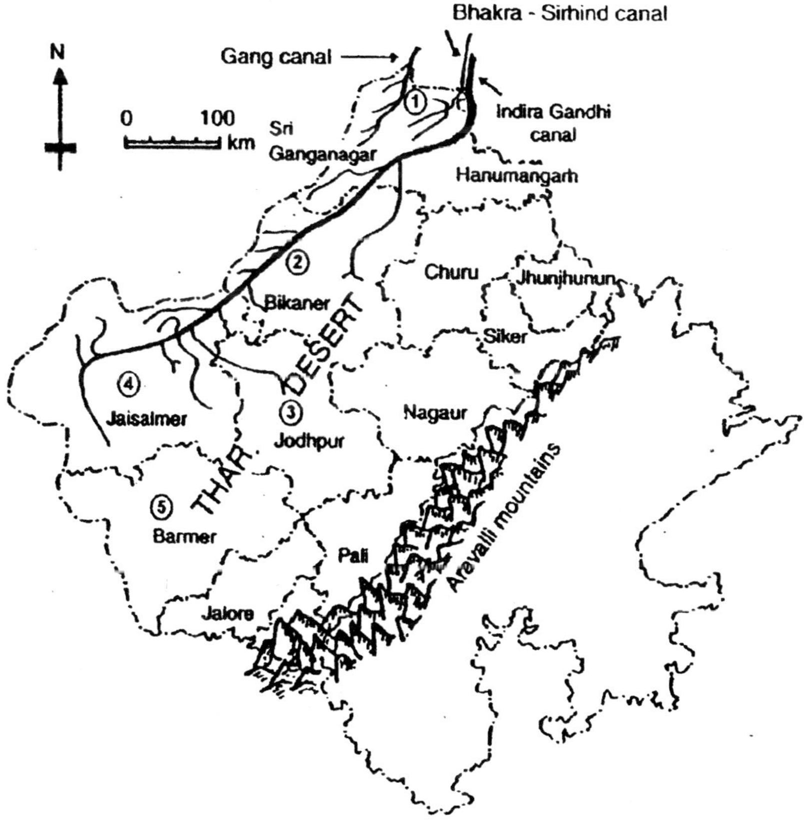

Fig. 1. Twelve desert districts in the Thar Desert. Those districts depicted by encircled numerals have experienced repeated *P. falciparum* dominated malaria epidemics; all except Barmer falling under canal based irrigation, particularly by the IGNP.

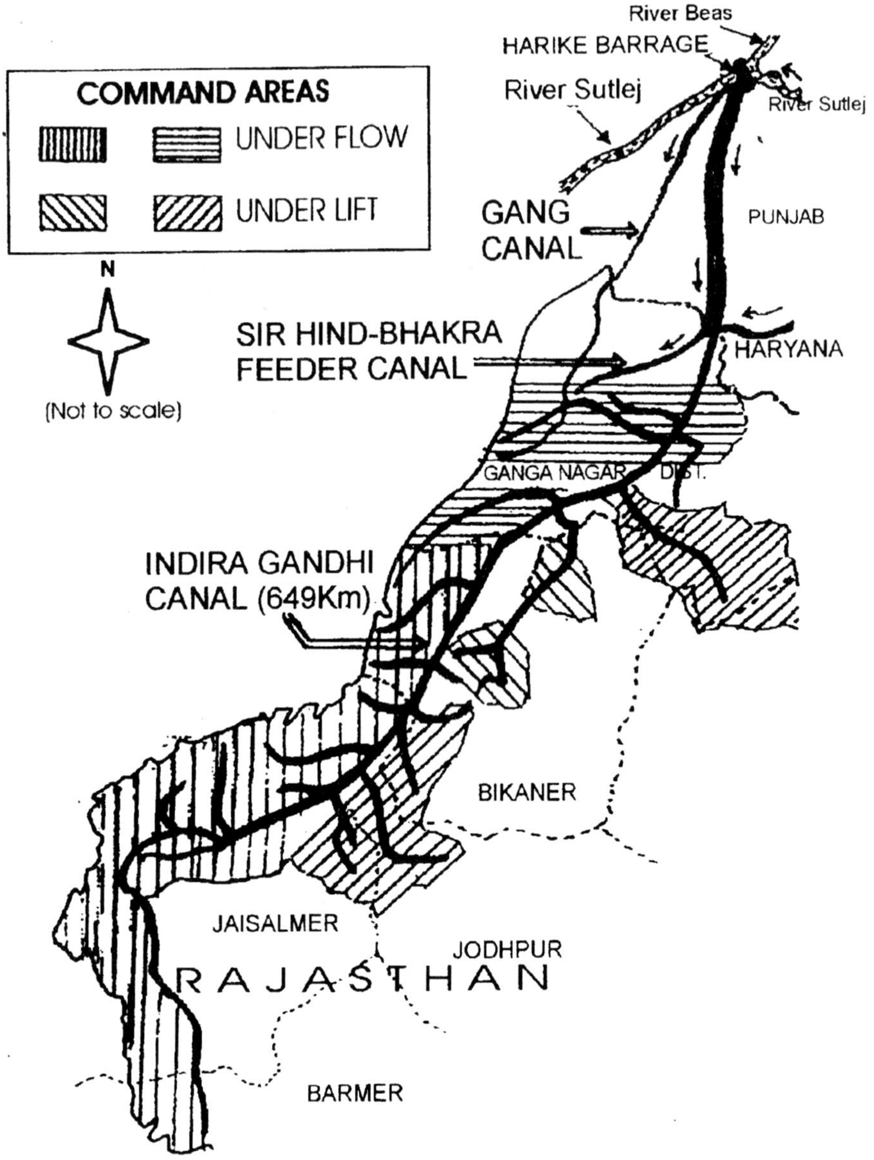

Fig. 2. Extent and types of water flow under the Indira Gandhi Nahar Pariyojana (IGNP) in the Thar Desert, Western Rajasthan

Besides the surplus irrigation water that accumulates on irrigated land, vast areas along the course of the Indira Gandhi main canal have been engulfed by seepage water from the irrigation canals on a perennial basis. An estimated 8,600 ha in the Nalibed area, Sri Ganganagar district, is permanently inundated under the IGNP Stage I by the Ghaggar river floodwater while 1000 ha is converted into marshy land due to water logging and excessive seepage from the canal in Stage II

(Anon., 1992). Supply of canal water to the desert region over several decades has caused a rise of the water table and water-holding potenial of the soil and a plentiful growth of vegetation, particularly hydrophytic weeds. Such extensive breeding grounds are the preferred sites of many anophelines including *Anopheles culicifacies,* the most serious vector of malaria in the whole Indian subcontinent, and *An. stephensi,* another malaria vector of great importance, originally breeding exclusively in 'tanka' and 'beri' in desert environs (Tyagi & Yadav, 1996) but now also breeding in expansive surface waters of different sorts.

Pant and Hingane (1988) studied trends in rainfall and temperature between 1901 and 1982 and showed an increasing trend in mean annual rainfall (141.3 mm per 100 years), and a decreasing trend in ambient temperature (-0.52°C for 100 years). Ramakrishna and Rao (1991) found an increase in the mean decadal rainfall in the Sri Ganganagar district which has been subjected to canal-based irrigation for a considerable period. As to the air temperature, Rao (1996) has demonstrated a decreasing trend of 0.039°C/year at Bikaner and 0.009°C/year at Jaisalmer. He also recorded 5-10% higher humidity measurements in cropped surfaces at 1m height in the irrigated villages compared to the un-irrigated cropped surfaces. Before the IGNP had been commissioned in the early 1970s, the groundwater table was between 40 and 51 meters below ground level (mbgl), which increased to 5 – 27 mbgl by 1981 – 82.

Irrigation-intensive cropping in the Thar Desert and its possible correlation with intensification of a vector-borne disease like malaria, *the numero uno*

Years of sustained canal-water supply to the northern desert area of the Thar desert has changed its cropping pattern a great deal. Till the turn of the first half of the 20th century Sorghum or Jwar (*Sorghum vulgare),* bajra or pearl millet (*Pennisetum typhoides)* etc. were the main desert crops completely dependent on the regular monsoon-fed rains. However, with the advent of extensive canal irrigation in the 1960s crops which required surface moisture for long periods have been introduced in the Thar, particularly in Sri Ganganagar district. Consequently, crops like cotton, wheat, mustard, groundnut and paddy are grown extensively in Sri Gnaganagar district and the process is continuously extending to other districts downstream with the southward advancement of the IGNP. The paddy cultivation is amongst the top ten crops grown in Sri Ganganagar district.

Irrigation-dependent cropping pattern, though not directly responsible for disease exacerbation is nevertheless a significant development in defining epidemiology of the disease in the Thar Desert. Until few years, paddy was grown only in Sri Ganganagar district, but soon it was extended to Bikaner, Sikar and certain areas in Jodhpur and Jaisalmer districts. According to an estimate (1994a), out of the 0.142 million ha under paddy cultivation in Rajasthan at 1993 level, 0.03

million ha (23%) is only from the Thar Desert area. Ironically, of the total output of 0.129 million tonnes annually from the State, the desert districts, lead by Sri Ganganagar, contribute more than 0.034 million metric tonnes. The productivity level of 2,600 kg/ha of paddy is also the highest in Sri Ganganagar distsrict when compared to the State average of 1,234 kg/ha. Sri Ganganagar district, where paddy cultilvation was introduced only in the mid-1970s, had paddy cultivation grown on about 6655 ha land with a yield of 27.50 Quintals/ha. A decade later, however, the paddy cultivation increased to 10,391 ha of land, with nearly double the yield in the same period. At present, the annual agricultural producton in Sri Ganganagar, the most irrigated desert district, has exceeded US$ 130 millions, having pady produced in over 17,000 ha land and making it by far the most rice-producing district in the whole State.

This radical change in the crop pattern is attributed to the introduction of canal irrigation in the desert. The initiation of cultivation in the seventies and enormous paddy output during the mid-eightiles assume great significance from the fact that it was during this period of 1983-84 when Sri Ganganagar district had experienced the first ever local outbreak of malaria, with *P. falciparum* in prominence. In ten most affected IGNP command villages in Sangaria Primary Health Centre under the district the *P. falciparum* proportion in 1983 and 1984 was 83.27% and 41.50%, respectively, compared to that of only 1.2% in the district during the nation-wide 1975 epidemic (Tyagi,1996). However, Tyagi and Yadav (2000) have cautioned that paddy cultilvation *per se* may not be directly responsible for exacerbation of *P. falciparum*-pronounced malaria but combined with rise in water, extensive water-logging and high prevalence of *A. culicifacies* it certainly exhibited a positive correlation between the two.

1. Malaria

Since the early 1990s *Plasmodium falciparum*-dominated malaria epidemics have more or less been occurring annually in the Thar Desert. This phenomenon is plausible with the cognizance of the fact that nearly all malaria epidemics in the Thar desert have come about with the progression in canal-irrigation work, particularly the massive Indira Gandhi Nahar Pariyojanà (IGNP). Before canalized irrigation had been initiated only *Anopheles stephensi,* breeding exclusively in household and/or community-based underground water reservoirs ('tanka' and 'beri'), and transmitting malaria at a low level, was prevalent in the interiors of the Thar Desert. Availability of canal waters has not only altered the desert physiography, but also the vector preponderance, distribution, victorial capacity, on one hand, but also triggered emergence of *P. falciparum*-dominated malaria in the virgin levees of the Thar Desert, on the other. Recently, Tyagi *et al.,* (1995) and Tyagi (2002, 2003, 2004) have propounded a new malaria-based classification of the

Thar region based on distributional patterns of the malaria vector species, *Anopheles stephensi* and *An. culicifacies* as well as the corresponding distribution of malaria parasites, *Plasmodium falciparum* and *P. vivax*.

Table 1. Frequency of malaria epidemics in the Thar desert districts (1975 – 2001)

District	Years of epidemics	Dominant malaria species at the epicentre of epidemic	Possible reason associated with epidemic	Deaths till 2001
Sri Ganganagar (incl. Hanumangarh)	1983, 1984, 1985	*P. falciparum*	Increased vector activity with increase in varied breeding sites	0
Bikaner	1975, 1976, 1990 1992, 1993, 1994	*P. falciparum*	Increased vector activity with increase in varied breeding sites	113
Churu	1975, 1976	*P. falciparum*	Inter-State population migration	5
Jhunjunu	1975, 1976	*P. falciparum*	Inter-State population migration	6
Nagor	1975, 1976, 1990	*P. vivax*	Inter-State population migration	2
Jodhpur	1990, 1992-1994, 2001- 2004	*P. vivax and P. falciparum*	Increased vector activity with increase in varied breeding sites; excessive quarry-mining brooding perennial vector breeding	72
Jaisalmer	1975, 1990, 1993 1996 1999 – 2001, 2002	*P. vivax and P. falciparum*	Increased vector activity with increase in varied breeding sites; non-immune human population	72
Pali	1975, 1976, 1990	*P. vivax and P. falciparum*	Inundative vectorism	3
Barmer	1975, 1990, 1994	*P. vivax and P. falciparum*	Inundative vectorism	178
Jalor	1975, 1976, 1990	*P. falciparum*	Inundative vectorism	20
Sikar	1975, 1976	*P. falciparum*	Flooding	10

Resurgence of *P. falcilparum*-pronounced malaria and canal irrigation in the Thar Desert

It is highly interesting to know as to how the introduction of canalized irrigation in a virgin land like the Thar Desert triggered not only an increase in vector density but also helped establish into the interiors of the desert the types of vector species which were earllier unknown for the region (Tyagi, 1996). It is now well established that *Anopheles stephensi,* chronollogically the oldest amongst all the presently available species in the desert, reigned supreme before the first canal waters were drawn in the northern most Sri Ganaganagar district through the neighbouring Punjap State, in 1927 (Zahar, 1990; Tyagi & Chaudhary, 1997; Tyagi, 1994b). The temporal advancement of both the anopheline fauna and the incidence of malaria synchronous to the availability of irrigation in the desert has also been explained by Tyagi (2002) (Table 2). From a status of mere four species, two primary (*An. culicifacies, An. stephensi*) and two secondary (*An. subpictus, An. annularis*), recorded till the early thirties (Covell, 1927; Christophers, 1933), the number of species rose to seven (44% of the entire anophelines recorded from Rajasthan State), from 5 of the 12 desert districts, by the mid-fifties, just at the time when second major canal, the Bhakra-Sirhind feeder, had been initiated (Puri, 1955, 1960). Following the mega IGNP irrigation project in the early sixties, even this number of anopheline fauna in the Thar Desert soared to eleven (68% of the total species ocurring in Rajasthan), with the inclusion of *An. nigermus* from Sri Ganganagar district, and *An. splendidus* and *An. d'thali,* new to Rajasthan, from Jaisalmer district. Recently, even *An. fluviatilis* has been reported occurring in Jaisalmer, raising the total number of *Anopheles* species to 12 that are prevalent in the Thar Desert. Shri Prakash *et al.* (*cf.* Chaper 7 in this book) even added one more, *An. jeyporiensis*, to the wealth of the anophelines from the Thar Desert raising the total to 13 species. While both *An. culicifacies* and *An. stephensi* have been repeatedly incriminated with malarial parasites, a third species, *An. subpictus,* too, had been occasionally found infected (Tyagi & Yadav, 2001). *Anopheles fluviatilis* which had been cited by a single specimen, had not been incriminated in the Thar, though it is an important vector of malaria in the southern mountainous and sylvan districts in southern Rajasthan.

Table 2. Possible evolutionary correlation amongst the anopheline fauna, malaria incidence and the extent of canal irrigation in the Thar Desert.

Period	Mosquitoes including vector species (*primary vector, ** secondary vector, *** known vector yet to be incriminated in Thar)	Cases reported		Irrigation ($x10^6$ ha)
		P. vivax	*P. falciparum* (%)	
Pre-1930	*An. stephensi**	-	-	-
1930-34	*An. stephensi*	-	-	-
1935-39	*An. stephensi,An. culicifacies*, An. subpictus**, An. annularis, An. vagus, A. splendidus*	-	-	0.24
1940-44	*An. stephensi,An. culicifacies, An. subpictus, An. annularis, An. vagus, A. splendidus*	-	-	0.24
1945-49	*An. stephensi,An. culicifacies, An. subpictus, An. annularis, An. vagus, An. splendidus, An. barbirostris*	-	-	0.24
1950-54	*An. stephensi, An. culicifacies, An. subpictus, An. annularis, An. vagus, An. splendidus, An. barbirostris*	-	-	0.24
1955-59	*An. stephensi,An. culicifacies, An. subpictus, An. annularis, An. vagus, An. splendidus, An. barbirostris*	-	-	0.24
1960-64	*An. stephensi,An. culicifacies, An. subpictus, An. annularis, An. vagus, An. splendidus, An. barbirostris*	-	-	0.32
1965-69	*An. stephensi,An. culicifacies, An. subpictus, An. annularis, An. vagus, An. splendidus, An. barbirostris*	-	-	0.37
1970-74	*An. stephensi,An. culicifacies, An. subpictus, An. annularis, An. vagus, An. splendidus, An. barbirostris*	-	-	0.71
1975-79	*An. stephensi,An. culicifacies, An. subpictus, An. annularis, An. vagus, An. splendidus, An. barbirostris*	639,754	21,649 (4.9%)	1.22
1980-84	*An. stephensi,An. culicifacies, An. subpictus, An. annularis, An. vagus, An. splendidus,An. barbirostris*	98,242	12,422 (12.6%)	1.23
1985-89	*An. stephensi,An. culicifacies, An. subpictus, An. annularis, An. vagus, An. splendidus,An. barbirostris*	11,746	16,533 (14.9%)	1.33
1990-94	*An. stephensi,An. culicifacies, An. subpictus, An. annularis, An. vagus, An. splendidus,An. barbirostris, An. fluviatilis****	386,208	106,963 (29.0%)	2.17
1995-2000	*An. stephensi,An. culicifacies, An. subpictus, An. annularis, An. vagus, An. splendidus, An. barbirostris, An. fluviatilis, An. nigerrimus, An. pulcherrimus, An. d'thali, An. jeyporiensis*	2,45,960	30,077 (10.9%)	2.17

Anopheles culicifacies, widely recognized as a major vector of malaria in India, is rampant in the plains as well as the desert areas currently under extensive irrigation-based development. Tyagi (2002) conducted three long term studies between 1989 and 1999 to corroborate this hypothesis (*vide infra*). It is noteworthy, however, that *An. culicifacies,* which exists in a "sibling species complex" of five species (A,B,C,D and E) in the Indian peninsula, in addition to most of the other parts of the mainland, is distinctively distributed in both the desert and non-desert districts, with the former harbouring the more vectorial species (Subbarao *et al.*, 1999; Tyagi, 2002).

Table 3. Sibling species complexes in malaria vectors prevalent in the Thar Desert.

Species	Sibling species in the complex	Type of sibling species	Sibling species type in Thar Desert
An. culicifacies	5	A,B,C,D and E	A, B and C
An. fluviatilis	3	S, T and U	Undetermined yet
An. subpictus	4	A, B, C and D	Undetermined yet
An. annularis	2	A and B	Undetermined yet

The *An. culicifacies* sibling species complex is estimated to be responsible for the transmission of 60-70% of the 2-3 million malaria cases reported annually. *Anopheles culicifacies* is also considered an important vector in some countries west of India and Sri Lanka (Sharma, 1999; Sharma and Mehrotra, 1986; Konradsen *et al.*, 2000). While in Sri Lanka, it is *An. culicifacies* 'B' which is credited with the malaria transmission in India the most important vector sibling species are *An. culicifacies* 'A' 'C' and 'D' whereas 'B' is considered comparatively a weaker or insignificant vector (Subbarao & Sharma, 1997). In the Thar Desert only *An. culicifacies* 'A' 'B' and 'C' have so far been sampled. Of these sibliIng species while the sympatrics 'A' and 'B'were common in only north-eastern areas of the Thar Desert, but extending to the whole of Punjab and Haryana as will as other northern states, the siblings 'B' and 'C' were distributed well over two-third parts of the Thar encompassing nearly entirely the western, central and southern areas of Rajasthan State. This implies that *An. culicifacies* 'B', which is possibly the most weaker vector amongst all sibling species, is contiguously present throughout the length of the Thar Desert and, while sibling species 'A' is restricted at the entry point of Indira Gandhi Canal in the erstwhile Sri Ganganagar district only, and sibling species 'C' occurs prodigiously at the fag end of the desert, in and around Barmer district. In view of this information, it can be safely comprehended as to why epidemics in recent years were invariably confined to the central southern areas of the Thar Desert where both specied A and B are expected to occur sympatrically. Regarding species complexes relatig other vector species, however, very little is known in the Thar desert, although Tyagi *et al.* (1991) have indicated on the possible differential roles being played by isolated populations within the species, *An. stephensi,* in the Thar Desert.

(i) ***The dynamics of malaria vectors:*** Study # 1 in Sri Ganganagar district (1989-19990).

A total of 16 villages in Sri Ganganagar district were selecteld for determining spatial and temporal distribution of the vector species. Of these, 10 villages namely Kamarani, Jhambhar, Sherekan, Masani, Salemgarh, Tibbi, Talwada Jheel, Tandurwali, Basir and Saliwala were located in the highly irrigated area surrounding the Sangaria Primary Health Centre, while the remaining 6 villages namely Udaipura, Harsingpur, Maler, Lalgarhi, Sangeeta and Deedasari were present in the unirrigated area surrounding the Pilibanga Primary Health Centre. Tyagi and Verma (1991) sampled a total of seven species that included, besides *An. culicifacies* and *An. stephensi,* also *An. splendidus, An. subpictus, An. vagus, An. annularis* and *An. nigerrimus* which were all reported for the first time from the area, but were constantly prevalent in adjoining States of Punjab and Haryana (Macdonald, 1931). Tyagi & Singh (1991) mapped out the distribution of *An. culicifacies* and *An. stephensi* in Sri Ganganagar district and found that *An. culicifacies,* breeding in a variety of irrigation-water formed habitats (breeding index = 6.7), constituted the bulk (94.3%) of the two species in the Sangaria PHC area. However, exactly the reverse occurred in the Pilibanga PHC area where *An. stephensi,* breeding more or less exclusively in the 'tanka' and 'beri' (Tyagi and Yadav, 1996a), formed 95.2% of the composition. Only *An. cuilicifacies* was incriminated in Sri Ganganagar district.

(ii) *Correlation between irrigation and malaria conflagration:* Study # 2 in the desert villages of Jodhpur (1990- 1996).

A total of ten villages, *i.e.*, two villages in the highly irrigated northernmost Jaisalmer District (Madassar and Awai villages) and eight in the unirrigated Jodhpur district (Dholia, Mokheri, Kuslava, Dayankor, Bhed, Mandla, Kanasar and Noore ki Dhani), where *An. stephensi* was predominant, were selected for entomological and epidemiological investigations. In the irrigated villages, *An. culicifacies* occurs in more or less in the same density as does *An. stephensi,* and above the critical density (Rao, 1984). The critical density of a vector, a concept developed by Ross (1911), and exemplified by Russell and Rao (1942) in respect of Indian anophelines by the 'per man-hour' method of sampling adults, implies that it is not necesary to reduce the vector population to zero levels to obtain cessation of malaria transmission. In other words, critical density means the maximum number of female adults belonging to a particular vector species collected by 'per man-hour' method below which the species is not able to effect the disease transmission (Viswanathan *et al.*, 1944; Singh *et al.*, 1957). What is interesting to note is that there is a visible shift in the preference for breeding and feeding by *An. stephensi.* Originallly breeding exclusively in the underground water reservoirs in the desert environment (*e.g.*, 'tanka' and 'beri'), the species has, with availability of a mélange of surface water in canal-irrigated villages, has adaped to additionally breed in the open surface water bodies formed mainly by the seepage from the irrigation canals. Tyagi and Yadav (2001) have shown that both *An. stephensi* and *An. culicifacies* exhibited two feeding peaks, one early in the night and the other just before the dawn. *Anopheles culicifacies* fed earlier than *An. stephensi* and stayed in higher densities until just before dawn. *Anopheles stephensi* was found to be more anthropohilic (52.2%) compared to *An. culicifacies* (26.8%). Between the two species, *An. stephensi* was also the most numerous (38.8%), with an averaged relative density per man-hour of 17.1 as compared to 3.3 of *An culicifacies,* in the irrigated villages. No specimen of *An. culicifacies* was ever sampled in the unirrigated villages.

While Only *An. stephensi* was incriminated with the malarial parasite in the xeric Jodhpur villages, in the irrigated Jaisalmer villages, both *An. culicifacies* and *An. stephensi* were frequently confirmed transmitting the infection, in addition to occasional parasite infestation in *An. subpictus.*

Taygi *et al.* (1994) and Tyagi and Yadav (2001) have demonstrated a relatively higher slide positivity rate (32.3%) in the irrigated IGNP villages of Jaisalmer district (*i.e.*, Madassar and Awai villages, both located within one kilometer distance from the main Indira Gandhi canal) than in the unirrigated villages (25.5%) in Jodhpur district *(vide supra,* all of which being more than 25 km away from the main IG canal) The most significant feature was, however, the higher proportion of *P. falciparum* in the IGNP villages (76.6%) compared to that in the unirrigated villages (16.6%) Further, interestingly enough, gametocyte carriers were more in the irrigated villages (17.8%) compared to the unirrigated villages (8.3%), indicating that a more active transmission was continuing in the wet villages. The fact that altogether 73.7% of all the positive cases were detected in children of less than 14 years of age is pointing towards a low immune population in the deep interiors of the desert where the irrigation water has arrived.

(iii) *Pathway of succession of P. falciparum-dominated malaria epidemics in the Thar Desert:* Study # 3 in Jaisalmer and Jodhpur.

Although documentable outbreaks of malaria in the Thar Desert are accountable since 1975 (Table 1), there is however every possibility that a high morbidity might have also ocurred during the nationwide epidemic in 1908 and perhaps thereafter as well on such occasions (Najera *et al.,* 1998). These epidemics invariably occurred in areas that were still to be canal-irrigated under IGNP and were mostly flood-prone such as in the lower reaches of the Luni river basin, *i.e.*, Barmer, Pali, Jalor and a part of Jodhpur. With the onset of IGNP in the Thar Desert, a different niche was carved out which was highly conducive to the propagation of both the vector mosquitoes and the disease parasites. Therefore, including the first *P. falciparum*-dominated malaria epidemic in an IGNP command area, *i.e.* the erstwhile Hanumangarh Tehsil under Sri Ganganagar distrsict - the entry point of the canal in the Thar Desert, as many as 13 epidemic episodes are enumerable till 2002 (Tyagi, 2002). Save for the 1990 epidemic whose epicenter was in Barmer, all others essentially had their cradle in the IGNP area, which speaks about the volume and severity of the conflagrating malaria problem in the desert! However, except outbreaks of 1990, 1992 and 1994, all others were generally focal in nature covering often a part of one district or two. These epidemics seem to have a strong correlation with the progression of the IGNP irrigation activities, which is also demonstrable.

(i) Malaria epidemics of 1983-1984 : Sri Ganganagar experienced a focal *P. falciparum*-dominated malaria outbreak for the first time in 1983, with 4925 positive cases of which 43.4% were *P. falciparum* cases. In spite of the fact that in 1983 the annual parasite incidence (API) was but second lowest during the last few years, the malaria outbreak in the Nalibed area in the erstwhile Hanumangarh tehsil of Sri Ganganagar caused high morbidity. Two primary health centers (PHCs), Sangaria and Pilibanga, which were most severely affected, had jointly contributed 72.2% of the total positive cases in the district during the epidemic. No death was however reported during the epidemic.

(ii) Malaria epidemic of 1990: In 1990, a widespread epidemic hit western Rajasthan distrsicts, but particularly the Barmer district claiming 47 lives (Mathur *et al.*,1992). A total of 19,322 positove cases (*P. falciparum 66%)* were recorded in Barmer district which was 2.5 times higher than what was reported in the preceding year. It was advocated that the epidemic in Barmer district occurred as a sequel to widespread floods in the region. It was demonstrated that the rise in the incidence of malaria was associated with the increase in the rainfall. During 1977, when annual rainfall was 13% above the normal, the annual parasite incidence (API) showed a conspicuous rise at 5.3 and annual *P. falciparum* incidence (AFI) between 0.03 and 0.1. However, 1985 was a year of exception when there was a noticeable increase in cases, particularly *P. falciparum,* despite a low rainfall. This increase was, however, attributed to the migratory cases, since the cases did not proliferate locally due to non-conducive transmission conditions. This was substantiated by a significant decline in the incidence in the next 2-3 years.

Both rural urban areas were equally and simultaneously affected by the epidemic. Though all the eight PHCs presented a rising trend, the PHC Baitu was the worst affected with API touching a peak of 55.3 and AFI being 19.6. The highest increase in AFI was noted in Mandli (4.5 times), Gudamalani (4 times), Ramasar (2.8 times), Ganga (2.3 times), Sindhari (1.75 times) and Chohtan (1.6 times). The Barmer town also recorded a 4-fold rise in API and 3-fold rise in AFI. The district's overall API and AFI recorded 2.5 and 3-fold increase, respectively. The mass blood survey carried out during the epidemic showed that all age groups were affected. The slide poitivity rate (SPR) was as high as 61 per cent. More than 93% of these had *P. falciparum* infection. A total of 103 infant positives were from Baitu PHC only. About 88% of the positsive infants were detected during September-November alone.

Baitu, along with Ramsar, also recorded maximum deaths (13), followed by Sindhari (6), Mandli (4), Ganga (3), Chohtan (2) and Gudamalani (2). Maximum number of deaths (21) were recorded during October while mortality figures in September and November were 12 and 14, respectively.

(iii) Malaria epidemics of 1992-1993: During late 1992 and early 1993, two consecutive epidemics of malaria broke out simultaneously in several districts, *i.e.*, Bikaner (Bajju and Kolayat PHCs), Jodhpur (Baap and Phalodi PHCs) and Jaisalmer (Pokaran, Nachna and Sangad PHCs). In Jaisalmer district alone a total of 2128 positive cases (*P. falciparum* 73.6%) were reported during the epidemic as against a combined total of 1809 cases (*P. falciparum* 27.6%) in the two preceding years, *i.e.*, 1990 and 1991. If the malariometric data from the three PHCs were reviewed it became clear that the malaria scenario has drastically changed over the years. A total of 2128 positive cases (*P. falciparum* 27.6%), clearly indicating a shift towards an ascending order.

In a pilot survey carried out in two desert villages near the Indira Gandhi canal, Madassar and Awai, during early 1993, the SPR was as high as 85.0% and the *P. falciparum* proportion as 88% (Tyagi *et al.*, 1994).

(iv) Malaria epidemic in 1994: In mid-1994 a major epidemic of malaria broke out in whole of Rajasthan State taking a toll of 452 lives. The worst affected three districts included Barmer, Jaisalmer and Jodhpur, contributing nearly two-thirds of the total mortality. *Plasmodium falciparum* accounted for over 60% of cases in the desert and was considered as the main cause of death (Sharma *et al.*, 1995; Tyagi *et al.*, 1995).

In a year long study during 1993-94, some very important observations were made. Following a rapid mass blood survey immdediately after the epidemic in the late autumn of 1992 in the two IGNP command villages (*P. vivax* 11.1%, *P. falciparum* 88.7%) and two unirrigated villages (*P. vivax* 17.1%, *P. falciparum* 82.7%), fever surveys were conducted for one full year between April 1993 and March 1994 which had shown a slightly higher slide positivity rate in the IGNP villages (32.3%) than in the unirrigated villages (25.5%). The most important feature was, however, the higher proportion of *P. falciparum* in the IGNP villages (76.6%) compared to that in the unirrigated villages (16,6%). Influenced by the high percentage of *P. falciparum* in the study villages, the slide positivity rate maintained a high profile throughout. *Plasmodium falciparum* exhibited two distinct peaks; first, during the summer months of April-May and second during July-August. Interestingly, more gametocyte carriers were present in the IGNP villages (16#, 17.8%) compared to the unirrigated villages (2#, 8.3%). Cumulatively, 73.7% of all the positive cases were found in children of less than 14 years of age. The male population (68.4%) suffered more than the females (31.5%).

In another short-term epidemiological investigation involving villages in three districts (Jodhpur, Barmer and Jaisalmer) with different canal-based developmental status, a rather high profile results were obtained (SPR 50.2%, SFR 28.5%). *Plasmodium falciparum* infection contributed 56.0% of the total cases. The high spleen rate (59.7%) was an indication of a serious situation of malaria in the study villages.

(v) Malaria epidemic in 1995 – 1996: Severe malaria epidemics occurred during 1995 and 1996, particularly the latter year when 45 deaths (60.70% of the total deaths in Rajasthan) were reported in Thar Desert region alone.

(vi) Malaria epidemic of 1999: A focal outbreak of malaria occurred in the villages situated close to the main Indira Gandhi canal near Ramgarh in Jaisalmer district, western Rajasthan. Stagnation of water over a month's period in the main canal as will as long standing rain water in the form of large lake-like bodies near these villages formed vast breeding grounds for the vectors like *An. culicifacies,* along with *An. stephensi* already breeding in the 'tanka'. and 'beri' in the epidemic-hit villages. A combined entomological-epidemiological investigation was undertaken in four of the 10 epidemic-hit villages, namely Seuva, Raghava, Raimala and Sadhana), all within a distance of 2 km from the main Indira Gandhi Canal. During mass blood survey in the four study villages, both thick and thin blood films from a total of 380 persons with a history of fever in last one month were prepared for examination. Of these 49 (12%) were *P. vivax* and 40 (10.5%) were *P. falciparum,* while 8 (2.1%) had mixed parasitism. Results of the mass blood survey pointed out towards a high slide positivity rate (25.5%) and *P. falciparum* proportion (49.5%) in the villages hit by the epidemic. While Raghwa village had presented the highest SPR (30.4%), Raimala showed the lowest SPR (13.1%). Interstingly, however, while on one hand Raimala singled out itself with the highest % *P. falciparum* (60%), Seuva offered the lowest *P. falciparum* proportion (41. 0%), on the other. Maximum *P. falciparum* cases nevertheless originated from Sadhna village. Only *An. culicifacies* was incriminated with the malaria parasite (Table 4).

Table 4. Malariometric data collected during malaria epidemic from the four desert villages in vicinity to the main Indira Gandhi canal in the newly irrigated Jaisalmer district

Village	Population	BSC/ BSE	*P. vivax*	*P. falciparum*	Mix	SPR	% Pf
Seuwa	505	143	23	11	5	27.2	41.0
Raghwa	600	69	9	10	2	30.4	57.8
Raimala	659	38	2	3	0	13.1	60.0
Sadhna	1100	130	15	16	1	24.6	55.5
TOTAL	*2864*	*380*	*49*	*40*	*8*	*25.5*	*49.5*

The highlight of this malaria outbreak was that the epidemic was restricted only to those villages which were in close vicinity of the main Indira Gandhi canal and where monsoon floodwaters had formed large-sized lakes in low-lying areas prodigiously breeding *An. culicifacies*. Another vector species, *An. stephensi,* was collected only from 'tanka' and 'beri' underground reservoirs (Tyagi & Yadav, 1996a). It appears that *An. culicifacies* has only recently entered the area (Tyagi *et al.*, 2001).

Tyagi (2002) has presented the annual parasite incidence and % *P. falciparum* to relate the two parameters to each other and to the period of availability of canal water for irrigation in the four desert districts under the IGNP command area (*cf.* Table 2). Generally, it would appear that more malaria cases occurred in the early seventies. However, it is significant to note an appreciable decline in malaria cases since 1977 on one hand, and a significant rise in% *P. Falciparum,* on the other. It is noteworthy here that Jaisalmer has been inexorably experiencing *P. falciparum*-dominated malaria outbreaks on an yearly basis since 1997, save for the period 1997-98. Two blocks of 1994-96 and 1999-2002 featuring unmistakable epidemic episodes are clearly identifiable.

Thus, taking cognizance of the chronological development of various epidemics since 1983 as well as their foci in the desert region, it appears that *P. falciparum* dominated malaria epidemics have advanced from northern most Sri Ganganagar district to the western and southern districts in the interior of the Thar Desert, more or less along the course of the Indilra Gandhi canal.

2. Cutaneous leishmaniasis

Leishmaniasis, as presently regarded by the World Health Organization, is one of the world's most complex diseases with several forms, some of which are self-curing (*e.g.,* cutaneous leishmaniasis), others hideously disfiguring (*e.g.*, muco-cutaneous and/or post-kala azar dermal leishmaniasis), still others fatal (*e.g.*, visceral leishmaniasis) (Tyagi, 1994a). Leishmaniasis is found in five continents and is endemic in the tropical and subtropical regions of 88 countries: 16 are developed countries, 72 are developing countries and 13 of them are among the least developed. There is an overall prevalence of 12 million cases worldwide. Globally the yearly incidence is believed to be 1.5 to 2 million new cases of cutaneous leishmaniasis (CL) and 500,000 new cases of visceral leishmaniasis (VL) (WHO, 1993). In recent years, epidemics in Sudan and India have claimed the lives of thousands of people. Geographically, cutaneous leishmaniasis has a distribution in both Old and New World countries, but particularly the former, where it prevails as a *Leishmania tropica* infection specifically in Sub-Saharan and North Africa, the Mediterranean Basin, the Middle East, Southern Asia and China. Rarely another form of infection, diffuse cutaneous leishmaniasis, is reported as caused by *L. aethiopica*. On the other hand, *L. tropica* can be the cause of chronic relapsing cutaneous leishmaniasis (CRCL) in some cases.

Of the two major types of leishmaniasis found in India, the dermal type – cutaneous leishmaniasis - is endemic in the Thar Desert, although present insignificantly also in Punjab, Delhi, Haryana and Gujarat where several epidemics have been reported in the past (Shahi, 1941; Jaswant Singh, 1933; Lysenko, 1971; Pathak *et al.,* 1990). Presence of this vector-borne disease in the Thar has assumed greater significance in recent years due mostly to two reasons: (i) the infection seems to be spreading in the Indira Gandhi canal region where incidently a correlation of the Indira Gandhi Nahar Pariyojana (IGNP) and malaria epidemics in past one-and-a-half decade was highlighted by the author (Tyagi & Chaudhury, 1997; Tyagi, 2002), and (ii) recent discovery of some correlation between cutaneous leishmaniasis and the HIV infection in some parts of the world (WHO, 2000). It is noteworthy here that while, on one hand, both the incidence of disease and the prevalence of sandfly vectors have regularly been reported from south-western areas in the IG canal region far away from the original epidemic site in Sri Ganganagar district in early 1970s (Bhardwaj & Joshi, 1971), the HIV and AIDS cases, on the other, have also increased in the region over past one decade.

Cutaneous leishmaniasis, caused by *Leishmania tropica,* is also popularly known as Tropical sore, Oriental sore, Delhi boil, Bay sore, Lahore boil, Ambala sore, Karnal sore, Gujarat sore, Cambay sore, Date sore *etc.,* indicating towards widespread distribution or the indelible nature of infection. It is further classified into "Dry Type" and "Moist Type", denoting largely to the type of ulceration at the site of infection. It is a disease that disfigures the infected person and in many a instance a social stigma, too. Sandflies, *Phlebotomus papatasi, P. salehi* and *P. sergenti,* transmit the disease which occurs in the Thar Desert both as anthroponotic cutaneous leishmaniasis (Dry type), restricted to urban areas, and zoonotic cutaneous leishmaniasis, restricted to rural (Kalra, 1996). The "dry" type of cutaneous leismaniasis occurred in the central, western and north-western parts of the Indian Subcontinent and the prominent places of infection included Lahore, Multan, Delhi, Dera-Ismailkhan etc., although reports of its occurrence in Hyderabad, the North-West Frontier Province and the United Provinces in the undivided India were also available (Daner & Ahmed, 1943).

Orientation of the three major canal systems, especially the IGNP, has changed the physiography of the Thar Desert a very great deal. Resultantly, the behaviour and distribution of the zoonotic reservoir, *Meriones hurrianae,* a desert gerbil, and the disease refractory *Tatera indica,* common in agricultural zone, have also altered significantly (Prakash, 1981). It is noteworthy here that, in past, *T. indica* formed a natural ecological barrier between villages developing agriculture and gerbil infested desert fields (Prakash *et al.,* 1971). However, due to extensive and rapid development of agricultural practices in the IGNP areas, the rodent reservoir has apparently shifted its habitats and begun to dwell close to the villages, thereby also increasing the threat of spread of the disease (Prakash, 1962). The changed disease scenario in the fast transforming Thar from desert to verdure has urgently called for an evaluation of the extent and intensity of prevalence and distribution of cutaneous leishmaniasis so that necessary control measures can be enforced before it turns out to be an uncontrollable serious public health problem.

Epidemics of cutaneous leishmaniasis in the Thar Desert

Cutaneous leishmaniasis is principally an endemic disease in certain specific climatic zones with varying degree of incidence in different ecological and seasonal conditions. Since the discovery of the parasite, *L. tropica*, in 1903, epidemics had been recorded from various parts of the pre-independent undivided India and that of today. The Jodhpur epidemic of 1950 was possibly due to displacement of people expatriated from Pakistan. Most significantly, the epidemics of both Sri Ganganagar and Bikaner involved development of agricultural practices in the hitherto arid lands now brought under the aegis of the IGNP (Table 5).

Table 5. Some notable outbreaks of cutaneous Leishmaniasis in the Thar desert

City	Year	No. of cases
Jodhpur	1950	?
Sri Ganganagar	1971	?
Bikaner	1971	828

The vectors of Cutaneous leishmaniasis have for long been highly susceptible to DDT and the indoor spraying @ 1-2 g (ai)/ sq.m. for control of vector mosquitoes of malaria controlled also the sandfly vectors simultaneously (Kaul *et al.*, 1978). Consequently many such areas where the disease had a history of sporadic occurrence no cases of cutaneous leishmaniasis surfaced after full prescribed rounds of DDT were employed, although sporadic cases continued to occur in several towns in north-western States including the Thar Desert region of Rajasthan (Table 6).

Table 6. Insecticide spray impacting incidence (and/or epidemics) of cutaneous leishmaniasis in the Thar Desert

District	Year of spray	No. of rounds of DDT/HCH	Cases / epidemic
Sri Ganganagar	1959-61	2 DDT	—
	1962-65	1	—
	1966	2	—
	1967-72	1	Several acive cases occurred in the IG canal area during 1972-73
Bikaner (urban)	1958-64	2 DDT	—
	1965-70		—
	1971-72		Epidemic recorded in 1971
Bikaner (rural)	1958-72	2 DDT	Some cases were detected in 1972 at Nokha
Jodhpur (urban)	1958-64	2 DDT	—
	1964-71	Nil	—
	1972	Focal spray HCH	Active cases detected in 1972

Prevalence of cutaneous leishmaniasis in the Thar Desert

In the Thar Desert which is the natural and most notable abode of cutaneous leishmaniasis in India, no systematic and long-term investigation or surveillance for cutaneous leishmaniasis has ever been conducted and, therefore, the information available is highly scattered and infrequent, and often insufficient to comprehend the overall impact of the disease burden in the Thar's rural population, albeit its incapacitating and disabling nature often casting a social stigma on the person of the infected ones (Sharma, *et al.*, 1973a,b; Purohit *et al.*, 1982; Verma *et al.*, 1979; Agarwal, 1980). The hospital records from both Bikaner and Jodhpur also indicate towards the magnitude of seriousness of disease incidence. Between 1997-1999, while in Bikaner 71, 139 and 113 cases were recorded consecutively, in Jodhpur between 1995-1998, cases of cutaneous leishmaniasis registered were 8,1,0 and 13, respectively. Even though these records seem underscored, nevertheless, it is evident that the disease incidence has been on a rising trend in last more than one and-a-half decade, which is quite alarming. In dogs, a major reservoir in Bikaner, an incidence of 6.8% was recorded during 1985 (Nirban, 1980).

Gender bias in nature and intensity of the cutaneous leishmaniasis infection has been duly recognized. This phenomenon has been more effectively demonstrated in the Thar Desert where the DALYs are proportionately more for men than for women, and for the rural population compared with the urban population (Table 7). Males received more infection than females and the age group most affected was 15-44 years, although a significant increase in infection rate of younger people below age 15 as well was recorded. It is noteworthy that a case of cutaneous leishmaniasis occurring in as early age as that of 18 months old boy had been reported from the Thar Desert.

Table 7. Ratio of infection by *L. tropica* in males and females in some studies in Bikaner between 1973 and 1999

Year of study	Ratio of infection in males and females
1973	1.8:1
1980	8:1
1985	7:1
1993	1.1:1
1999	2.2:1

Studies so far conducted have nevertheless substantiated some degree of gender preference of males for infection by *L. tropica* in the Thar Desert. In the Thar region cutaneous leishmaniasis infection was reported to be highest in the 5-40 years age group, albeit a recent investigation crediting a total of 91.1% to the 1-40 years age group. However, *L. tropica* infection could be found in as fragile age of 6 months as the senile age of 60 years. Possible risk factors identified to associate with cutaneous leishmaniasis in the Thar Desert are presented in Table 8.

Table 8. Risk factors identified to associate with cutaneous leishmaniasis in the Thar Desert

Risk factors	Attributes
Age	1-40 years; including economically active population in 15-40 years age group
Sex	Male
Occupation	Miner, colonizer, farm worker, army personnel, kiln worker, road worker
Environment	Hot climate, presence of deep rodent (*M. hurrianae*) burrows, scanty and sparse vegetation, availability of micro-climatic moisture and new or secondary forest galleries (as provided by IGNP in the Thar).
Economic factors	Reclamation of desert land for agriculture in the wake of new and extensive irrigational avenues under IGNP, new road construction, enhanced re-forestation activities, agriculture and mining.
Social factors	Migration, urbanization with encroachment over agricultural lands approximating the desert fringe, settlements of work-force or labourers in the work sites.
Zoonosis & anthropological or ethical factors	Shift in habitat of the desert gerbil, *M. hurrianae*, the known zoonotic host for *L. tropica*, as well as that of *T. indica*, refractory to the protozoan infection; rats are worshipped (Deshnoke Temple where thousands of rats wander with uninhibited freedom); stray dogs are socially protected and fed (like in Bikaner and also in Jodhpur)

The disease in the Thar has traditionally been prevailing in the hot and dry climatic zone around Bikaner, though with the advent of major canal systems like the IGNP an altogether new sylvan verdure is ushered in the Thar Desert in the form of social forestry and vast agricultural lands, most of which grow crops requiring water stagnation for long time. Yet, inevitably some risk factors related to cutaneous leishmaniasis in the Thar Desert can be comprehended which are subject to further investigation. Obviously, therefore, cutaneous leishmaniasis can be wisely viewed as an occupational hazard, particularly so for men owing to their nature of work and less covered body.

Distribution of Lesions

Cutaneous lesions can occur on any part of the body, but these are present more frequently on face, fore- and hind limbs and abdomen as shown in Table 9. These lesions may occur singly or plurally (upto 9-12), but a recent study indicated presence of solitary lesion in 56.68% cases while multiple lesions (2-9) were present in 43.32% cases. Of the various types of skin lesion of cutaneous leishmaniasis, the cutaneous and nodular ulcerative types (92.23%) had been found to be most commonly occurring in human-beings in Bikaner.

Table 9. Distribution of cutaneous lesions (%) on different body parts during different years

Part of body infected	1973	1980	1985	1995	1999
Face	18.6	44.44	31.25	22.22	22.22
Upper limbs	38.6	19.44	15.62	25.01	22.75
Lower limbs	22.80	8.33	15.62	11.20	13.33
Abdomen	7.10	—	6.25	5.55	0.55

Sandfly fauna and vectors of cutaneous leishmaniasis

A total of eight species belonging to the genera *Phlebotomus* and *Sergentomiya* have so far been reported from the Thar Desert (Lewis, 1978). However, only three species have been incriminated with the parasite *L. tropica* (Kaul *et al.*, 1973; Mohan & Suri, 1975). In a recent study relative abundance of five species, viz., *P. papatasi, S. punjabensis, S. babu, S. clydei* and *P. baghdadis* was determined and it was demonstrated that the vector species, *P. papatasi*, had dominated the fauna (Table 10).

Table 10. Relative abundance and per-cent distribution of sandflies during different seasons in Bikaner

Species	# Sampled	% Relative abundance	Monsoon & post-monsoon (July-Oct)	Winter (Nov.-Feb.)	Spring (Mar.-Apr.)	Summer (May-June)
P. papatasi	1664	71.1	47.6	3.4	32.2	16.8
S. punjabensis	445	19.0	60.9	1.8	24.3	13.0
S. babu	172	7.3	54.7	2.3	27.9	15.1
S. clydei	42	1.8	54.8	4.8	30.9	9.5
S. baghdadis	18	0.8	50.0	11.1	27.8	11.1

When compared with a non-desert district, like Jaipur, the per-man-hour density of the vector sandflies was found much higher in the desert districts such as Bikaner and Jodhpur. Sandflies are prevalent in almost all climatic zones in Rajasthan, although some biotic and abiotic factors are more helpful to distribution and abundance of certain species (Singh, 1999). In Table 11, a comparative data on the per-man-hour density of seven sandfly species, including two vector species, is given to evidence the dominance of the sandfly fauna in the Thar Desert.

Table 11. The per-man-hour (PMH) density comparision in seven sandflies in three different districts from Rajasthan

Species	Jodhpur		Bikaner		Jaipur	
	No. collected	Mean PMH	No. collected	Mean PMH	No. collected	Mean PMH
P. papatasi	47	1.0	1274	16.5	178	0.8
P. sergenti	0	0.0	0	0.0	2	0.005
S. punjabensis	194	1.8	137	1.6	114	1.1
S. babu	3	0.005	32	0.4	36	0.1
S. clydei	2	0.04	7	0.07	2	0.01
S. christophersi	0	0.0	0	0.0	7	0.07
S. baghdadis	0	0.0	0	0.0	7	0.03

Zoonotic reservoirs of cutaneous leishmaniasis

Both dogs and rodents serve as the zoonotic reservoirs for cutaneous leishmaniasis in the Thar Desert. There have been several studies on dogs as zoonosis for cutaneous leishmaniasis in the Thar but the most recent and significant investigation centred on the diagnosis and treatment trials of cutaneous leishmaniasis in dogs and antigenic characterization of fractions of *L. tropica* (Ahuja *et al.*, 1993). In Bikaner, cutaneous leishmaniasis was reported from both domestic and wild dogs, where cutaneous ulcerative type of cutaneous leishmaniasis was more dominant compared to other two types, muco-cutaneous ulcerative and muco-cutaneous-cum-cutaneous ulcerative type, respectively. Also, breed-wise predilection of infection has been found and the disease occurred more in short-haired breed than the lon-haired breeds. Rabbits and rats could not ever be substantiated as zoonosis in the Thar Desert.

Control of vectors and disease

In India, vectors of both cutaneous and visceral leishmaniasis had been effectively made by indoor spraying of DDT @ 1-2 gm (a.i.) during the anti-malaria vector campaigns (Sen, 1959). Although, reports are brought on record about *P. argentipes* having developed resistance against DDT, there exists no substantial proof that *P. paptasi* is not susceptible to the adulticide dosage (Mukhopadhyay *et al.*, 1992). However, in Bihar resurgence of *P. papatasi* was noted following DDT spraying (Mukhopadhyay *et al.*, 1987). Therefore, in general, recurrence of high vector population and subsequently the disease epidemic had been a result of cessation or withdrawal of DDT spraying under the National Malaria Eradication Programme in the leishmaniasis infested States, like Bihar and perhaps also Rajasthan. So far only adult sandflies have been targeted for control in the nature, as identification of sites of sandfly larvae and their control is generally very difficult and cost-prohibitive. Some juvenile hormone and insect development inhibiting compounds have shown promise against the larval instars of sandflies.

In human beings, cutaneous leishmaniasis has been controlled traditionally with 10% sodium antimony gluconate, 4-8 intralesional injections at weekly intervals. Alternatively, a new salt, 2% berberine sulphate, 4-8 intralesional injections at weekly intervals also have proved very effective (Saxena, 1970). Relative percentage efficacy between these two anti-leishmanials has been to the order of 97.77% for berberine sulphate compared to 92.85% for sodium antimony gluconate (Ahuja, 1999). These results were adequately substantiated when a comparative trial of metronidazole, co-trimoxazole and berberic disodium citrate in the treatment of cutaneous leishmaniasis was conducted in Bikaner (Singh, 1980). Besides, different treatment methods including surgical, electrodesiccation, superficial X-ray therapy, freezing with carbondioxide snow and carbondioxide laser-beam for cutaneous leishmaniasis have been tried with varying degree of success (Gerald, 1974). It is noteworthy that dogs suffering with cutaneous leishmaniasis lesions of various types and magnitude are also treated as regularly as human beings due to the fear of transmission of the infection if remained untreated for long. The tendency of nurturing or protecting stray dogs in many parts of the Thar region by the localfolk without caring healthy interaction with them is being considered one of the major constraints in treating infected dogs.

Impact of IGNP initiative

Until the initiation of canal-based irrigation in the Thar Desert, cutaneous leishmaniasis which occurred in both anthroponotic as well as zoonotic types in urban and rural set-ups, respectively, had a clear demarcation of its distribution in a given ecosystem with different climatic characteristics. While the former, the "dry" type (anthroponotic cutaneous leishmaniasis, ACL) confined to towns and urban agglomeration, the "wet" type (zoonotic cutaneous leishmaniasis, ZCL) was prodigiously prevalent in the desert lands recently brought under cultivation and irrigated under the IGNP. Therefore, anthroponotic cutaneous leishmaniasis was widespread in the highly irrigated Sri Ganganagar district, whereas zoonotic cutaneous leishmaniasis was restricted to more drier Bikaner district, during the early seventies (Sharma *et al.*, 1973a,b). It is noteworthy that over the years, the IGNP has brought nearly 12% of the Thar Desert area under its coverage, and is presently covering, besides Sri Ganganagar, also Bikaner, Jodhpur and Jaisalmer districts, while Barmer is also planned to be covered in the third phase of development of the IGNP already on the verge of completion of its second phase. If Bikaner could be momentarily considered the central focal site of concentration of the infection before the dawn of canal systems in the Thar Desert, it would become quite plausible to draw pathways of extention of cutaneous leishmaniasis, especially zoonotic type infection, more or less within the reach of the IGNP (Figure 3).

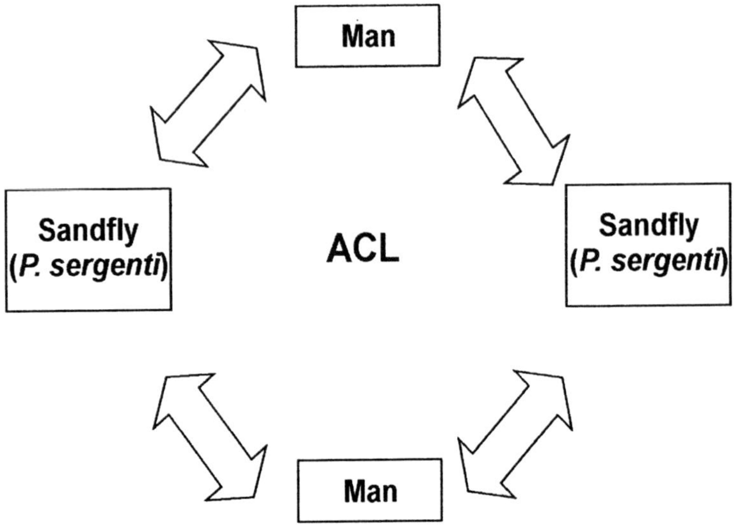

Figure 3a. Pathways of anthroponotic cutaneous leishmaniasis (ACL) transmission by *P. sergenti* (non-endemic phase).

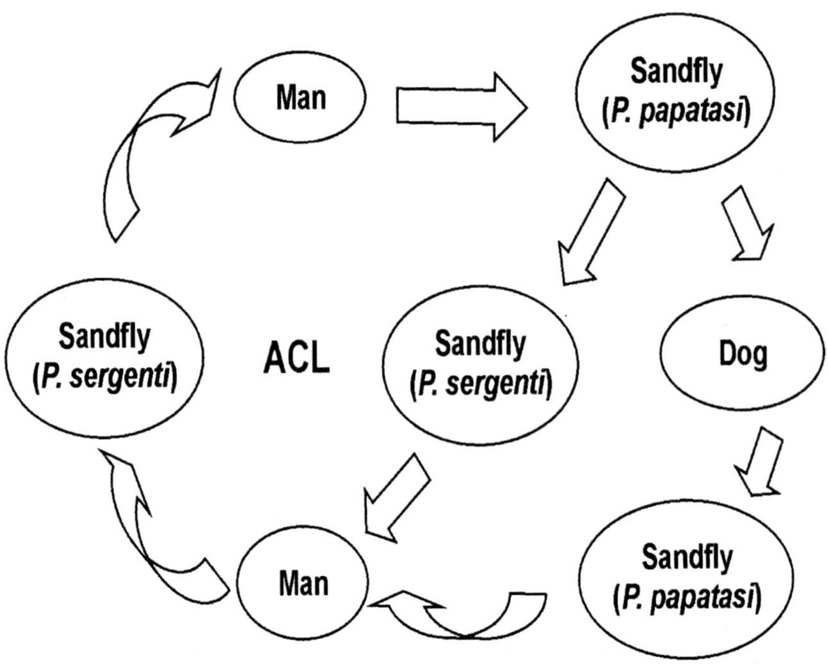

Figure 3b. Pathways of anthrponotic/zoonotic cutaneous leishmaniasis (ACL/ZCL) transmission by sandflies, *P. sergenti* and *P. papatasi*, through dogs as zoonotic reservoirs (epidemic phase).

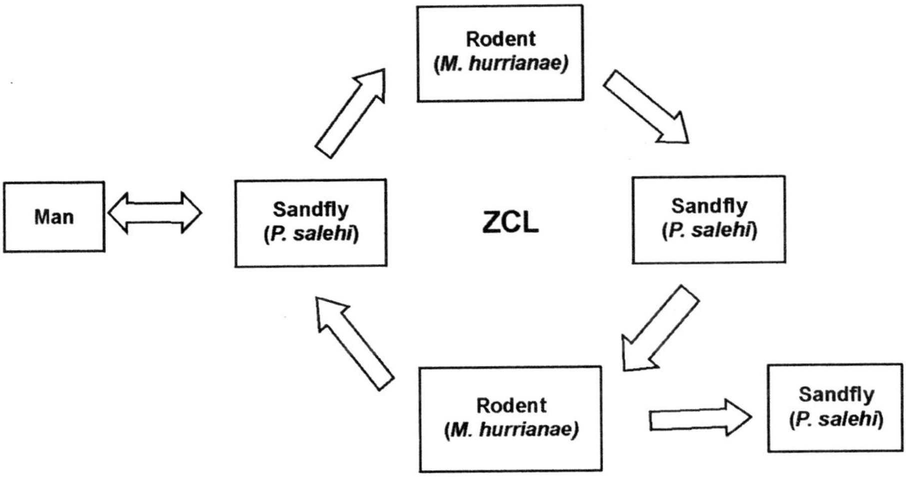

Figure 3c. Zoonotic cutaneous leishmaniasis (ZCL) in the IGNP area where rodent to rodent transmission is maintained by *P. salehi* and the infection to humans is a case of occupational hazard.

3. Dracunculiasis

Dracunculiasis is no more in India today. India was amongst the very few countries where the infection was last reported some years ago and after its eradication from this country in 1998, the disease is recently certified in early 2000 to have been wiped out completely from the face of the country. This event of annihilation of as severely debilitating infection as dracunculiasis is truly a marvelous success story, second only to that of man's victory over small-pox.

Dracunculiasis, dracontiasis or as most commonly called as Guineaworm disease, has possibly been amongst the oldest infections known to have afflicted the man through centuries. Accounts of yore indicate its presence in the Red Sea area during the pre-Christian era. Besides India, the disease had until recently been occurring in Pakistan, Afghanistan, Saudi Arabia, Iran, Africa, Turkistan, Nigeria, West Indies and South America (Lindberg, 1936; Muller, 1971). Autochthonous cases were reported from Indonesia. Stoll (1947) was among the first to have compiled the global prevalence of the disease; 0.1 million infections in Central and South America, 15 million in Africa, 3.3 million in the earstwhile USSR and 30 million in Asia. Nearly 5-25 million cases were thought to have occurred in *pro rata* India (Singh & Raghavan, 1957, Stoll, 1947).

The disease has been widely known in India from early times (Maplestone and Rao, 1966) and in some of the vernacular languages, especially in the Thar Desert, the worm is given the name of 'naru', 'valu' or 'bala' – referable to its thread-like appearance. The disease was caused by the invasion of the connective tissue by a long, thin worm,

Dracunculus medinensis, the female of which comes to the surface of the skin where it causes a blister, usually on parts of the body which come in contact with water. Dracunculiasis had been prevalent in many rural tracts and its prevalence usually was very high, particularly till about mid-1900. Although the infection was not directly fatal, nonetheless, it caused much suffering and deformities to the affected people. In fact those who were inflicted by the worm also got incapacitated for several months, but at times deformed more or less permanently. In principle, the Guineaworm disease was a rural health problem, though in some towns where sanitation including water supply was very poor the disease occurred with varying degree of severity.

Rajasthan amongst the worst affected endemic states in the pre-and post-independence India

In the first quarter of 20th Century, dracunculiasis was endemically prevalent in seven states, namely, Rajasthan, Andhra Pradesh, Tamil Nadu, Gujarat, Maharashtra, Madhya Pradesh and Karnataka, although it occurred in very low frequency in couple of other States, too (Table 12).

The disease was practically confined to villages which did not generally have provision of protected and safe water supply, and was completely absent in big towns and cities with modern conduit system of water supply. The disease showed a marked increase during the dry months preceding the monsoon, even though infection prevailed throughout the year. It was common observation that the infection depleted considerably during the years with very heavy monsoon, and the *vice versa.* Deep step wells were often associated with the sites of infected Cyclops.

Table 12. Minimum roughly estimated population at risk in different States in the *pro rata* India

Name of the state	Population at risk (in million)
Tamil Nadu	0.9
Andhra Pradesh	1.0
Karnataka	0.03
Madhya Pradesh	0.4
Rajasthan	2.3
Maharashtra	0.2
Gujarat	0.21
Punjab	0.3
Jammu & Kashmir	0.1

The guineaworm disease cases were reduced from nearly 40,000 in 12, 840 villages in 89 districts of seven endemic states (Tamil Nadu, Andhra Pradesh, Gujarat, Karnataka, Madhya Pradesh, Maharashtra and Rajasthan) in year 1984 to only 755 cases (98% reduction) in 803 villages in 23 districts in four states (Karnataka, Madhya Pradesh, Maharashtra and Rajasthan) in the year 1993 (Table 13 & 14).

Table 13. Situation of dracunculiasis in the recently identified seven endemic states, as on January 1, 1984.

States	# Districts	# PHC	# Villages	Population at risk
Andhra Pradesh	6	54	1160	15,66,218
Gujarat	13	56	444	10,58,012
Karnataka	8	73	991	16,66,123
Madhya Pradesh	21	131	3647	27,23,934
Maharashtra	15	99	1213	10,58,452
Rajasthan	23	146	5376	48,49,340
Tamil Nadu	3	5	9	1,00,48
Total	89	564	12,840	12,932,127

Table 14. State-wise guineaworm endemicity as on December, 1994

States	Districts	PHC	Villages	Population	Cases
Andhra Pradesh	3	17	26	122,648	Nil
Gujarat	Deleted on Dec. 31, 1993				
Karnataka	3	20	51	113,547	10
Madhya Pradesh	5	13	59	56,615	13
Maharashtra	Deleted on Dec. 31, 1993				
Rajasthan	12			10,048	
Tamil Nadu	Deleted since 1984-1985				348
Total	23	104	803	1,110,880	371
% decline since 1984	74%	82%	94%	91%	99%

Prevalence of Guinea-worm disease in the Thar Desert, Rajasthan (1984-1999)

In Rajasthan, the disease was prevalent much before the formation of the state. The State has been highly endemic contributing to more than 50% cases in the country during the 1980s (Johnson & Joshi, 1982; Tyagi, 1996a). Since the inception of GWEP there was downward trend and decline in cases. Due to the concerted efforts the State achieved zero guinea-worm disease status in 1997, with the last 9 cases recorded in July 1996 from three villages under Peelwa PHC in Jodhpur district. The year-wise number of affected districts, PHCs, villages and cases since 1984 in the State is given in Table 15.

It is noteworthy to mention that at the beginning of the GWEP in 1984, the State of Rajasthan reported an annual incidence of 15,210 cases from 6776 villages belonging to 147 PHCs under 23 out of 28 districts (Basu & Kaul, 1985; Sharma & Bora, 1999). Therefore, it is indeed a remarkable scientific achievement to now bring the State to a zero guineaworm disease level.

Table 15. Year-wise endemicity in Rajasthan State (1984-1999)

Year	Districts	PHCs	Villages	No. of cases
1984	23	147	6776	15210
1985	21	131	3349	11644
1986	19	118	3276	10500
1987	17	90	2755	7896
1988	17	85	2254	5619
1989	16	76	2009	4872
1990	15	70	1564	3376
1991	13	64	1310	971
1992	14	62	956	761
1993	12	54	667	545
1994	12	34	300	348
1995	8	19	160	60
1996	4	9	59	9
1997	3	7	21	0
1998	1	1	3	0
1999	1	1	3	0

Thar Desert in Rajasthan as the last site of epidemics of dracunculiasis before eradication

Two major epidemics occurred before the disease came to be completely annihilated from India by 1999 (Sharma and Bora, 1999; Kaul *et al.*, 1991). Of the two outbreaks one, and the last ever reported outbreak, occurred in the Thar Desert in Rajasthan in 1992. An epidemic of guinea-worm disease occurred in 20 villages of six out of ten Primary Health Centres in Jodhpur district alone. A total of 271 cases were recorded from the district, of which 113 (41.7%) were recorded in village Bapini under Mathania PHC which reported totally 75.76% in the district (Table 16).

Table 16. Guinea-worm cases (# 271) recorded from Jodhpur during 1992

Name of PHC	Name of village	No. of cases
Banar	Salwa Kala	01
Bap	Bap	01
	Jamba	01
Mathania	Nimbon Ka Talab	30
	Bapini	113
	Nosar	28
	Jakhan	22
	Kadwa	08
	Matora	01
	Palli	03
Peelwa	Kerla	11

Continue...

Name of PHC	Name of village	No. of cases
	Denok	12
	Chadi	10
	Paleena	09
	Khara	02
Salawas	Kalijal	01
	Kankelao	01
	Badliya	10
	Mogra Kala	03
Somesar	Bhandu Charni	04

Sites of extrusion of the worm

When dracunculiasis infection in India was on its zenith, about 90% of the cases presented sites of worm's exodus on the lower extremities, since probably these are the parts of the body that came in contact with water. In an investigation involving study on the infection in 880 cases, it was found that lower extremities were as the most preferred sites of worm's expulsion (Table 17).

Table 17. Relative per cent appearance of the worm from different sites on the body

Site of worm's expulsion	% Appearance per expulsion site
Feet & Ankles	50.1
Legs	27.1
Knees	4.9
Thighs	10.4
Upper Limbs	4.5
Trunk	2.7
Buttocks	0.3

The severity of infection, however, increases manifold when the worm-expulsion occurs concurrently on many sites, i.e., multiple infection. Therefore, a patient could manifest worm's expulsion on the same or different parts of the body either synchronously or at short intervals. The sight of as many as 15 worms from one individual had been often reported, although from a resident of Rajasthan as many as 22 worms were also removed during only one year. There is indirect evidence of immunity development in the patients suffering with multiple infections.

After-effects of the disease

As soon as the blister occurs the infected person begins to get somewhat incapacitated and debilitated from attending to his work. If the worm is completely removed without secondary infections following the infected person could be soon fit to work after the ulcer is healed. However, if the partly protruding adult worm is either retracted or withdrawn into the host's tissue or broken half way out the disease might lead to serious complications or even permanent deformities. These complications are resulted after secondary infections by certain bacteria leading to cellulitis, arthritis, synovitis. The patients develop fixed joints (ankle or knee) due to prolonged immobilization attributable to inflammation and suppuration associated with the disease, often leading to a crippled state.

Seasonal vicissitudes played a key role in occurrence of the disease. It varied from place to place. While in Madras (Chennai) the incidence of clinical dracunculiasis occurred in the months of march to June, in Colaba district of former Bombay State the disease occurred in February to May, and in Osmanabad district in Andhra Pradesh between October and July. In Jodhpur district of Rajasthan the incidence was more during May to September with peak in July during which tome more than 85% of the cases surfaced.

Prevention and Control Measures

Since it is absolutely clear that the infection to an individual could have occurred only if the water containing the cyclops infected with the guinea-worm larvae was drunk, the method for prevention from the disease also happened to be rather very simple; just to prevent from drinking such water or drink only safe water after properly filtering and/or boiling. Also, the drinking water sources should be protected from infection with guinea-worms.

Cyclopiciding with the help of temephos at the rate of 0.5-1ppm has been a very effective methodology. Doing so, not only the water gets rid of Cyclops, but the water remains to be good for drinking. Some have suggested different columns of water feeding fishes such as *Barbus puckelli* and *Gambusia* spp.

3. Dengue

Dengue, also commonly called as breakbone fever, and dengue haemorrhagic fever (DHF) have emerged as a major public health problem in India in recent years, with more and more new foci being formed in both urban and rural areas. Four viruses, dengue 1 to 4, classified in an antigenic complex of the flavivirus genus, under family flaviviridae, are the aetiological agents of this entity. These agents of cubic symmetry of 40-50 nm in diameter have a lipid envelope and a positive single stranded RNA. The viral genome of approximately 11000 base pairs in length encodes three structural proteins (capsid, C, membrane protein, M, and envelope glycoprotein, E) and seven non-structural proteins (NS1, NS2a, NS2b, NS3, NS4a, NS4b, and NS5). The 5' and 3' non-coding regions are important for regulating viral replication. The main biological properties of the viruses are located in the E protein, including receptor binding, haemagglutination of erythrocytes, neutralizing antibody induction and protective immune response.

The four serotypes are antigenically very similar to each other but different enough to elicit only partial cross protection, after infection by one of them. Recovering victims are generally immune to future infections, but only from the strain they were infected with earlier. A person can potentially experience all four different strains of dengue. In man each of the serotypes has been associated with classical dengue and dengue haemorrhagic fever. It is generally understood that DEN-2 has a strong correlation with the onset of the dengue shock syndrome (DSS).

The primary vector mosquito is *Aedes aegypti,* although *Ae. albopictus* has also been recently incriminated in nature by Tyagi and his team (*cf.* Tyagi, 2003, 2004). In urban environment, the urbanization process, which has left many without adequate water, sewer systems or waste management, and created new breeding grounds for the vector, has hastened the spread of disease. Dengue is a notifiable disease and is included in the list of diseases under priority research by the TDR/WHO.

Dengue is endemic in 100 countries throughout the Americas, South-east Asia, the Western Pacific islands, Africa and Eastern Mediterranean. Almost 40% of the world's population is at the risk of the disease. Presently, however, an estimated 50 million infections are believed to occur annually worldwide, of which about 40,000 cases are those of DHF. It is, therefore, not very overwhelming to observe that in the South-eastern Asian region dengue (including DHF and DSS), identified for the first time as a new disease in 1953 outside India, has by now become a leading cause of hospitalization and deaths among children during last 10-15 years, ranking after only the diarrhoeal disease and acute respiratory infections. The estimated quantum of annual cases in the Region is between 2030 million and DHF cases about 2,00,000. The case fatality rate is higher in the endemic countries, being around 3.5%. The disease has also formed new foci in lands that were hitherto almost a *terra incognito* for the infection, such as the interior Thar Desert and Kerala state.

In as far as Rajasthan State is concerned, but outside the Thar Desert, dengue epidemics had been recorded only twice. During an epidemic in Ajmer district in 1969, among 97 cases studied, only two cases of clinical encephalitis were seen. Dengue virus serotype 3 was isolated from cerebrospinal fluid of one, but no shock syndrome was seen. The second epidemic of dengue fever was recorded in Jaipur in 1971. So far as the Thar Desert is concerned, of the two epidemics the latest occurred in 2001 in Jodhpur district, while the former came about in 1985 in Jalore district (Chouhan *et al.*, 1991). The 1985 epidemic in Jalore district was reportedly associated with the dengue haemorrhagic fever (DHF), a total of 110 cases (80 males and 30 females) were reported, of which five had died. Involvement of central nervous sytem was found in at least in 3 patients who were detected to suffer with encephalitis with tonic spasm and delirium. Recently, *Aedes aegypti* was suggested to transmit the infection through the trans-ovarial route (Joshi *et al.*, 1996). The author has found prodigious breeding of *Ae. aegypti* in the "tanka" in Thar rural environs, and correlated this new breeding behaviour of the vector with the possible spread of the vector and the disease in the Thar desert region (Tyagi & Hiriyan, 2003).

4. Japanese encephalitis

There are a group of important arthropod-borne viral infections that attack the central nervous system and cause inflammation of the brain. There are several types of encephalitides, such as the Japanese encephalitis, St. Luis encephalitis etc. These diseases are normally transmitted between animal populations and occasionally to man. In Asia, including India, it is Japanese encephalitis which is a major concern since it affects mostly children below 12 years of age.

It is a zoonotic disease caused by a group B virus belonging to the family Flaviviridae, and transmitted mainly by culicine mosquitoes. The disease acquires its name from the fact that it repeatedly occurred in the form of severe epidemics in Japan between 1873 and 1968 (Mather and That, 1984). The disease is presently widely spread over particularly in the Southeast Asia with an annual incidence of about 45,000 cases, mostly in children under five years (as in Singapore) or 12 years (as in India). It is fatal in 7-28% cases, while 50% of the survivors tend to develop neurological handicaps. Japanese encephalitis is now a vaccine preventable disease.

Japanese encephalitis, an arbovirus (=**ar**thropod **bo**rne **virus**) infection, which, though like other virus infections, can be transmitted by blood taken for transfusion in the period of viraemia, is nevertheless particularly carried in Southeast Asia by the culicine mosquitoes belonging to the *Culex vishnui* group, particularly *Culex tritaeniorhynchus* breeding in rice fields. The incubation period is 6-8 days and the onset may be acute or gradual, with high fever, rigidity and exaggerated reflexes, headache dizziness, convulsions or even delusions and later coma. Meningeal symptoms may predominate. Many vertebrates including pigs, a sparrow, herons, certain lizards and possibly bats are naturally infected and maintain a continuous chain of infection.

Japanese encephalitis (JE) is prevalent in most of the Southeast Asian countries. In India Japanese encephalitis was first recorded in 1955, although the first serologically confirmed epidemic of the disease had occurred in West Bengal in 1973 when 763 cases including 325 deaths were reported. Since then the disease had been occurring now and then in many States particularly southern and northeastern States in the country. As far as the Thar Desert is considered, Sri Ganganagar district, which is also the most irrigated desert district, supercedes all other districts of the State in the production of paddy as well as is reported to breed JE vectors in abundance. Even so, Japanese encephalitis has never been reported from the Thar desert, although extensive surveys conducted by the National Insitute of Virology, Pune in India in recent past have indicated presence of specific neutralizing antibodies to JE virus in Rajasthan.

5. Filariasis

Filariasis (Filum=thread-like) is a group of various different kinds of infections caused by any of the filarial worms, but tradionally the term mainly refers to lymphatic filariasis. Filariasis is widespread through many tropical and subtropical parts of the world, and, in fact, it has earned a reputation of global problem. Even though the death is rarely associated with lymphatic filariasis infection, nevertheless the disease is a major social and economic scourge in large areas of warmer Africa, Asia, Western Pacific and parts of the Americas. While 1.1 billion people in 80 countries live in areas where there is a risk of contracting infection, over 120 million people living in 73 countries are estimated to be infected with or without clinical manifestations all over the world. In south-eastern Asian region alone, according to an estimate, about 600 million (60%) people in eight countries are constantly at the risk of contracting infection; India being the most endemic and holding 70% of the Asian burden. In India, lymphatic filariasis, particularly bancroftian type of filariasis, is a major public health problem. At present, it is estimated that about 420 million (109 million urban and 311 million rural) people are living in fialriasis endemic areas. While nearly 45 million people are presently detected to have one or more chronic filarial lesions, it is estimated that there occur at least 6 million attacks of acute filarial disease annually. As part of a global effort India is seeking to put lymphatic filariasis on the list of eliminated diseases before 20 years from now, or much sooner, if possible, as declared recently in the second meeting of the Global Alliance for the Elimination of Lymphatic Filariasis (GAELF) held under the aegis of World Health Organization.

The Thar Desert, or for that matter the Rajasthan State, was never considered a filariasis prone State. However, in 1985, a few cases were detected in the migrated labour population in Alwar district, without a mosquito being incriminated. In the Thar Desert region, the author conducted a short-term entomo-epidemiological investigation in both mobile (such as those in the Indian Army) and residual populations (such as those who did not ever go to hyperendemic States in the country), in the early 1990s, and had found several cases within the Army personnels and their families suffering with the infection of lymphatic filariasis, while the known vector species, *Cx. quinquefasciatus* was also incriminated (Tyagi, 1996a).

6. Plague

Plague is caused by the bacillus, *Yersinia pestis*, which occurs in great profusion in buboes, in addition to spleen, intestines, lungs, kidneys, liver and other viscera as well as in blood, though in small number only. The bacillus is also present in sputum when the disease is moved to pneumonic type. The bacterium is a short, thick coccobacillus measuring 1.5-2 x 0.5 x 0.7 μm, with rounded ends. It is non-motile, Gram-negative, indole positive and gives nitrite reaction with sulphanilic acid and α-naphthalamine.

In principle, plague is a disease of rodents, incidentally transmitted to man through the bite of infected rat flea (*Xenopsylla cheopsis*). Plague or black death has been swamping enormous human populations all over the world for more than 3000 years, although since mid-20th century the frequency of its scourges has been on the decline, sporadic and less frequent. In India, plague was first scientifically documented in early 17th century, and the most significant last events of epidemic occurred in 1994 first in the form of bubonic plague in the Beed district (Maharashtra State) and secondly in the form of pneumonic plague in the Surat city (Gujarat State) and, finally more recently during early 2002 in Jabbal area (which incidentally has been struck by the infection for the third time; the first one being in 1983) in Himachal Pradesh.

Plague is yet another disease potent to spread in a pandemic form just like cholera, influenza, and even AIDS. Plague had been occurring in both the developing and developed countries, and is considered not to be easily eradicated due to zoonosis in wild rodents of variety and their ectoparasite, the flea. As a result some countries continue to report cases every year. Over the last 20 years, the mean annual global plague case fatality rate has been 9% (range 6 to 14%). A pandemic disease like plague, which has already created havocs during 1994 and more recently during early 2002 in the country, bears no documented epidemic history in the Thar Desert in recent past, but old records of the first quarter of the 20th century do substantiate catastrophic epidemics in the Thar Desert. However, the recent high alert enforced by the State health department leaves but little space for any complacence especially in the light of presence of a far more conducive environment harbouring rodent and flea species, known to be susceptible to *Yersinia pestis,* in some parts of the Thar Desert.

DISCUSSION ON THAR'S VECTOR-BORNE DISEASES

The major objective of bringing the Himalayan waters to the xeric environment of the Thar was to transform it into verdure through growing irrigation intensive crops, like paddy, groundnut, cotton, mustard, wheat, sugarcane *etc.*, besides providing drinking water to the desert dwellers. The change in crop pattern, retaining high moisture on surface, and excessive channelization rife with mismanagement of irrigation waters attracted several anophelines such as *An. culicifacies,* which were earlier unknown for the interior part of the desert. Thus *An. culicifacies* made its headway into the interiors of the Thar Desert along with the irrigation and got settled in vast areas covered under the Indira

Gandhi Nahar Pariyojana (IGNP) command. The distribution of *P. falciparum*-dominated malaria in the Thar deset is more or less synchronous to the dissemination of IGNP-related irrigated agriculture and of *An. culicifacies*.

The disease is presently directly linked with impediments in human development and national economic progress (Anon, 2000). Malaria is one of the most important infections worldwide, while the parasite *Plasmodium falciparum* is the main cause of severe clinical malaria and death. In some areas, severe malaria problems developed following major ecological or social changes, particularly the agricultural or the exploitation of forests or socio-political interests (Nichols, 1907; Hackenitz, 2002; Mohapatra *et al.,* 1998). Malaria has also often been closely linked to irrigation and agricultural development in several countries such as Sudan, Tanzania, Burkina Faso, Sri Lanka, Zimbabve and India (EI Gaddal *et al.,* 1985; Oomen *et al.,* 1994; Ijumba *et al.,* 2002; Carnevale and Robert, 1987; Konradsen *et al.,* 2002; Senzanje *et al.,* 2002; Pant & Gratz, 1979). Yet, perhaps more than any crop, malaria has been associated more intensely with paddy (Najera, 1987). It has been difficult to determine how much of the malaria transmission occurred due to vectors that bred in ricefields, though one thing is certain that ricefield-breeding *Anopheles* do account for a great deal of the malaria transmission in rice-growing areas of the world (Gratz, 1988; Bang and Pant, 1983). Ijumba & Lindsay (2001) have extensively dealt with this phenomenon, the paddy paradox, of better health achieved as result of increased economy in the irrigated rice schemes and relatively low malaria incidence, implying that the benefits from increased irrigation is often increased wealth which in turn, means improved health care, better housing, more resources available for anti-mosquito measures, etc.

The great Indian Thar Desert in northwestern Rajasthan State is a unique ecosystem which, notwithstanding a persistant low and variable malaria endemicity with potential for sporadic outbreaks has been almost a *terra incognito* for cyclic annual epidemics till recent years (Anon., 1987a, b, 1988; Sharma, 1986a,b; Tyagi, 1994a). Before malaria epidemics began to occur regularly in the desert region in the mid-eighties onwards, most of the positive cases were reported from the hyperendemic hilly and forested terrain of Udaipur in the southern part of Rajasthan (Sharma, 1986a,b). Thus, while in 1986 the Udaipur zone had contributed 44.1% positive cases and 67.7% *P. falciparum* cases as against 24.4% positives and 11.6% *P. falciparum* cases in the desert region (Anon., 1987a), in 1994 only 14.1% positives (*P. falciparum* 26.6%) hailed from the Udaipur zone compared to 53.3% positives from desert districts with a whopping 62.5% *P. falciparum* cases (Sharma *et al.,* 1995). Amongst 12 desert districts, Barmer has mostly been the most malarious, contributing highest 31.3% positive cases and 35.8% *P. falciparum* cases during 1994 (Mathur *et al.,* 1992; Tyagi *et al,* 1995; Tyagi and Chaudhary, 1997). Malaria scenario in recent years, however, has completely reversed as malaria epidemics are being experienced invariably every year, particularly in the canalized irrigation areas under the Indira Gandhi Nahar Pariyojana (IGNP) (Table 1). During past seven decades, three major canal systems have been introduced into the desert regions of Rajasthan State, namely the Gang canal, the Bhakra-Sirhind feeder canal and the Indira Gandhi canal, running over 10,000 km in length of distributaries (Fig.1). Consequently, the physiography of the land has changed a great deal in its soil composition, landscape, climate, potential for water retention and, most important of all, the cropping and crop patterns. As a result some areas, especially those in the vicinity of the canalized irrigation, have become perennial breeding sites for vectors and are regularly experiencing outbreaks of malaria, particularly that caused by *Plasmodium falciparum*. Periodic malaria epidemics between 1983 and 2000 claiming hundreds of deaths are ample proof of the changing pattern of the disease epidemiology in the Thar desert region. It is to be emphasized here that almost all these epidemics after 1983 were characterized by a high *P. falciparum* incidence and deaths, which were notably wanting in all the earlier instances of epidemics, including those of 1975 and 1976. The only exception, however, was the 1990 epidemic in the flood-prone Barmer district, which was dominated by *P. vivax* with some deaths. The progression of these epidemics is clearly demonstrable as having taken place from northern most Sri Ganganagar district downstream to southwestern districts, more or less along the IGNP command area in the Thar desert. As a result, the Thar region now appears to provide an excellent model for the study of impact of a fast changing xeric environment on the (1) exacerbation of *P. falciparum*-dominated malaria. (2) pathways of malaria epidemics, (3) succession in anopheline fauna, (4) changes in preferences of vector breeding and feeding and, most importantly, (5) how water and environment management can help reduce the disease burden in a newly canal-irrigated desert ecosysytem.

Malaria resurgence is generally associated with the irrigation schemes, although it is more due to untidy and defective engineering in canal construction and irrigation practices (Russell, 1938; Singh, 1948). To this, the Indira Gandhi Nahar Pariyojana (IGNP) is no exception and is contrarily rife with both natural and man-made vagaries (Tyagi & Chaudhary, 1977; Tyagi & Yadav, 1996 b). Some of the major deficiencies which have rendered the canal waters highly suitable for vector breeding include (i) cessation of flow due to heavy sedimentation and profuse hydrophytic plants, (ii) excessive peeling-off of the canal lining allowing water percolation through embankments, (iii) vast collections of seepage water along the main Indira Gandhi Canal, (iv) pilferage of canal water leading to short or long-term pool formations, and (v) undue procrastination in construction activity of the main canal and distributaries resulting into stagnation of canal water for long periods (the IGNP was envisaged to be completed by early 1980s, but due mainly to shortage of funds and other reasons, the construction actilvities are unduly delayed and taken up in peace-meal!).

The resultant rise in water table became a perennial source of still water in which malaria vectors bred (Sharma, 1996). Irrigation is also responsible for increasing transmission season of malaria, and as noticed in some parts of India, it has changed areas from epidemic to endemic malarious regions (Sharma *et al.*, 1994). Thus, the Thar Desert seems to be heading towards becoming a hotbed for endemic malaria, although this appears to take still many decades before the Thar Desert would have experienced a spate of fulminating *P. falciparum*-dominated malaria epidemics.

The situation of water management in both irigated agriculture and other vital needs of the rural folks in Thar Desert is really very grave. With more and more areas being brought under the IGNP, the faulty and untidy canal irrigaation, along with perennial seepage collections, is leading to formation of innumerable groundwater sources conducive to vector breeding. The only vector control method employed in the Thar Desert is the indoor application of residual insecticides (Tyagi, 2002), and absolutely nothing is practiced to check mosquito breeding in the irrigated ricefields. In the Thar Desert, however, there is hardly any awareness at agricultural level about the correlation between irrigated ricefields and malaria mosquito breeding.

From the foregoing review of malaria emergence in the Thar Desert in the wake of irrigated agriculture, it becomes more than clear that *P. falciparum* - dominated malaria epidemics started to occur after the establishment of canalized irrigation, particularly under the IGNP. However, in view of the fact that the IGNP presently covers but 11% of the entire desert region in Rajasthan, the current situation seemingly presents only the tip of malarial iceberg.

Cutaneous leishmaniasis, by virtue of both its causative agent *L. tropica* and the vector species, seem to be affected a great deal under the changing Thar environment, and is now spreading to wider areas with changed epidemiology (Bhardwaj and Joshi, 1971).

Dracunculiasis has been eradicated from India, but the vector continues to prevail. Therefore, absolute vigilance has to be exercised in future not to allow recurrence of the disease, as what was seen in case of plague, a disease caused by a bacterium through fleas and rodents, in some parts of the country.

Filariasis in the Thar has but shown up haphazardly, still in an alarming way, and has to be carefully surveyed. Both the vector, *C. quinquefasciatus* and the pathogen worms circulate in the environment. Japanese encephalitis *per se* has never been reported from the Thar but in view of occurrence of the vector species and the changing ecology of the region with all the possible disease constituents likely to interact every effort should be made to follow up vigilantly in future.

Dengue is one serious vector-borne infection that has to be very meticulously examined in future in the Thar Desert. Its vectors are fast spreading in the rural environs and the virus is already widespread, In a non-immune population like that of the Thar Desert it would be catastrophic if the disease gets settled.

REFERENCES

Agarwal, S.K. 1980. A study of prevalence of human cutaneous leishmaniasis in Bikaner city. M.D. Thesis, Rajasthan Univ., Jaipur.

Ahuja, A. 1999. Diagnosis and treatment trials of cutaneous leishmaniasis in dogs and antigenic characterization of fractions of *Leishmania tropica*. Ph.D. Thesis, Rajasthan Agricultural Univ., Bikaner.

Ahuja, A., S.K. Purohit, J.S. Yadav and Netra, 1993. Cutaneous leishmaniasis in domestic dogs. *Indian J. Pub. Hlth.* 37: 29,

Anonymous, 1987a. Rajasthan: The health scenario. Indian Institute of Health Management Research, Jaipur. 197pp.

Anonymous, 1987b. Indira Gandhi Nahar Project. New Age Printing Press, Jaipur. 6pp.

Anonymous, 1988. Climate of Rajasthan State. Controller of Publications. Government of India, Delhi. 137pp.

Anonymous, 1992. Profile of Indira Gandhi Nahar Project. Indira Gandhi Nahar Board. Government of Rajasthan, Jaipur. 38pp.

Anonymous, 1994a. Resource atlas of Rajasthan. Department of Science and Technology. Government of Rajasthan, Jaipur. 231pp.

Anonymous, 1994b. A great venture of miraculaous changes in desert: Indira Gandhi Nahar. Indira Gandhi Nahar Board, Jaipur. 4pp.

Anonymous, 2000. Roll Back Malaria – Advocacy Guide WHO/CDS/RBM/2000. 26.

Bang, T.H. and C.P. Pant, 1983. A review on disease vectors breeding in rice fields in tropical Asia and research needs. *J. Comm. Dis. 15: 268-279.*

Basu, R.N. and S.M. Kaul, 1985. Independent Appraisal of Guinea-Worm Eradication Programme. *Journal of Communicable Diseases* 17:131-139.

Bhardwaj, T.P. and Joshi, K.R., 1971. Preliminary report on the study of endemic focus of tropical leishmaniasis with a note on the existence of canine reservoir. *J. Com. Dis.* 3: 92.

Blanford, W.T., 1876. Physical geography of the Great Indian desert. *Journal of Asiatic Society of Bengal* 45: 86-103.

Bradly, D.J., 1988. The epidemiology of ricefield-associated diseases. In vector-borne disease control in humans through rice agroecosystems management. Proceedings of the workshop on research and training needs in the field of integrated vector-borne disease control in riceland agroecosystems of developing countries. The International Rice Research Institute, Philippines.

Carnevale, P. and V. Robert, 1987. Introduction of irrigation in Burkina Faso and its effect on malaria transmission. Edited version of the working papers presented to the 7th Annual Meeting of the Joint WHO/FAO/UNEP Panel of Experts on Environmental Management for vector control, 7–11 September, 1987. *Effects of Agricultural Development on Vector-Borne Diseases.* (AGL/MISC/12/87). Food and Agriculture Organization of the United Nations, Rome.

Chouhan, G.S., F.M. Rodgigues, B.H. Sheikh, M.V. Ilkal, S.S. Kangaro, K.N. Mathur and K.R. Joshi, 1991. Clinical and virological study of dengue fever in Jalore city, Rajasthan, 1985. *Indian Journal of Medical Reearch* 91: 414-419.

Christophers, S.R., 1933. The fauna of British India. Diptera IV. Family Culicidae, Tribe Anophelini. Taylor and Francis, London. 371pp.

Covell, G. 1927. A critical review of the data recorded regarding transmission of malaria by different species of Anopheles, with notes on distribution, habits and breeding places. *Indian Medical Research Memoirs* 7: 87-99.

Daner, M.B. and S.S. Ahmed, 1943. The occurrence of Oriental sore in Hyderabad State. *Ind. Med. Gaz.* 73: 296.

Dossou-yovo, J., J. Doannio, F. Riviere and J. Duval, 1994. Rice cultivation and malaria transmission in Bouake city (Cote d'Ivoire). *Acta Tropica* 57: 91-94.

El Gaddal, A.A., A.A.M. Haridi, F.T. Hassan and H. Hussein, 1985. Malaria control in the Gezira-Managil irrigated scheme of the Sudan. *Journal of Tropical Medicine & Hygiene* 88: 153-159.

Gratz., N.G., 1988. The impact of rice production on vector-borne disease problems in developing countries. Proceedings of the workshop on research and training needs in the field of integrated vector-borne disease control in Riceland agroecosystems of developing countries, 9 – 14 March, 1987. International Rice Research Institute. WHO/FAO/UNEP Panel of Experts on Environmental Management for Vector Control pp. 7-12.

Gerald, D. D., 1974. Treatment of cutaneous leishmaniasis by freezing. J.A.M.A. 27: 658.

Gupta, R.K., 1975. The landscape and land forms. In: Gupta, R.K. & Prakash, I. (Eds). Environment Analysis of the Thar Desert. Pp. 1-21. English Book Depo, Dehradun. 484 pp.

Hackenitz, E.A., 2002. Malaria and agriculture in the Netherlands: Land reclamation, water management and risk assessment of the society concept. Malaria in irrigated agriculture IWMI Working Paper 47. SIMA Document 2: 25-58.

Heron, A.M., 1938. On the physiography of Rajaputana. *Abstract Paper Indian Science Congress Association,* Calcutta. p.8

Hill, R.B. and F.J.C. Cambournac, 1941. Intermittent irrigation in rice cultivation, and its effect on yield, water consumption and Anopheles production. *Am. J. Trop. Med.* 21: 123-144.

Ijumba, J.N. and S.W. Lindsay, 2001. Impact of irrigation on malaria in Africa: Paddies paradox. *Medical and Veterinary Entomology* 15: 1-11.

Ijumba, J.N., F.W. Mosha and S.W. Lindsay, 2002. Malaria transmission risk variations derived from different agricultural practices in an irrigated area of northern Tanzania. *Medical and Veterinary Entomology* 16: 28-38.

Jaswant Singh, S.A.S., 1933. Some observations on the mosquitoes and sandflies of Rajputana Records of Malaria Survey of India 3: 579-581.

Johnson, S. and V. Joshi, 1982. Dracontiasis in Western Rajasthan, India. *Transactions of Royal Society of Tropical Medicine & Hygiene* 76:36-40.

Joshi, V. et al., 1996. Transovarial transmission of dengue 3 virus by *Aedes aegypti. Trans. Roy. Soc. Trop. Med. & Hyg.* 90: 643-644.

Kalra, N.L. (1996). Epidemiology of cutaneous leishmaniasis and its control in India. *Proc. Nat. Acad. Sci. India* 66: 189.

Kaul, H.N., V. Dhanda, and G.B. Modi, 1973. The Phlebotomine sandflies (Diptera: Psychodidae) from Rajasthan, India, with description of *Sergentomiya* (*Sintonius*) *sirohi* sp. nov. *Indian J Med. Res.* 61:528.

Kaul, S.M., B.L. Wattal, V.N. Bhatnagar, and K.K. Mathur, 1978. Preliminary observations on the susceptibility status of *Phlebotomus argentipes* and *Ph. papatasi* to DDT in two districts of North Bihar (India). *J. Com. Dis.* 10: 208.

Kaul, S.M., Ashok Kumar, S.R. Belambe, and T. Verghese, 1991. Outbreak of dracontiasis in Bhiwandi Town in Maharashtra: A report. *Journal of Communicable Diseases* 23: 22-28.

Konradson, F., W. van der hoek, F.P. Amerasinghe, C. Mutero and E. Boelee, 2002. Engineering and malaria control: Learning from the past 100 years (in: E. Boelee, F. Konradsen W. van der Hoek, eds., *Malaria in irrigated Agriculture).* Papers and abstracts for the SIMA. Special Seminar at the ICID 18th International Congress on Irrigation and Drainage, Montreal, 23rd July, 2002. pp. 5-14.

Lewis, D.J., 1978. The Phlebotomine sandflies (Diptera: Psychodidae) of the Oriental Region. *Bull. British Museum (Nat. Hist.),Entomology,* 37: 217.

Lindberg, K., 1936. Dracunculose en Iran. *Arch. Schiffs-u Tropenhgy.* 40: 330.

Luh, Pao-ling, 1984. The wet irrigation method of mosquito control in rice fields: an experiment in intermittent irrigation in China. FAO Irrg. Drain Pap. 41: 133-136.

Lysenko, A.J.A. (1971). Distribution of leishmaniasis in the old world. *Bull. Wld. Hlth. Org.* 44: 515.

Macdonald, G., 1931. Report on malaria survey in Bikaner State. *Records of the Malaria Survey of India* 2: 603-617.

Maplestone, P.A. and S. Sundar Rao, 1966. Dracontiasis or Guinea–worm disease. *Health Bull. No. 7, Govt. of India Press, Faridabad,* pp. 25.

Mathur, K.K., G. Harpalani, N.L. Kalra, G.G.K. Murhy and M.V.V.L. Narsimhan, 1992. Epidemic of malaria in Barmer (Thar desert) Rajasthan during 1990. *Indian J. Malariol.* 29:1-10

Mahapatra, P.K., A. Prakash, D.R. Bhattacharayya and J. Mahanta, 1998. Malaria situation in north- eastern region of India. *ICMR Bulletin* 28: 21-30.

Mather, T.H. and T.T. That 1984. Environmental management for vector control of Japanese encephalitis vectors in Taiwan by chemical means. *J. Formosan Med. Assoc* 73295-312.

Mohan, K. and J.C. Suri 1975. Studies on cutaneous leishmaniasis in India. V. Isolation of *L. tropica* from gerbils, sand flies and human. *J. Com. Dis.* 7: 353.

Mukhopadhyay, A.K., A.K. Chakravarty, V.R. Kureel, and Shivraj 1987. Resurgence of *Phlebotomus argentipes* and *Ph. Papatasi* in parts of Bihar (India) after DDT praying. *Indian J. Med. Res.* 85:158.

Mukhopadhyay, N.B.L. Saxena and M.V.V.L. Narasimham 1992. Susceptibility status of *Phlebotomus argentipes* to DDT in some Kala-azar endemic districts of Bihar, India. *WHO/CTD/VBC/*92.995.

Muller, R. 1971. Dracunculus and Dracunculiasis. *Advances in Parasitology* 9: 71- 151.

Najera, J.A. 1987. Malaria and rice: strategies for control. Vector-borne disease control in human through rice agroecosystem management. Proceedings of the workshop on research and training needs in the field of integrated vector-borne disease control in Riceland agroecosystems of developing countries, 9-14 March 1987. International Rice Research Institute. WHO/FAO/UNEP Panel of Experts on Environmental Management for Vector Control. pp. 123-132.

Najera, J.A, R.L. Konznetzson and C. Delacollette 1998. Malaria epidemics: Detection and control, Forecasting and prevention. WHO/MAL/98. 1084.pp.1.81.

Nichols, F.P. 1907. The efects of large drainage works on the prevalence of malaria. *Journal of the Rolyal Army Medical Corps 8: 343-360*

Nirban, V.K. 1980. Incidence of *Leishmania tropica* infection in and around Bikaner, and preliminary trials of its cultivation on artificial media. M.V.Sc. Thesis, Sukhadia Univ., Udaipur.

Oomen, J.M.V., J. de Wolf and W.R. Jobin 1994. Health and irrigation: Incorporation of disease-control measures in irrigation, a multi- faceted task in design, construction and operation. Vol. 2, No. 2. International Institute for Land Reclamation and Improvement, The Netherlands. 119pp.

Pant, C. P. and N. Gratz 1979. Malaria and agricultural development. *Outlook agric.* 10:11-115.

Pant, G.B. and L.S. Hingane 1988. Climatic changes in and around the Rajasthan desert during the 20th Century. *Journal of Climatology* 8: 391-401.

Pathak, K.M.L., D. Kochar and M. Kapoor, 1990. Cutaneous Leishmaniasis in Rajasthan, India: *Indian J. Animal Hlth.* 29:187.

Prakash, I., 1962. Ecology of gerbils of the Rajasthan desert India. *Mammalia* 26: 311-331.

Prakash, I., R.K. Gupta, A.P. Jain, B.D. Rana, and B.K. Dutta 1971. Ecological evaluation of rodent populations in the desert biome of Rajasthan. *Mammalia* 35:384-423.

Prakash, I. 1981. Ecology of the Indian desert gerbil, *Meriones hurrianae*. Central Arid Zone Research Institute, ICAR, Jodhpur, pp. 88.

Purohit, S.K., D.K. Kochar, B.B. Lal, and P.R. Jatkar, 1982. Cultivation of *Leishmania tropica* from untreated cases of oriental sore. *Indian J. Pub. Hlth.* 26: 34.

Puri, I.M., 1955. The Distribution of *Anopheles* mosquitoes in India. Health Bull. No. 17; Malaria Bureau No. 7. Government of India 104pp.

Puri, I.M., 1960. Synoptic keys for the identifcation of the full grown larvae of the Indian anopheline mosquitoes. Health Bulletin No. 16. Malaria Bureau 7, GOI Press, Calcutta, pp. 104.

Raghavan, N.G.S., 1958. Dracontiasis in India, its public health importance. Bulletin of National Society of India for Malaria & other Mosquito-Borne Diseases.

Ramakrishna, Y.S. and A.S Rao, 1991 Climaatic features of the Indira Gandhi Canal region. (In: Prospects of Indira Gandhi Canal Project. I.P. Abrol and J. Venkateshvaralu), ICAR, New Delhi, 1-10 pp.

Rao, A.S., 1996. Climatic changes in the irrigated tracts of Indira Gandhi Canal region of arid western Rajasthan, India. Annals of Arid Zone 35:111-116.

Rao, T.R., 1984. The anophelines of India. Malaria Research Centre, Delhi, 518pp.

Ross, R., 1911. *The prevention of malaria* . Johan Muray, London, 667 pp.

Russell, P.F. 1938. Malaria due to defective and untidy irrigation: A preliminary discussion. *J. Mal. Inst. India* 1: 339-349.

Russell, P.F. and T.R. Rao 1940. The *Anopheles* of rice fields in south-eastern Madras. *J. Mal. Inst. India 3:* 427-446.

Russell, P.F. and T.R. Rao, 1942. A study of the density of *Anopheles culicifacies* in relation to malaria endemicity. *Am . J. Trop. Med.* 22: 535-558.

Saxena, H.C., V.N.S. Tomar and M.R. Songra, 1970. Efficacy of a new salt of berberine (Uni Berberine) in Oriental sore. *Current Med. Practice* 14: 247.

Sen, P. 1959. Studies on insecticide resistance of public health importance in West Bengal, India. *Indian J. Malariol.* 13:19.

Senzanje, A., E. Haackenitsz and M. Chitima, 2002. Malaria and irrigated agriculture in Zimbabwe: Impact assessment, costing and quantification under field condition. (In: E. Boelee, F. Konradsen W. van der Hoek; Eds., *Malaria in Irrigated Agriculture)* Papers and abstracts for the SIMA Special Seminar at the ICID 18th International Congress on Irrigation and Drainage, Montreal, 23rd July, 2002. pp. 15-23.

Sharma, G.K., 1986a. Malaria and its Control India. 1. Directorate of National Malaria Eradication Programme, Delhi. 347pp.

Sharma, G.K., 1986b. Malaria and its Control in India. III. Directorate of National Malaria Eradication Programme, Delhi. 618pp.

Sharma, R.C. and D. Bora (1999). National Guinea Worm Eradication Programme: Report of the Seventh Independent Evaluation, April 1999.

Helminthology Division, National Institute of Communicable Diseases, Directorate of Health Services, Delhi, pp. 157.

Sharma, R., *et al.*, 1995. Expert committee on malaria: Report. Medical & Health Department, Government of Rajasthan, Jaipur. 132pp.

Shahi, M.M., 1940. Report on the epidemic of Oriental sore in Delhi. *Indian Med. Gaz.* 76: 449.

Sharma, M.I.D., J.C. Suri, N.L. Kalra, and Krishna Mohan, 1973a. Studies on cutaneous leishmaniasis in India. I. A note on the current status of cutaneous leishmaniasis in north-western India as determined during 1973. *J. Com. Dis.* 5: 73.

Sharma, M.I.D., J.C. Suri, N.L. Kalra, and Mohan Krishna, 1973b. Studies on cutaneous leishmaniasis in India. III. Detection of a zoonotic form of cutaneous leishmaniasis in Rajasthan. *J. Com.Dis.* 5: 149.

Sharma, V.P. and K.N. Mehrotra, 1986. Malaria resurgence in India: A critical study, *Social Science & Medicine* 22: 835-845.

Sharma, V.P. 1996. Re-emergence of malaria in India. *Journal of Medical Research* 103: 26-45.

Sharma, V.P. 1999. Current scenario of malaria in India. *Parasitologica* 41: 349-353.

Sharma, V.P., A. Srivastava, and B.N. Nagpal, 1994. A study of the relationship of rice cultivation and annual parasite incidence of malaria in India. *Soc. Sci. & Med.* 38: 165-178.

Singh, J., 1948. Malaria incidental to engineering construction. *Health Bulletin* 32, Malaria Bureau 12. 50pp.

Singh, J., B.A. Rao and S.P. Ramakrishnan, 1957. National Malaria Control Programme in India, 1953-1956. *Ind,J. Malariol.*11: 419.

Singh, J. and N.G.S. Raghavan, 1957. Dracontiasis in India, its public health importance. *Bulletin of National Society of India for Malaria & other Mosquito-Borne Diseases* 5: 3.

Singh, K.V., 1999. Climatological factors in the distribution of *Phlebotomous* Sand flies (Diptera: Psychodidae) in semi-arid areas of Rajasthan, India. *J. Arid Environments* 42:43.

Singh, P.A 1980. Comparative trial of metronidazole, co-trimoxazole and berberic disodium citrate in the treatment of cutaneous leishmaniasis – a clinical study conducted in Bikaner. M.D. Thesis, Rajasthan Univ., Jaipur.

Singhvi, A.K. and Kar, A., 1992. Thar Desert in Rajasthan: Land, Man and environment. *Geological Survey of India, Bangalore,* 191pp.

Stoll, N.R., 1947. This wormy world. *Journal of Parasitology* 32:1-28.

Subbarao, S.K., N. Nanda and K. Raghavendra, 1999. Malariological stratification of India using *Anopheles culicifacies* sibling species prevalence. *ICMR Bulletin* 29: 75-80.

Subbarao, S.K. and V.P. Sharma, 1997. Anopheline species complexes and malaria control. *Indian Journal of Medical Research* 106: 164-173.

Tyagi, B.K., 1992. Control of malaria vectors in India. *Indian Rev. Life Sci.* 12: 211-238.

Tyagi, B.K., 1994a. Distribution of arthropod vector-bone communicable diseases and control of their vectors in India. *Indian Rev. Life Sci.* 14: 223-243.

Tyagi, B.K., 1994b. Distribution of malaria transmitting mosquitoes (*Culicildae: Anophelilnae; Anophelini*) in the Thar desert. Abstr. Pap. Meeting of Experts on Gaps in Research on the Faunal Diversity in the Thar Desert, Jodhpur, p. 37.

Tyagi, B.K., 1996a. Major vector-borne diseases in the Thar Desert, India. *Annals of Medical Entomology* 5: 13-19.

Tyagi, B.K., 1996b. Chronological and spatial distribution of anopheline vectors of malaria in the Thar desert. In: Ghosh, A.K., Baqri, Q.H. & Prakash, I. (Eds), Faunal Diversity in the Thar Desert: Gaps in research, pp. 227-239. Scientific Publishers (India), Jodhpur. 410pp.

Tyagi, B.K., 2002. Malaria in the Thar Desert: Facts, Figures and Future. Agrobios (India), Jodhpur. 165pp.

Tyagi, B.K., 2003. Medical Entomology: A handbook of medically important insects and other arthropods. Scientific Publishers (India), Jodhpur, pp. 262.

Tyagi, B.K., 2004. The Invincible Deadly Mosquitoes: India's health and economy enemy No. 1. Scientific Publishers (India), Jodhpur, pp. 276.

Tyagi, B.K., S.K. Bansal, and K.V.S. Verma, 1991. Intsraspecific morphological variations in two ecotypes of *Anopheles stephensi* Liston in northwestern Rajasthan (Diptera: Culicidae). *Ann. Entomol.* 9: 85-89.

Tyagi, B.K. and R.C. Chaudhary, 1997. Outbreak of *falciparum* malaria in the Thar Desert with particular emphasis on physiographic changes brought about by extensive canalization and their impact on vector density and dissemination. *J. Arid Environments* 36: 541-555.

Tyagi B.K., R.C. Chaudhary and S.P. Yadav, 1995. Epidemic malaria in Thar desert, India. *Lancet* 346: 634-635.

Tyagi, B. K., K.V. Singh, S. K. Bansal and S. P. Yadav, 1994. Malaria epidemic in some villages of north-western desert Rajasthan. *J. Appl. zool. Res* 5:152-155.

Tyagi, B.K. and K.V.S. Verma, 1991. Anopheline mosquitoes of Sri Ganganagar district (Rajasthan) transmitting malarial parasite. *J. Appl. zool Res.* 2: 85-91.

Tyagi, B.K. and S.P. Yadav, 1994. Impact of agricultural patterns and water management on the distribution of disease transmitting mosquitoes. Abstr. Pap. Meeting of Experts on Gaps in Researches in the Thar Desert, Jodhpur. 36.

Tyagi, B.K. and S.P. Yadav, 1996a. Malariological and sociological significance of 'tanka' and 'beri' In the Thar Desert, Western Rajasthan, India. *J. Arid Environments* 33: 497-501.

Tyagi, B.K. and S.P. Yadav, 1996b. Correlation of irrigation and flood water management with malaria in the Thar Desert. *WHOSEARO Regional Forum* 2: 5-11.

Tyagi, B.K. and S.P. Yadav, 1998. Impact of paddy cultivation on the prevalence of malaria and vector mosquitoes in the Thar Desert- a possible forecasting indicator for future epidemics. Abstr. Pap. Joint Annual Conference of Indian Society of Malaria & other Communicable Diseases and Indian Association of Epidemiology, Delhi, P.11.

Tyagi, B.K. and S.P. Yadav, 2000. Paddy cultivation and prospect of malaria in theThar Desert, India: A review. *Journal Applied zoological Researches* 11: 41-47.

Tyagi, B.K. and S.P. Yadav, 2001. Bionomics of malaria vectors in two physiograhically different areas of the epidemic prone Thar Desert, north westren Rajasthan (India). *J. Arid Environments* 437: 161-172.

Tyagi, B.K. and J. Hiriyan, 2003. A brief note on the breeding of dengue vector *Aedes aegypti* (Linnaeus) in the rural Thar Desert, north-western Rajasthan (India). *Dengue Bulletin* 28.

Tyagi, B.K., S.P. Yadav, P.K. Dam and R. Sachdev, 2001. Malaria outbreak in the IGNP command area in Jaisalmer district, Thar Desert, India. *J. Comm. Dis* 33: 8-95.

Verma, K.C., N.C. Bhargava, and R.K. Joshi, 1979. Cutaneous leishmaniasis. *Indian J. Dermatol. Venerol. and Lepreol.* 45: 341.

Viswanathan, D.K., T.R. Rao and T.S. Rama Rao, 1944. The behaviour of *Anopheles fluviatilis*. Part II. Experiments on the behaviour of gravid females. *J. Mal. Inst. India* 6:243-245.

Yadav, S.P. and B.K. Tyagi, 2000. Responses of women with malaria history in relation to family management and support in the rural Thar Desert, India under constant threat of increasing mosquitogenic conditions. *Annals of Medical Entomology*: 32-41.

Zahar, A. R., 1990. Vector bionomics in the epidemiology and control of malaria. The WHO European Region and The WHO Eastern Mediterranean Region. Vol. 11. Applied Field Studies. WHO/VBC/90. 1. Mal/90.1.90pp.

WHO. The Sand Killer (Video) (1993). World Health Organization/UNDP, Geneva,

W.H.O. (2000). Leishmania/HIV co-infection: South-western Europe, 1990-1998. Department of Communicable Disease Surveillance and Response. *WHO/ Leish*/2000.42.